CAMBRIDGE STUDIES IN
ADVANCED MATHEMATICS 66

HOLOMORPHIC DYNAMICS

Already published

1 W.M.L. Holcombe *Algebraic automata theory*
2 K. Petersen *Ergodic theory*
3 P.T. Johnstone *Stone spaces*
4 W.H. Schikhof *Ultrametric calculus*
5 J.-P. Kahane *Some random series of functions, 2nd edition*
6 H. Cohn *Introduction to the construction of class fields*
7 J. Lambek & P.J. Scott *Introduction to higher-order categorical logic*
8 H. Matsumura *Commutative ring theory*
9 C.B. Thomas *Characteristic classes and the cohomology of finite groups*
10 M. Aschbacher *Finite group theory*
11 J.L. Alperin *Local representation theory*
12 P. Koosis *The logarithmic integral I*
13 A. Pietsch *Eigenvalues and s-numbers*
14 S.J. Patterson *An introduction to the theory of the Riemann zeta-function*
15 H.J. Baues *Algebraic homotopy*
16 V.S. Varadarajan *Introduction to harmonic analysis on semisimple Lie groups*
17 W. Dicks & M. Dunwoody *Groups acting on graphs*
18 L.J. Corwin & F.P. Greenleaf *Representations of nilpotent Lie groups and their applications*
19 R. Fritsch & R. Piccinini *Cellular structures in topology*
20 H. Klingen *Introductory lectures on Siegel modular forms*
21 P. Koosis *The logarithmic integral II*
22 M.J. Collins *Representations and characters of finite groups*
24 H. Kunita *Stochastic flows and stochastic differential equations*
25 P. Wojtaszczyk *Banach spaces for analysts*
26 J.E. Gilbert & M.A.M. Murray *Clifford algebras and Dirac operators in harmonic analysis*
27 A. Frohlich & M.J. Taylor *Algebraic number theory*
28 K. Goebel & W.A. Kirk *Topics in metric fixed point theory*
29 J.F. Humphreys *Reflection groups and Coxeter groups*
30 D.J. Benson *Representations and cohomology I*
31 D.J. Benson *Representations and cohomology II*
32 C. Allday & V. Puppe *Cohomological methods in transformation groups*
33 C. Soulé et al *Lectures on Arakelov geometry*
34 A. Ambrosetti & G. Prodi *A primer of nonlinear analysis*
35 J. Palis & F. Takens *Hyperbolicity and sensitive chaotic dynamics at homoclinic bifurcations*
36 M. Auslander, I. Reiten & S. Smalo *Representation theory of Artin algebras*
37 Y. Meyer *Wavelets and operators*
38 C. Weibel *An introduction to homological algebra*
39 W. Bruns & J. Herzog *Cohen–Macaulay rings*
40 V. Snaith *Explicit Brauer induction*
41 G. Laumon *Cohomology of Drinfeld modular varieties I*
42 E.B. Davies *Spectral theory and differential operators*
43 J. Diestel, H. Jarchow & A. Tonge *Absolutely summing operators*
44 P. Mattila *Geometry of sets and measures in Euclidean spaces*
45 R. Pinsky *Positive harmonic functions and diffusion*
46 G. Tenenbaum *Introduction to analytic and probabilistic number theory*
47 C. Peskine *An algebraic introduction to complex projective geometry I*
48 Y. Meyer & R. Coifman *Wavelets and operators II*
49 R. Stanley *Enumerative combinatorics*
50 I. Porteous *Clifford algebras and the classical groups*
51 M. Audin *Spinning tops*
52 V. Jurdjevic *Geometric control theory*
53 H. Voelklein *Groups as Galois groups*
54 J. Le Potier *Lectures on vector bundles*
55 D. Bump *Automorphic forms*
56 G. Laumon *Cohomology of Drinfeld modular varieties II*
59 P. Taylor *Practical foundations of mathematics*
60 M. Brodmann & R. Sharp *Local cohomology*
64 J. Jost & X. Li-Jost *Calculus of variations*
66 S. Morosawa et al *Holomorphic dynamics*

S. Morosawa, Y. Nishimura, M. Taniguchi, T. Ueda
Kochi University, Osaka Medical College, Kyoto University, Kyoto University

Holomorphic Dynamics

CAMBRIDGE
UNIVERSITY PRESS

PUBLISHED BY THE PRESS SYNDICATE OF THE UNIVERSITY OF CAMBRIDGE
The Pitt Building, Trumpington Street, Cambridge, United Kingdom

CAMBRIDGE UNIVERSITY PRESS
The Edinburgh Building, Cambridge CB2 2RU, UK www.cup.cam.ac.uk
40 West 20th Street, New York, NY 10011-4211, USA www.cup.org
10 Stamford Road, Oakleigh, Melbourne 3166, Australia

Originally published in Japanese as Fukuso-rikigakukei josetsu
by S. Morosawa, M. Taniguchi and T. Ueda by Baifukan, Tokyo

This revised English edition by S. Morosawa, Y. Nishimura, M. Taniguchi, T. Ueda
First published by Cambridge University Press

Printed in the United Kingdom at the University Press, Cambridge

Typeface Times 10/13pt. *System* LaTeX 2_ε [DBD]

A catalogue record of this book is available from the British Library

Library of Congress Cataloguing in Publication data

Holomorphic dynamics / S. Morosawa, Y. Nishimura, M. Taniguchi, T. Ueda.
p. cm.
Includes bibliographic references.
ISBN 0 521 66258 3
1. Differentiable dynamical systems. 2. Holomorphic mappings.
3. Functions of complex variables. I. Morosawa, S. (Shunsuke),
1958– .
QA614.8.H66 1999
512'.352–dc21 99-14417 CIP

ISBN 0 521 66258 3 hardback

Contents

Preface

Recently, various phenomena caused by complex dynamics have drawn much attention not only from mathematicians but also from scientists over wide areas. Here we mean by complex dynamics vaguely the actions induced by iteration of analytic maps of complex number spaces. The researches on them are developing rapidly and relevant papers have become enormous in number. Nowadays, there are several excellent introductions, especially to iteration of rational functions, which is one of the central topics of complex dynamics, such as those written by Milnor, Beardon, Carleson and Gamelin, Steinmetz, and McMullen.

This book is a more comprehensive introduction to holomorphic dynamics: complex dynamics induced by holomorphic maps, including the cases of entire functions, Kleinian groups and polynomial functions of several complex variables such as complex Hénon maps, as well as the case of rational functions. Also, we emphasize the substantial value of the classical theory of complex analysis in research in holomorphic dynamics.

First, in Chapter 1, we start with several typical fractal sets appearing under iteration of polynomials, and then summarize fundamental facts on the behavior of polynomials near fixed points under iteration. We also explain the Mandelbrot set for quadratic polynomials and some results concerning Hausdorff measures and the logarithmic capacity. Finally, we introduce the concept of polynomial-like maps, one of the new fundamental concepts in this field.

In Chapter 2, we give a dichotomy of the Riemann sphere or the complex plane, respectively, with respect to dynamics of a given rational or entire function, from the viewpoint of the normal family. On one part called the Fatou set, the action is tame, but it is chaotic on the other called the Julia set. After summing up classical results on value distribution theory, we explain some fundamental properties of these sets valid for both cases, of rational functions and of entire functions, including denseness of repelling periodic points in Julia sets, and a general version of the non-wandering domains theorem, after introducing the Teichmüller spaces of holomorphic dynamics.

In Chapter 3, we explain the dynamics of entire functions, considered as holomorphic maps of the complex plane into itself, and give several examples of wandering domains and so-called Baker domains, which never appear in the case of rational functions. Also we discuss fundamental features of the exponential family and other tame families of entire functions, including the Cantor-bouquet structure and topological completeness.

Chapter 4 starts with a survey of recent topics about the classical Newton method. Next, we give a self-contained exposition (fairly independent of Chapter 2) of the standard theory due to Fatou, Julia and Sullivan on iteration of rational functions. Also we include a survey of Shishikura's fundamental theorems, and discuss closely so-called hyperbolic rational functions.

Now, we may consider the group action of Kleinian groups (discrete subgroups of the Lie group consisting of linear fractional transformations) as dynamical systems, which are nowadays called conformal dynamics. In this case, the Julia sets and the Fatou sets are conventionally called the limit sets and the ordinary sets, respectively. But it is classically recognized that properties of these sets resemble each other. Having gathered up such analogies, we now obtain a fairly long list, which is called Sullivan's dictionary, some of which we discuss in Chapter 5.

Thus Chapters 1–5 are devoted to the case of holomorphic dynamics on domains in the Riemann sphere. The theory in this case has a long history and is rather mature, as is seen from the fact that there are many excellent books on it. Hence we intend also to include a modern survey of recent progress in these chapters.

On the other hand, Chapters 6–9 treat the case of two complex variables, which is still a frontier in research on holomorphic dynamics. As a consequence, there are few guide books which give comprehensive introductions to such topics. Thus we intend to give a self-contained exposition of holomorphic dynamics of two variables. Thus, these chapters can be read almost independently of the preceding ones, though many motivations and many ways of thinking are in the same flavor.

Now, Chapter 6 starts with a survey of fundamental facts of the complex analysis of several variables. Next, after explaining the behavior near fixed points of a holomorphic automorphism of the multi-dimensional complex number space, we construct several examples of Fatou–Bieberbach domains, namely, proper domains biholomorphic to the whole space. Finally we introduce stable and unstable submanifolds near saddle (fixed) points as a local theory.

In Chapter 7, we first classify polynomial automorphisms of two variables and show that complex Hénon maps are of essential importance. Next, we concentrate on complex Hénon maps in the remainder of this chapter. We divide the whole space into the tame part consisting of escaping points and the chaotic one consisting of non-escaping points as in Chapter 1 and explain structures of each part. In particular, we represent the chaotic part by using the theory of symbolic dynamics.

Chapter 8 is an introduction devoted to the pluripotential theory, which is crucial for the measure theoretic treatment of the dynamics of several complex variables. First, we introduce the notion of currents and some kind of Lelong theorem on the solution of the $\partial\bar{\partial}$-problem. We discuss fundamental properties of the Green functions and the stable and the unstable currents, and after proving the fundamental convergence theorem we explain some of the pluripotential theoretic properties of polynomial automorphisms.

Finally, in Chapter 9, we explain, as the interaction between the pluripotential theory and the dynamics of polynomial automorphisms, fundamental global properties of important invariant sets, such as the basins of Siegel disks or Herman rings, when the absolute value of the Jacobian is less than 1. Finally, we discuss hyperbolic polynomial automorphisms.

Chapters 1–5 are written by Shunsuke Morosawa and Masahiko Taniguchi, Chapters 6 and 7 are by Tetsuo Ueda, and Chapters 8 and 9 are by Yasuichiro Nishimura. Chapters 1–7 are a revised, updated, and extended translation of a book written by Morosawa, Taniguchi and Ueda in Japanese, published by Baifukan. We express heartfelt gratitude to Shigehiro Ushiki and Yasuhiro Gotou for allowing us to use computer graphics which they produced, and Mitsuhiro Shishikura for his valuable comments.

February, 1998
The authors

1

Dynamics of polynomial maps

A simple mechanism can make a chaos. The case of iteration of polynomial maps is no exception. Indeed an extremely simple map, $z \to z^2 + c$, can create deep confusion. Understanding of, or, more modestly, an attempt to understand, such phenomena is the theme of this book.

More precisely, consider a monic polynomial

$$P(z) = z^k + a_1 z^{k-1} + \cdots + a_{k-1} z + a_k$$

of degree k. Then P can be considered as a self-map of $\mathbb{C}$, and the issue is to discuss what happens when we apply P to a given point z iteratively. Set

$$z_0 = z, z_1 = P(z), \ldots, z_n = P(z_{n-1}), \ldots,$$

and consider the behavior of the sequence $\{z_n\}_{n=1}^{\infty}$ or the orbit of z. Here, one might take some stable dependence on the initial value z for granted. But this is not true in general, and the set of unstable initial values is, almost always, incredibly complicated, and nowadays is called a fractal. Moreover, for almost every such initial value, the behavior is chaotic enough.

First, in §1.1, we give several typical examples of fractal sets caused by polynomial maps, and explain basic properties of them. Then §1.2 summarizes local theory near a fixed point of a general holomorphic map.

Turning to quadratic polynomial maps as above, we introduce the Mandelbrot set, which in itself is a fractal and represents the very interesting family of quadratic polynomial maps, in §1.3.

As the fundamental tools to measure how complicated the set considered is, we introduce in §1.4 the Hausdorff measures and the logarithmic capacity, and give some basic facts about them.

Finally, in §1.5, we define a polynomial-like map, which is one of the fundamental concepts in the modern theory of complex dynamics, and explain some basic results.

In this chapter, we explain not only basic facts on the dynamics of polynomials, but also several central issues about complex dynamics. In particular, we include many important results without proofs. As general references for those results, we cite books by Beardon (1991), Carleson and Gamelin (1993), Milnor (1990), and Steinmetz (1993).

1.1 The set of escaping points for a polynomial

1.1.1 Typical examples of polynomials

Fix a monic polynomial

$$P(z) = z^k + a_1 z^{k-1} + \cdots + a_{k-1} z + a_k$$

of degree $k \geq 2$, and consider the sequence of $\{P^n\}_{n=0}^{\infty}$, where and in what follows, we set

$$P^0(z) = z,\ P^1(z) = P(z), \ldots, P^n(z) = P(P^{n-1}(z)).$$

Then for every z with sufficiently large $|z|$, we have

$$|z| < |P(z)| < |P^2(z)| < \cdots.$$

In fact, choose an $R > 2$ so large that $|P(z)| > |z|^k/2$ for every z with $|z| > R$. Then inductively, we see that

$$R < |z| < |z|^k/2 < |P(z)| < \cdots < |P^{n-1}(z)| < |P^{n-1}(z)|^k/2 < |P^n(z)|.$$

Thus for every such z, $P^n(z)$ tends to ∞ in the Riemann sphere $\widehat{\mathbb{C}} = \mathbb{C} \cup \{\infty\}$.

Definition We set

$$I_P = \left\{ z \in \mathbb{C} \,\middle|\, \lim_{n\to\infty} P^n(z) = \infty \right\},$$

and call it the *set of escaping points* for P. Also we set

$$K_P = \mathbb{C} - I_P$$

and call it the *filled Julia set* of P. The boundary $J_P = \partial K_P$ of K_P is called the *Julia set* of P.

Remark Since $k \geq 2$, any polynomial

$$P(z) = a_0 z^k + a_1 z^{k-1} + \cdots + a_{k-1} z + a_k \quad (a_0 \neq 0)$$

is affine conjugate to a monic polynomial of degree k. More precisely, set $A(z) = \omega z$, where ω is a $(k-1)$th root of a_0, and we see that $A \circ P \circ A^{-1}(z)$ is a monic polynomial of degree k.

The following example is clear.

Example 1.1.1 *Let $P(z) = z^2$. Then we have*

$$I_P = \{|z| > 1\}, \quad K_P = \{|z| \le 1\}, \quad J_P = \{|z| = 1\}.$$

Example 1.1.2 *Let $P(z) = z^2 - 2$. Then we have*

$$I_P = \mathbb{C} - [-2, 2], \quad K_P = J_P = [-2, 2],$$

where $[-2, 2]$ denotes the interval $\{x \in \mathbb{R} \mid -2 \le x \le 2\}$.

Verification Set $B(z) = z + 1/z$, and we can take the single valued branch of B^{-1} from $\mathbb{C} - [-2, 2]$ onto $\{|z| > 1\}$. A simple calculation shows that $P(z) = B \circ P_0 \circ B^{-1}(z)$ with $P_0(z) = z^2$. Hence by Example 1.1.1, we see that I_P contains $\mathbb{C} - [-2, 2]$. On the other hand, it is clear that $P([-2, 2]) = [-2, 2]$, and hence the assertions follow.

Recall that P restricted to $[-2, 2]$ is a typical unimodal map which represents an action that expands the interval to twice its length and then folds it into two.

In general, J_P has a very complicated shape. Actually, the above examples are exceptions. We show the variety of the Julia sets by examples.

Remark The reasons why the Julia sets in the examples below have such shape are not easy, and need several results discussed later. But for the sake of convenience, we include brief explanations.

We start by clarifying what we consider as fractal sets. There are several ways to define fractal sets. Self-similarity relates closely to fractal sets. But there exist so many complicated sets, for which self-similarity is not so clear or seems to be absent. From a quantitative viewpoint, we may call every simple curve with Hausdorff dimension, defined in §1.4, greater than 1 a fractal curve. But we take a looser definition (also see §4.1).

Definition Let E be a closed subset of $\mathbb{C}$ without interior. We say that E is a *fractal set* if E cannot be represented as a countable union of rectifiable curves.

We say that a domain D in $\widehat{\mathbb{C}}$ is a *fractal domain* if the boundary ∂D is a fractal set.

As to fractal sets in the multi-dimensional complex number space, we will discuss these in Chapters 6–9.

Even in the case that the Julia set J_P is a simple closed curve, J_P is terribly complicated except for trivial cases.

To give such examples, we perturb $P(z) = z^2$ in Example 1.1.1. That is, consider $P_c(z) = z^2 + c$ with sufficiently small c. Then simple computations show that there is a fixed point z_c, a solution of $P(z) = z$, near $z = 0$ and hence the value of the derivative at z_c is also near to 0. Moreover, we will show that the structure of J_{P_c} is circle-like.

Example 1.1.3 *When $|c|$ is sufficiently small, the Julia set J_{P_c} of $P_c(z) = z^2+c$ is a simple closed curve, but has a very complicated shape, called a fractal circle. See Figure 1.1(a), the Julia set of $z^2 + c$, $c = 0.59\cdots + i0.43\cdots$, and Figure 1.1(b), that of $z^2 + c$, $c = 0.33\cdots + i0.07\cdots$.*

Verification By the corollary to Theorem 1.5.1, each Julia set is a quasicircle. In particular, it is a simple closed curve. Theorem 1.4.7 implies that it is a fractal set.

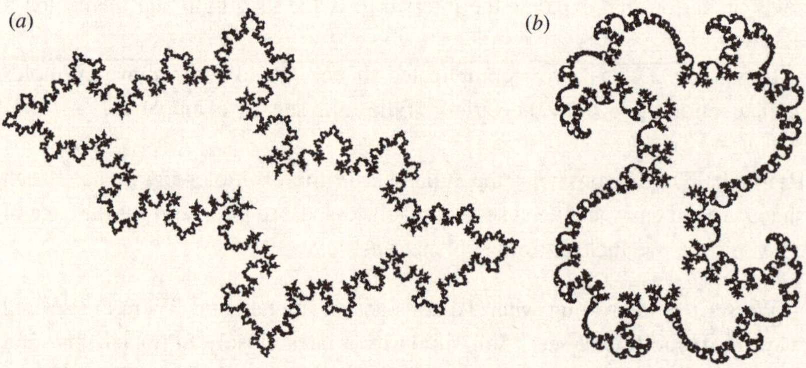

Fig. 1.1. Jordan curves as Julia sets

Definition We say that a compact set K in $\widehat{\mathbb{C}}$ is *locally connected* if, for every $z_0 \in K$ and every neighborhood U of z_0, $U \cap K$ contains a connected neighborhood of z_0 in K.

A connected and locally connected compact set is arcwise connected. For some conditions on local connectedness, see §4.4.

Example 1.1.4 *Set*

$$E = \left\{ x + iy \,\middle|\, 0 < x < 1,\, y = \sin \frac{1}{x} \right\},$$

and the closure of E is not locally connected.

Definition We call a compact set K a *dendrite* if K is a connected and locally connected compact set without interior whose complement $\mathbb{C} - K$ is connected.

Example 1.1.5 *A typical example of a dendrite is the (filled) Julia set of* $z^2 + i$. *As other examples, we cite in Figure 1.2 the (filled) Julia sets of*

(a) $z^2 + c, \quad c \in \mathbb{R},\ c^3 + 2c^2 + 2c + 2 = 0,$
(b) $z^3 + c, \quad c = \sqrt{\omega - 1},\ \omega^2 + \omega + 1 = 0.$

Verification These Julia sets are connected by Theorem 1.1.4, for the orbit of the unique critical value 0 is preperiodic and hence bounded. They have no interior points, which can be shown as in the proof of Theorem 4.2.18. Further, since they are subhyperbolic, defined in §4.4, the Julia sets are locally connected.

When the Julia set is a dendrite, it coincides with the filled Julia set. There is another typical case that these are coincident with each other. It is the case that the Julia set of a polynomial is a Cantor set. We will define and discuss Cantor sets in §1.4. Here we temporarily call a set E a Cantor set if E has no isolated points, and every connected component of E consists of a single point. Then it is easy to see that such an E has uncountably many connected components.

By definition, if a closed set without interior has uncountably many connected components, then it is a fractal set. We give in Figure 1.3 the Julia sets of

(a) $z^2 + c, \quad c = 0.38\cdots + i0.22\cdots,$
(b) $z^2 + c, \quad c = -0.82\cdots + i0.33\cdots.$

These Julia sets are Cantor sets by Theorem 1.1.6, for the c are outside of the Mandelbrot set defined in §1.3.

In the case of polynomials P of degree greater than 2, the filled Julia set may have interior points even if it has uncountably many connected components. We give the following example.

Example 1.1.6 *The Julia set of* $P(z) = z^3 + az$ *with* $a = 0.038 + i1.95$ *consists of uncountably many components and countably many components have interior points. See Figure 1.4.*

(*a*)

(*b*)

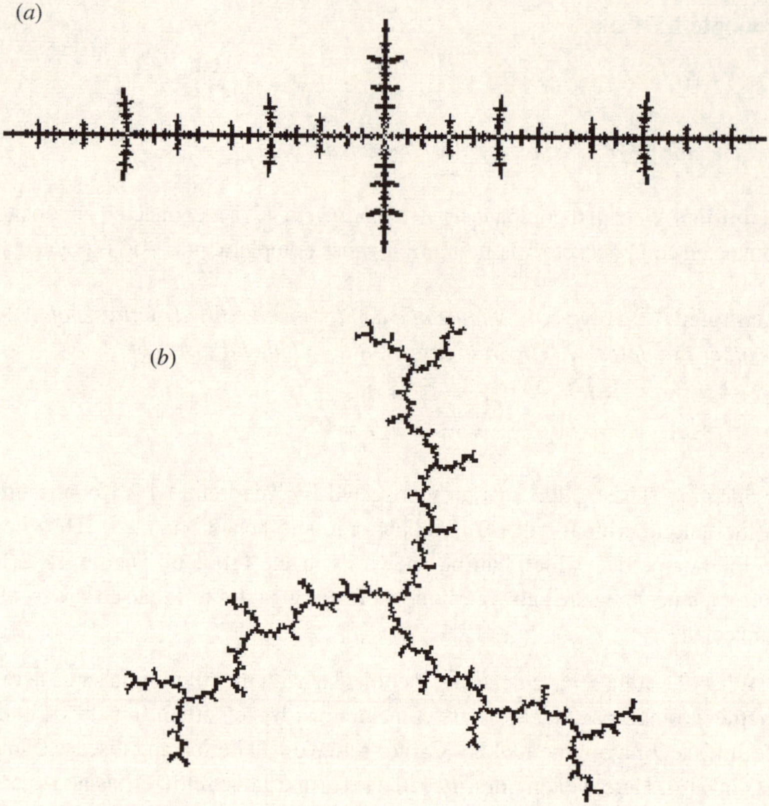

Fig. 1.2. Dendrites as Julia sets

Verification Since Theorem 2.3.6 shows that Julia sets are perfect sets, a disconnected Julia set consists of uncountably many components. To see the filled Julia set of P has countably many connected components which have interior points, we need only pursue the orbit of the finite critical points $\pm\sqrt{-a/3}$, as is seen from the fact proved in Chapter 2. We remark that the shape of each component depends on the polynomial induced by a suitable renormalization, which is defined in §1.5.

(a)

(b)

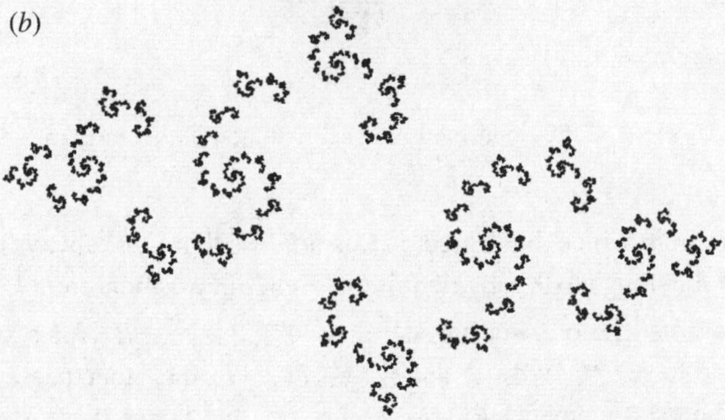

Fig. 1.3. Cantor sets as Julia sets

1.1.2 Basic properties of the filled Julia set

We have seen that the Julia set may have incredibly varied shapes even restricted to simple polynomial maps. One purpose of this book is to discuss such variety of shapes and characters of the Julia sets, though we stress explanation of the unified nature which a fairly large class of complex dynamics are equipped with.

We start with the following elementary fact.

Theorem 1.1.1 *Let $P(z)$ be a polynomial of degree not less than 2. Then the filled Julia set K_P is a non-empty compact set.*

Moreover, I_P is a domain, that is a connected open set.

Remark When $P(z) = z + c$ $(c \neq 0)$, it is clear that K_P is empty. The case of a general linear map is left to the reader.

Fig. 1.4. Disconnected Julia set with non-empty interior

Proof of theorem Since the degree of P is not less than 2, P always has a fixed point, which belongs to K_P by definition. Hence K_P is non-empty.

Next, as noted before, setting $V = \{z \in \mathbb{C} \,|\, |z| > R\}$ for a sufficiently large R, we have $P(V) \subset V$ and $V \subset I_P$. On the other hand, for every $z_0 \in I_P$, the orbit elements $P^n(z_0)$ tend to ∞, and hence there is an N such that $P^N(z_0) \in V$. Thus we conclude that

$$I_P = \bigcup_{n=1}^{\infty} (P^n)^{-1}(V), \tag{1.1}$$

which in particular implies that I_P is an open set and that K_P is a bounded closed set.

Finally, if there exists a connected component of $(P^n)^{-1}(V)$ disjoint from V, then it should be bounded, for $(P^n)^{-1}(V)$ contains V. This contradicts the maximal principle, and hence we conclude that $(P^n)^{-1}(V)$ is a domain containing V. Thus the above equation shows the assertion. □

The equation (1.1) also implies the following.

Theorem 1.1.2 (Complete invariance) *Let $P(z)$ be a polynomial of degree not less than 2. Then the sets I_P, K_P and J_P are completely invariant under the action by P, i.e., letting E be one of these sets, we have*

$$P(E) = P^{-1}(E) = E.$$

Proof By (1.1), we see that $P^{-1}(I_P) = I_P$. Since $P((P^n)^{-1}(V)) = (P^{n-1})^{-1}(V)$, we also have that $P(I_P) = I_P$. Thus I_P is completely invariant, and so is the complement K_P. Since P is a continuous open map, J_P is also completely invariant. □

Theorem 1.1.2 implies that every point of K_P has a bounded orbit. Hence we can characterize K_P as the set of points with bounded orbits.

Definition A closed set E in $\widehat{\mathbb{C}}$ is called *perfect* if E has no isolated points.

A domain D in $\widehat{\mathbb{C}}$ is *simply connected* if the complement $\widehat{\mathbb{C}} - D$ is connected.

Theorem 1.1.3 *Let $P(z)$ be a polynomial of degree not less than* 2*. Then the filled Julia set K_P is a perfect set.*

Moreover, every connected component of the interior of K_P is simply connected.

Proof Suppose that there is an isolated point z_0 of K_P. Then we can draw a simple closed curve C in I_P such that the intersection of the interior W of C in $\mathbb{C}$ and K_P consists only of the point z_0.

Now take such a V as in the proof of Theorem 1.1.1. Then by (1.1), we can find N such that $P^N(C) \subset V$. Since $P^N(z_0)$ is not contained in V, $P^N(W)$ should contain $\mathbb{C} - V$. Then complete invariance of K_P means that $K_P = \{z_0\}$. In particular, $P^{-1}(z_0) = \{z_0\}$, and hence $P(z)$ should have the form $a(z - z_0)^k + z_0$. But then z_0 should belong to the interior of K_P, which is a contradiction.

Next, if there were a connected component of the interior of K_P which were not simply connected, then $I_P \cup J_P = \overline{I_P}$ would not be connected, which contradicts Theorem 1.1.1. □

Remark For every $z \in I_P$, the closure of $\bigcup_{n=1}^{\infty}(P^n)^{-1}(z)$ contains J_P, and for every $z \in J_P$, we have

$$\overline{\bigcup_{n=1}^{\infty}(P^n)^{-1}(z)} = J_P.$$

These facts follows from the non-normality of $\{P^n\}$ on the Julia set (see Theorem 2.3.4), and give one of the standard ways to draw the Julia set.

Definition The solutions of $P'(z) = 0$ are called *critical points* of $P(z)$. We denote by C_P the set of all critical points of $P(z)$, and call it the *critical set* of

$P(z)$:

$$C_P = \{z \in \mathbb{C} \mid P'(z) = 0\}.$$

Now whether $I_P \cup \{\infty\}$ is simply connected or not depends on the location of the critical set.

Theorem 1.1.4 *Let $P(z)$ be a polynomial of degree not less than 2. Then the filled Julia set K_P is connected if and only if $I_P \cap C_P = \emptyset$.*

Proof As in the proof of Theorem 1.1.1, fix R so that $V = \{|z| > R\}$ is contained in I_P.

First, suppose that $I_P \cap C_P$ is the empty set. Then P^n gives a smooth k^n-sheeted covering of $V_n = (P^n)^{-1}(V)$ onto V, where k is the degree of P. Since $V \cup \{\infty\}$ is simply connected, so is every $V_n \cup \{\infty\}$. Hence the union $I_P \cup \{\infty\}$ of increasing domains $V_n \cup \{\infty\}$ is simply connected.

Next suppose that $I_P \cap C_P$ is not empty, and let N be the minimum of n such that $V_n \cap C_P \neq \emptyset$. Then applying Lemma 1.1.5 below to the proper holomorphic map $P : (V_N - \overline{V_{N-1}}) \to (V_{N-1} - \overline{V_{N-2}})$, we see that V_N is not simply connected. Hence K_P is not connected. □

Lemma 1.1.5 (Riemann–Hurwitz formula for domains) *Let D_1 and D_2 be domains in $\widehat{\mathbb{C}}$ whose boundaries consist of a finite number of simple closed curves. Let $f(z)$ be a proper holomorphic map of D_1 onto D_2. Then:*

(i) *Every $z \in D_2$ has the same number k of preimages including multiplicity.*
(ii) *Denote by N the number of critical points of f in D_1 including multiplicity. Then*

$$(2 - d_1) = k(2 - d_2) - N,$$

where d_j is the number the boundary components of D_j.

In particular, when both the D_j are simply connected, f has at most $k - 1$ critical points.

The number k in the lemma is called the degree of f.

Proof The first assertion follows since f is a proper open map. The second assertion follows by applying Euler's formula to suitable triangulations of D_j. □

A typical example of disconnected K_P is those such as in Figure 1.3. We say that a compact set E is *totally disconnected* if every connected component of E consists of a single point.

Theorem 1.1.6 *Let $P(z)$ be a polynomial of degree not less than 2. If the set I_P of escaping points contains C_P, then K_P is totally disconnected. In particular, $K_P = J_P$.*

Proof The assumption implies that the forward orbit

$$C^+(P) = \bigcup_{n=1}^{\infty} P^n(C_P)$$

of C_P accumulates only to ∞. Hence there is a simple closed curve C such that the interior W of C contains K_P and the exterior $V = \mathbb{C} - \overline{W}$ contains $C^+(P)$. As in the proof of Theorem 1.1.1, we see that $I_P = \bigcup_{n=1}^{\infty}(P^n)^{-1}(V)$, and hence $P^N(V \cup C_P) \subset V$ for a sufficiently large N. This implies that $(P^N)^{-1}$ has k^N single valued holomorphic branches on W, which we denote by $\{h_j\}_{j=1}^{k^N}$. Then the elements of $\mathcal{U}_1 = \{U_j = h_j(W)\}$ are pairwise disjoint, and each U_j contains k^N images $\mathcal{U}_2 = \{U_{jk} = h_j(U_k)\}_{k=1}^{k^N}$ of them. Inductively, we can define the set $\mathcal{U}_\ell$ of domains for every positive integer ℓ, and $K_P = \bigcap_{\ell=1}^{\infty} \mathcal{U}_\ell$.

Thus it suffices to show that, for any sequence of domains $U(\ell) \in \mathcal{U}_\ell$ such that $U(\ell) \subset U(\ell - 1)$, the intersection $E = \bigcap_{\ell=1}^{\infty} U(\ell)$ consists of a single point. For this purpose, let $R_j = \{1 < |z| < m_j\}$ be the ring domain conformally equivalent to $W - \overline{U_j}$ for every j. We call the quantity $\log m_j$ the *modulus* of R_j. Set $m = \min_j m_j$. Then since every $U(\ell - 1) - \overline{U(\ell)}$ is conformally equivalent to one of the R_j, it contains a ring domain conformally equivalent to $\{1 < |z| < m\}$.

Here the crucial fact is the following subadditivity of the moduli of ring domains. Suppose that a ring domain R contains disjoint ring domains S_1 and S_2 essentially, i.e., each S_j separates $\widehat{\mathbb{C}} - R$. Let $\log m$ and $\log m_j$ be the moduli of R and S_j. Then

$$\log m \geq \log m_1 + \log m_2.$$

This is a direct consequence of the extremal length characterization of the modulus.

Thus $W - \overline{U(\ell)}$ contains essentially a ring domain with modulus $\ell \log m$. Hence $W - E$ contains essentially a ring domain with arbitrarily large modulus. If E contains more than two points, this is impossible. □

Remark In the case that I_P contains C_P, the proof of Theorem 1.1.6 indicates how to describe the dynamics on $K_P = J_P$ symbolically.

Take the *space of sequences of k symbols*

$$\Sigma_k = \{0, 1, \ldots, k-1\}^{\mathbb{Z}^+} = \{(m_0, m_1, \ldots) \mid m_j \in \{0, 1, \ldots, k-1\}\},$$

where $\mathbb{Z}^+$ is the set of non-negative integers. We equip Σ_k with the product topology. In other words, we take the topology so that the injection

$$(m_0, m_1, \ldots) \to \sum_{j=0}^{\infty} 2m_j(2k-1)^{-j-1}$$

becomes a homeomorphism into $\mathbb{R}$.

Then letting k be the degree of P, we can identify J_P with Σ_k so that the dynamics of P on J_P equals the canonical *shift operator*

$$\sigma((m_0, m_1, \ldots)) = (m_1, m_2, \ldots)$$

on Σ_k. Further see §3.2.1 and §7.4.

1.2 Local behavior near a fixed point

1.2.1 Schröder equation for λ with $|\lambda| < 1$

The behavior of the dynamics of a polynomial $P(z)$ can be fairly well understood at least near a fixed point of $P(z)$. We gather in this subsection classical and basic facts on local behavior of a general holomorphic function near a fixed point. First, ∞ is a fixed point of $P(z)$ considered as a holomorphic endomorphism of $\widehat{\mathbb{C}}$, and has a distinguished character.

Theorem 1.2.1 (Böttcher) *Fix a monic polynomial $P(z)$ of degree $k \geq 2$. Then for a sufficiently large R, there exists a conformal map $\phi(z)$ of $V = \{|z| > R\}$ into $\mathbb{C}$ which has a form*

$$\phi(z) = bz + b_0 + \frac{b_1}{z} + \cdots$$

and satisfies

$$\phi(P(z)) = \{\phi(z)\}^k. \tag{1.2}$$

Moreover, such a $\phi(z)$ is uniquely determined up to multiplication by a $(k-1)$th root of 1.

We call such a function $\phi(z)$ as in the above theorem a *Böttcher function* for P at ∞. Also see §7.3.

Proof We may assume that $P(V) \subset V$, and hence there exists a single valued holomorphic branch $\psi(z)$ of

$$\log \frac{P(z)}{z^k}$$

such that $\lim_{z \to \infty} \psi(z) = 0$. Since $P(z) = z^k \exp \psi(z)$, we see inductively that

$$P^n(z) = z^{k^n} \exp\{k^{n-1}\psi(z) + \cdots + \psi(P^{n-1}(z))\}.$$

Hence we can take

$$\phi_n(z) = z \exp\left\{\frac{1}{k}\psi(z) + \cdots + \frac{1}{k^n}\psi(P^{n-1}(z))\right\}$$

as a branch of the k^nth root of $P^n(z)$.

Now we may assume that $\psi(z)$ is bounded on V. Then

$$\sum_{j=1}^{\infty} k^{-j}\psi(P^{j-1}(z))$$

converges uniformly on V, and hence so does the sequence $\{\phi_n(z)\}$. The limit $\phi(z)$ of $\phi_n(z)$ is holomorphic on V and satisfies equation (1.2), for $\phi_n(P(z)) = \{\phi_{n+1}(z)\}^k$. Since $\phi(z)$ is injective near ∞, we conclude with the first assertion. The second assertion follows from (1.2). □

In general, a Böttcher function cannot be continued analytically to the whole of I_P. Also see the remark below. But we have the following.

Proposition 1.2.2 *Let $\phi(z)$ be a Böttcher function for a monic polynomial $P(z)$ of degree $k \geq 2$, and set*

$$g(z) = \log|\phi(z)|.$$

Then $g(z)$ can be extended to a harmonic function on the whole of I_P.

Moreover, set $g(z) = 0$ for every $z \in K_P$. Then $g(z)$ is a continuous subharmonic function on $\mathbb{C}$.

The function $g(z)$ in the proposition is the Green function on $I_P \cup \{\infty\}$ with pole ∞, which will be defined in §1.4.3.

Proof Since $I_P = \bigcup_{n=1}^{\infty}(P^n)^{-1}(V)$, we extend $g(z)$ to the whole of I_P by setting

$$g(z) = \frac{1}{k^n} g(P^n(z))$$

if $z \in (P^n)^{-1}(V)$. This $g(z)$ is well defined by (1.2) and clearly harmonic.

Next, set $A = \max_{z \in \overline{V} - P(V)} g(z)$, and take any neighborhood W of K_P. Then

$$(P^N)^{-1}(V) \cap W = \varnothing$$

implies that

$$g(z) \leq k^{-N} A \qquad (z \in W).$$

Hence we have the second assertion. □

Remark It is known as Sibony's theorem that this continuous subharmonic function $g(z)$ on $\mathbb{C}$ is actually Hölder continuous.

When C_P is contained in K_P, then any Böttcher function $\phi(z)$ can be extended to a conformal map of I_P onto $\{|z| > 1\}$. In general, set

$$M = \max\{g(z_0) \mid z_0 \in C_P\};$$

then we can extend $\phi(z)$ to a conformal map of $\{z \in \mathbb{C} \mid g(z) > M\}$.

Indeed, if $r > M$, then $\{g(z) = r\}$ is a simple closed curve, and hence we can find a (multi-valued) conjugate harmonic function $g^*(z)$ of $g(z)$ such that $G(z) = \exp(g(z)+ig^*(z))$ is single valued on $\{|z| > r\}$. Choosing the additive constant of $g^*(z)$ so that $G(z) = \phi(z)$ near ∞, we conclude with the assertion. In particular, if $C_P \subset K_P$, Proposition 1.2.2 implies that $G(z)$ is a conformal map of I_P onto $\{|z| > 1\}$.

Now we turn to the case of a finite fixed point, which we assume to be 0 in the rest of this section. We can generalize Böttcher's theorem as follows.

Theorem 1.2.3 *Let $f(z)$ be a holomorphic function in a neighborhood of the origin with the Taylor expansion*

$$f(z) = c_0 z^k + c_1 z^{k+1} + \cdots \qquad (c_0 \neq 0).$$

Then there exist a neighborhood U of the origin and a conformal map ϕ : $U \to \mathbb{C}$ fixing the origin and satisfying

$$\phi(f(z)) = \{\phi(z)\}^k.$$

Moreover, such a ϕ is uniquely determined up to multiplication by a $(k-1)$th root of 1.

Again, we call such a $\phi(z)$ a *Böttcher function* for f (at 0). The proof is the same as that of Theorem 1.2.1 and hence omitted.

Remark We can derive Theorem 1.2.1 from Theorem 1.2.3 by taking the conjugate by $T(z) = 1/z$.

Note that Theorem 1.2.3 treats the case of a fixed point in C_P for a monic polynomial P. When a given fixed point does not belong to C_P, or more generally, when a holomorphic function $f(z)$ in a neighborhood of the origin fixes the origin and satisfies $f'(0) \neq 0$, we have the Taylor expansion

$$f(z) = \lambda z + c_2 z^2 + \cdots \quad (\lambda \neq 0).$$

We call this λ the *multiplier* of f at the fixed point 0. The case that $\lambda = 0$ has been treated in Theorem 1.2.3.

Theorem 1.2.4 (Koenigs) *Suppose that a function $f(z)$ holomorphic near the origin has the Taylor expansion*

$$f(z) = \lambda z + c_2 z^2 + \cdots \quad (0 < |\lambda| < 1).$$

Then there are a neighborhood U of the origin and a conformal map $\phi(z)$ of U such that $\phi(0) = 0$ and that

$$\phi \circ f(z) = \lambda \phi(z) \quad (z \in U). \tag{1.3}$$

Here, $\phi(z)$ is unique up to multiplicative constants.

Definition We call the equation (1.3) the *Schröder equation* for f. When the Schröder equation has a solution, then we say that f is *linearizable* at 0.

Proof of theorem Set

$$\phi_n(z) = \lambda^{-n} f^n(z)$$

for every n and we have

$$\phi_n \circ f(z) = \lambda^{-n} f^{n+1}(z) = \lambda \phi_{n+1}(z).$$

Fix an η such that $\eta^2 < |\lambda| < \eta < 1$, and we can choose a positive δ so small that

$$|f(z)| < \eta |z| \quad \text{and} \quad |f(z) - \lambda z| \leq c|z|^2$$

on $\{|z| < \delta\}$, where $c = |c_2| + 1$. Further, set $\rho = \eta^2/|\lambda|$ (< 1), and we obtain that

$$\begin{aligned} |\phi_{n+1}(z) - \phi_n(z)| &= |\lambda^{-n-1}\{f(f^n(z)) - \lambda f^n(z)\}| \leq |\lambda|^{-n-1} c |f^n(z)|^2 \\ &\leq c\rho^n |\lambda|^{-1} |z|^2. \end{aligned}$$

Thus $\phi_n(z)$ converges to

$$\phi(z) = \phi_1(z) + \sum_{n=1}^{\infty} \{\phi_{n+1}(z) - \phi_n(z)\}$$

uniformly on $\{|z| < \delta\}$, and it is clear that $\phi(z)$ satisfies the Schröder equation. The uniqueness follows by comparing the coefficients. □

Remark When $|\lambda| > 1$, we can show the same assertion by considering the inverse function.

1.2.2 Schröder equation for λ with $|\lambda| = 1$

In the case that $|\lambda| = 1$, the situation is very complicated. We gather several basic facts without proofs, which will be found in the books referred to before.

First, for almost every such λ, the corresponding Schröder equation has a solution.

Theorem 1.2.5 (Siegel) *For almost every λ on the unit circle $\{|z| = 1\}$, every function $f(z)$ represented by a convergent power series*

$$f(z) = \lambda z + c_2 z^2 + \cdots$$

is linearizable at 0, i.e., there exist a neighborhood U of the origin and a conformal map $\phi(z)$ such that

$$\phi(f(z)) = \lambda\phi(z) \quad (z \in U).$$

Remark Actually, Siegel showed that, if $t \in \mathbb{R} - \mathbb{Q}$ and is *Diophantine*, i.e., there are constants $c > 0$ and $b < +\infty$ such that

$$\left| t - \frac{p}{q} \right| \geq \frac{c}{q^b}$$

for every integer p and positive integer q, then $f(z)$ with $\lambda = e^{2\pi i t}$ is linearizable at 0.

Here, if we take $b > 2$, then the total length of the set of non-Diophantine numbers in $[0, 1]$ is not greater than $\sum_{q=N}^{\infty} 2cq^{1-b}$, which is the total length of intervals $\{t \in [0, 1] \,|\, |t - (p/q)| < cq^{-b}\}$ with $q \geq N$ for every N, and which tends to 0 as $N \to +\infty$. Hence almost every number is Diophantine.

On the other hand, at every *rationally indifferent* fixed point, that is every fixed point with a multiplier $\lambda = e^{2\pi i t}$ with $t \in \mathbb{Q}$, the function is not linearizable at 0. In particular, λs corresponding to the non-linearizable case are dense in the unit circle.

Theorem 1.2.6 (Petal theorem of Leau and Fatou) *For every function $f(z)$ represented by a convergent power series*

$$f(z) = z + c_{p+1}z^{p+1} + \cdots \quad (c_{p+1} \neq 0)$$

at 0, *there are* $2p$ *domains* $\{U_j, V_j\}_{j=1}^{p}$ *such that*

(i) *for* $U = \bigcup_{j=1}^{p} U_j$,

$$f(\overline{U}) \subset U \cup \{0\}, \quad \bigcap_{n=1}^{\infty} f^n(\overline{U}) = \{0\},$$

and for $V = \bigcup_{k=1}^{p} V_k$, $(f|_V)^{-1}$ *is univalent on* V, *and*

$$(f|_V)^{-1}(\overline{V}) \subset V \cup \{0\}, \quad \bigcap_{n=1}^{\infty}(f|_V)^{-n}(\overline{V}) = \{0\},$$

(ii) *domains* $\{U_j\}_{j=1}^{p}$ *are pairwise disjoint and so are* $\{V_j\}_{j=1}^{p}$, *and*

(iii) *the domain* $\{0\} \cup U \cup V$ *is a neighborhood of* 0.

Definition We call a component of U satisfying condition (i) an *attracting petal* of $f(z)$ at the origin. Also, an attracting petal of $(f|_V)^{-1}(z)$ such as V_k in Theorem 1.2.6 is called a *repelling petal* of f.

Remark When f has the form

$$f(z) = \lambda z + c_{p+1}z^{p+1} + \cdots \quad (c_{p+1} \neq 0)$$

with a primitive mth root of unity λ, consider $f^m(z)$, and we have a similar assertion as above. In particular, $f(z)$ has a family of attracting petals, divides it into invariant subfamilies and permutes the elements of each subfamily cyclically.

Thus, when $\lambda = e^{2\pi i t}$ with a rational t, the Schröder equation for f at the origin has no solution.

Example 1.2.1 *Let*

$$f(z) = z + c_2 z^2 + \cdots \qquad (c_2 \neq 0)$$

and define $T(z) = -1/(c_2 z)$. *Further we set*

$$\begin{aligned} \Omega_+ &= \{|\operatorname{Im} z| > 2c - \operatorname{Re} z\}, \\ \Omega_- &= \{|\operatorname{Im} z| > 2c + \operatorname{Re} z\}, \end{aligned}$$

with a sufficiently large c. *Then* $T(\Omega_+)$ *is an attracting petal of* $f(z)$ *at the origin and* $T(\Omega_-)$ *is a repelling petal of* $f(z)$ *at the origin.*

Verification By taking a conjugate by $T(z)$, we have

$$g(z) = T \circ f \circ T^{-1}(z) = z + 1 + \frac{a_1}{z} + \cdots.$$

We choose c so that

$$|g(z) - z - 1| < 1/\sqrt{2}$$

if $|z| \geq c$. Then $g(z)$ is nearly a translation near ∞. Hence a simple computation shows that $\overline{g(\Omega_+)} \subset \Omega_+ \cup \{\infty\}$, and $T(\Omega_+)$ is an attracting petal of $f(z)$ at the origin. Similarly, we see that $T(\Omega_-)$ is a repelling petal of $f(z)$ at the origin.

As examples of quadratic polynomials $z^2 + c$ having attracting petals, we show as Figure 1.5 the cases that (*a*) $c = 1/4$ and (*b*) $c = 0.31 \cdots + i0.03 \cdots$.

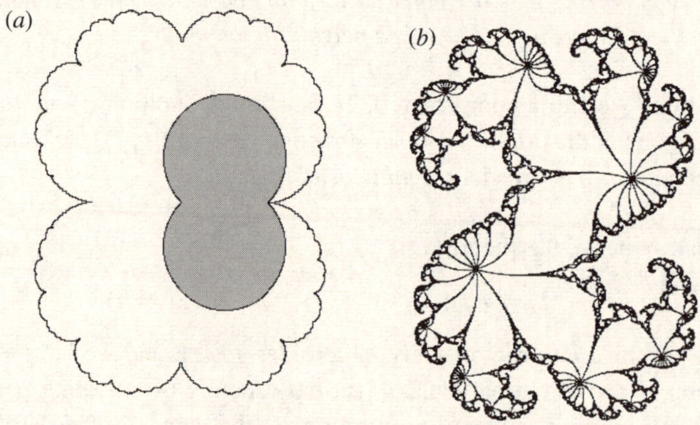

Fig. 1.5. Julia sets with parabolic basins

Remark The explanation in the previous example indicates that, for any attracting petal U of $f(z)$, there exists a conformal map $\phi(z)$ of U which satisfies the *Abel equation*

$$\phi(f(z)) = \phi(z) + 1 \quad (z \in U).$$

We call such a $\phi(z)$ a *Fatou function* for the attracting petal U.

As another typical case where the Schröder equation has no solutions, we cite the following.

Theorem 1.2.7 (Cremer) *For every* λ *with* $|\lambda| = 1$*, set*

$$Q_\lambda(z) = \lambda z + z^2.$$

Then the set of λ *such that the Schröder equation for* Q_λ *has no solutions is a generic set, that is one which can be represented as the intersection of at most countably many open dense subsets, in* $\{|\lambda| = 1\}$.

Definition For a function $f(z)$ holomorphic near z_0 which has the Taylor expansion

$$f(z) = z_0 + \lambda(z - z_0) + c_2(z - z_0)^2 + \cdots \quad (\lambda = e^{2\pi i t};\ t \in \mathbb{R} - \mathbb{Q}),$$

z_0 is called an *irrationally indifferent* fixed point. Further, we say that z_0 is a *Cremer point* of $f(z)$ if the Schröder equation for $f(z)$ at z_0 has no solutions.

Corollary *The set of* λ *such that* Q_λ *has a Cremer point is dense in the unit circle.*

Remark Suppose that 0 is a fixed point of a polynomial P with the multiplier λ with $|\lambda| = 1$. Then it is easy to see that, if P is linearizable at 0, then 0 does not belong to J_P. The converse is also true. See Theorem 2.1.9.

Finally, we note the relation between Cremer points and the continued fractional expansion in the case of quadratic polynomials.

Definition Let $t \in \mathbb{R} - \mathbb{Q}$, and consider the continued fractional expansion

$$t = \cfrac{1}{a_1 + \cfrac{1}{a_2 + \cfrac{1}{a_3 + \ddots}}},$$

which induces the sequence $\{q_n\}_{n=1}^{\infty}$ of the denominators of the rational approximation of t, i.e. $\{q_n\}$ are determined by the following equations inductively:

$$q_0 = 1, q_1 = a_1, \ldots, q_{n+1} = q_{n-1} + q_n a_{n+1}, \ldots.$$

We say that t is a *Brjuno number* if

$$\sum_{n=1}^{\infty} \frac{1}{q_n} \log q_{n+1} < \infty.$$

Theorem 1.2.8 (Brjuno–Yoccoz) *Let $\lambda = e^{2\pi i t}$ with $t \in \mathbb{R} - \mathbb{Q}$, and $Q_\lambda(z) = \lambda z + z^2$. Then the Schröder equation for Q_λ has a solution near the origin if and only if t is a Brjuno number.*

In general, t is a Brjuno number if and only if every function $f(z)$ represented by a convergent power series

$$f(z) = \lambda z + c_2 z^2 + \cdots$$

is linearizable at 0. For the details, see Yoccoz (1996).

1.3 Quadratic polynomials and the Mandelbrot set

In the case of quadratic polynomials $P_c(z) = z^2 + c$, Theorem 1.1.4 means that K_{P_c} is connected if and only if the orbit of 0 is bounded. So we define

$$\mathcal{M} = \{c \in \mathbb{C} \mid \{P_c^n(0)\}_{n=1}^{\infty} \text{ is a bounded sequence}\},$$

and call it the *Mandelbrot set*. See Figure 1.6.

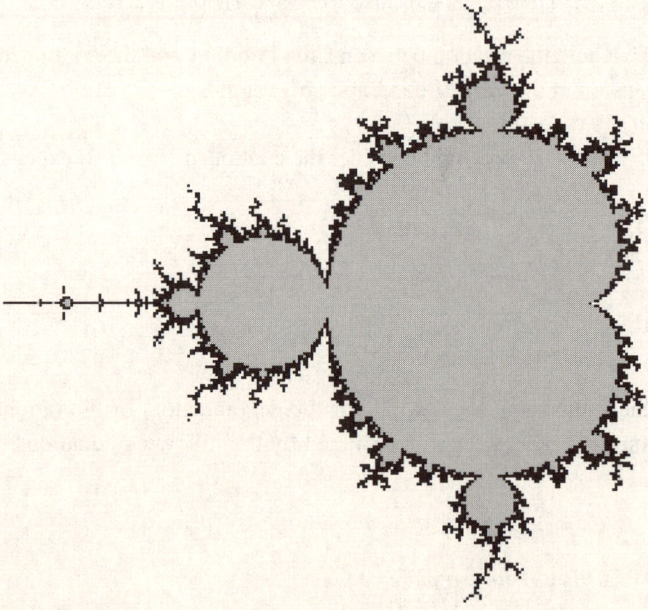

Fig. 1.6. Mandelbrot set

Thus, the Mandelbrot set consists of such c that K_{P_c} is connected. Moreover, if c does not belong to $\mathcal{M}$, then K_{P_c} is totally disconnected by Theorem 1.1.6.

Theorem 1.3.1 *The Mandelbrot set $\mathcal{M}$ is a compact set contained in* $\{|z| \leq 2\}$, *and $\mathbb{C} - \mathcal{M}$ is connected.*

The intersection of $\mathcal{M}$ with $\mathbb{R}$ equals $[-2, 1/4]$.

Proof When $|c| > 2$, $|P_c^2(0)| \geq |c|(|c| - 1)$ and inductively we have

$$|P_c^{n+1}(0)| \geq (|c|(|c|-1)^{2^{n-2}})^2 - |c| \geq |c|(|c|-1)^{2^{n-1}},$$

which tends to $+\infty$ as $n \to \infty$. Hence $\mathcal{M} \subset \{|c| \leq 2\}$.

Further, even if $|c| \leq 2$, we see that, when $|P_c^n(0)| > 2$ for some n, then c does not belong to $\mathcal{M}$. This implies that $c \in \mathcal{M}$ if and only if $|P_c^n(0)| \leq 2$ for every n. Hence $\mathcal{M}$ is a closed set in $\{|z| \leq 2\}$.

Further, apply the maximal principle to each $P_c^n(0)$, as a holomorphic function of c, and we conclude that $\mathbb{C} - \mathcal{M}$ should be connected.

In the case of real c, suppose that $c > 1/4$. Then $P_c(x) = x$ has no real solutions, and hence real $P_c^n(0)$ cannot accumulate in $\mathbb{R}$. If $c < -2$, then clearly c does not belong to $\mathcal{M}$. Finally, when $-2 \leq c \leq 1/4$ it is easy to show that

$$|P_c^n(0)| \leq a$$

for every n, with the not smaller solution a of the equation $x^2 - x + c = 0$. □

Theorem 1.3.2 (Douady–Hubbard) *The Mandelbrot set $\mathcal{M}$ is connected.*

To prove this fundamental result, first recall that the Böttcher function $\phi_c(z)$ for $P_c(z)$ in the proof of Theorem 1.2.1 has the following representation.

$$\begin{aligned}\phi_c(z) &= \lim_{n\to\infty} \{P_c^n(z)\}^{2^{-n}} = z\prod_{n=1}^{\infty} \frac{P_c^n(z)^{2^{-n}}}{P_c^{n-1}(z)^{2^{-(n-1)}}} \\ &= z\prod_{n=0}^{\infty}\left(1 + \frac{c}{P_c^n(z)^2}\right)^{2^{-n-1}}.\end{aligned}$$

Lemma 1.3.3 *Let $g_c(z)$ be the extension of* $\log|\phi_c(z)|$ *to $\mathbb{C}$ as in Proposition 1.2.2. Then*

$$g_c(z) = \lim_{n\to\infty} \frac{1}{2^n} \log^+ |P_c^n(z)|, \tag{1.4}$$

where $\log^+|w| = \max\{\log|w|, 0\}$.

Here the convergence is uniform on compact sets of $\mathbb{C}^2 = \mathbb{C} \times \mathbb{C}$, and hence $g_c(z)$, considered as a function of (z, c), is continuous on $\mathbb{C}^2$.

Proof First, when $z \in I_{P_c}$, it is clear by equation (1.3) that (1.4) holds. And when $z \in K_{P_c}$, the right hand side of (1.4) equals 0, for $\{P_c^n(z)\}$ is bounded. Hence (1.4) is valid.

Next, to show the continuity, note that each term $\frac{1}{2^n} \log^+ |P_c^n(z)|$ is a continuous function of (z, c). Assume that $|c| \le A$. Then

$$\left|\log^+ |P_c(z)| - 2\log^+ |z|\right| \le |P_c(z) - z^2| = |c| \le A$$

for every $z \in \mathbb{C}$. Hence

$$\left|\frac{1}{2^n} \log^+ |P_c^n(z)| - \frac{1}{2^{n-1}} \log^+ |P_c^{n-1}(z)|\right| \le \frac{A}{2^n},$$

which implies that the convergence of (1.5) is uniform on $\{(z, c) \mid z \in \mathbb{C}, |c| \le A\}$. □

Proof of Theorem 1.3.2 The second statement of Lemma 1.3.3 means that

$$\Omega = \{(z, c) \in \mathbb{C}^2 \mid g_c(z) > g_c(0)\}$$

is a domain, and that $\phi_c(z)$ is a holomorphic function of two variables z and c on Ω (cf. §6.1). In particular,

$$\Phi(c) = \phi_c(c) = c \prod_{n=0}^{\infty} \left(1 + \frac{c}{P_c^n(c)^2}\right)^{2^{-n-1}}$$

is a holomorphic function of c on $\mathbb{C} - \mathcal{M}$.

Moreover, we see that $\Phi(c)$ has a single simple pole at ∞. Since $\log |\Phi(c)| = g_c(c)$, $|\Phi(c)|$ converges to 1 as c tends to the boundary of $\mathbb{C} - \mathcal{M}$. Hence the argument principle implies that $\Phi(c)$ is a conformal map of $\mathbb{C} - \mathcal{M}$ onto $\mathbb{C} - \overline{\Delta}$. Thus $\widehat{\mathbb{C}} - \mathcal{M}$ is simply connected, which implies the assertion. □

Remark We call a conformal map of a simply connected domain G onto the unit disk Δ a *Riemann map* of G.

For instance, $1/\Phi$ with the above Φ is a Riemann map of $\widehat{\mathbb{C}} - \mathcal{M}$. A famous unsolved problems about the Mandelbrot set is whether or not the inverse $(1/\Phi)^{-1}$ of the Riemann map can be extended to a continuous map of $\overline{\Delta}$. By a classical theorem of Carathéodory (Theorem 4.4.12), this problem is equivalent to the following one of the central conjectures.

MLC conjecture The Mandelbrot set is locally connected.

Note that there exist quadratic polynomials with Julia sets not locally connected. For instance, the Julia set should be very complicated near a Cremer point, and the following fact can be imagined, though we omit the proof.

Theorem 1.3.4 (Douady–Sullivan) *If a quadratic polynomial P_c has a Cremer point, then the Julia set J_{P_c} is not locally connected.*

Now, the interior of $\mathcal{M}$ consists of countably many connected components. The component D_1 containing the origin is called the *main cardioid*, and corresponds to those quadratic polynomials having an attracting fixed point.

Example 1.3.1 *Set*

$$D_1 = \left\{ c = \frac{\lambda}{2} - \frac{\lambda^2}{4} \middle| |\lambda| < 1 \right\}.$$

Then P_c with $c = \frac{\lambda}{2} - \frac{\lambda^2}{4} \in D_1$ has a fixed point z_c with $P_c'(z_c) = \lambda$. Further, J_{P_c} is a simple closed curve for every $c \in D_1$.

Verification The last statement will be shown by the corollary to Theorem 1.5.1. Here, $c = 0$ corresponds to $\lambda = 0$, and is called the center of D_1.

Example 1.3.2 *Set*

$$D_2 = \{c \mid |4c + 4| < 1\}.$$

Then D_2 corresponds to those P_c such that P_c^2 has a fixed point z_c with $|(P_c^2)'(z_c)| < 1$ which satisfies $P_c(z_c) \neq z_c$.

The set D_2 is the connected component of the interior of $\mathcal{M}$, whose center is -1.

Definition For a general holomorphic map f of a domain into itself, we say that α is a *periodic point* of f if there is a positive integer N such that $f^N(\alpha) = \alpha$. Also we call the minimum N_0 of such N the *period* of α, and the set $\{\alpha, f(\alpha), \ldots, f^{N_0-1}(\alpha)\}$ the *cycle* of α.

We say that α is *preperiodic* if α is not periodic but there is a positive m such that $f^m(\alpha)$ is periodic.

Hence in our terminology, *periodic points are not preperiodic*, and a fixed point is a periodic point of period 1.

In the examples above, the critical point 0 of P_0 is a fixed point, and the critical point 0 of P_{-1} is a periodic point of period 2. For the solutions c of

$$c^3 + 2c^2 + c + 1 = 0,$$

0 is a periodic point of period 3 of P_c, and these c are the centers of pairwise disjoint connected components of the interior of $\mathcal{M}$. Inductively, we can determine components each of which corresponds to a quadratic polynomial having periodic critical point. We denote by E the set of the centers of all such components:

$$E = \{c \in \mathbb{C} \mid 0 \text{ is a periodic point of } P_c\}.$$

Theorem 1.3.5 *The set E of the centers are contained in the interior of $\mathcal{M}$, and the closure of E contains the boundary of $\mathcal{M}$.*

In particular, the interior of $\mathcal{M}$ is dense in $\mathcal{M}$.

Proof Fix a topological disk U not containing 0 such that $U \cap \partial\mathcal{M} \neq \varnothing$. Then we can take a single valued holomorphic branch $h(c)$ of $\sqrt{-c}$ on U.

Suppose that $\overline{E} \cap U = \varnothing$. Then every

$$f_n(c) = \frac{P_c^n(0)}{h(c)}$$

omits the values $0, 1$ and ∞. Hence by the three-point condition (Theorem 2.1.5), $\{f_n\}$ is a normal family. But for c not belonging to $\mathcal{M}$, $f_n(c)$ should tend to ∞, which is a contradiction. □

On the other hand, we do not know whether the union of all components containing points in E equals the interior of $\mathcal{M}$. This is a special version of the HD (hyperbolic dense) conjecture, which we will discuss in §1.4.2 (also see §4.4).

HD2 conjecture Every connected component of the interior of the Mandelbrot set contains a point of E.

Finally, since

$$\frac{1}{\Phi^{-1}(1/z)} \quad (z \in \Delta)$$

is a bounded holomorphic function on the unit disk, a classical theorem of Fatou implies that

$$\lim_{R \to 1} \Phi^{-1}(Re^{2\pi i\theta})$$

exists for almost every $\theta \in \mathbb{R}$.

Theorem 1.3.6 (Fatou) *Every bounded holomorphic function on Δ has a non-tangential limit almost everywhere on the boundary.*

In other words, the arc

$$\ell_\theta = \Phi^{-1}(\{Re^{2\pi i\theta} \mid R > 1\})$$

has a finite end point, which we denote by a_θ, for almost all θ. At least, this is the case when θ is a rational number. The arc ℓ_θ is called the *external ray* for θ. There may be several external rays ending at the same point a. In such a case, $\mathcal{M} - a$ is not connected. Further we know the following.

Theorem 1.3.7 *Let θ be a rational number. We write $\theta = p/q$ with mutually prime positive integers p and q.*

(i) *When q is odd, P_{a_θ} has a rationally indifferent periodic point. Moreover, 0 does not belong to the Julia set of P_{a_θ}.*
(ii) *When q is even, 0 is a preperiodic point of P_{a_θ} belonging to the Julia set, which in turn is a dendrite.*

We call such an $a_\theta \in \mathcal{M}$ as in case (ii) a *Misiurewicz point*. Conversely, all Misiurewicz points and all $a \in \mathcal{M}$ such that P_a has a rationally indifferent periodic point are end points of extremal rays ℓ_θ with rational θ.

Remark We can regard θ as an element of the space Σ_2 of sequences of two symbols $\{0, 1\}$ by representing $\theta = \sum_{j=0}^{\infty} \theta_j 2^{-j-1}$ and associating $(\theta_0, \theta_1, \ldots)$ with θ. Then θ is rational if and only if (θ_j) is either periodic or preperiodic with respect to the shift operator. More precisely, (θ_j) is periodic if and only if q is odd.

When θ is rational, the Julia set of P_{a_θ} is always locally connected, for P_{a_θ} either is subhyperbolic (cf. §4.4) or has a parabolic cycle. An argument similar to that in the proof of Theorem 1.3.5 shows that Misiurewicz points are dense in the boundary of $\mathcal{M}$.

A remarkable fact here is that the local configuration of the Mandelbrot set near a Misiurewicz point a_θ is quite similar to that of the Julia set of P_{a_θ} near the same $a_\theta = P_{a_\theta}(0)$ (cf. Tan Lei 1990). Also see Figure 1.7, which gives another kind of local similarity.

Remark For polynomials of degree greater than 2, we can consider similar subsets to the Mandelbrot set. Such a set is called the connected locus. But the situation is more complicated.

When the filled Julia set is connected as in the case of P_c with $c \in \mathcal{M}$, the union of the set of escaping points and ∞ is simply connected. Hence, we can consider external rays in the set of escaping points similarly as above.

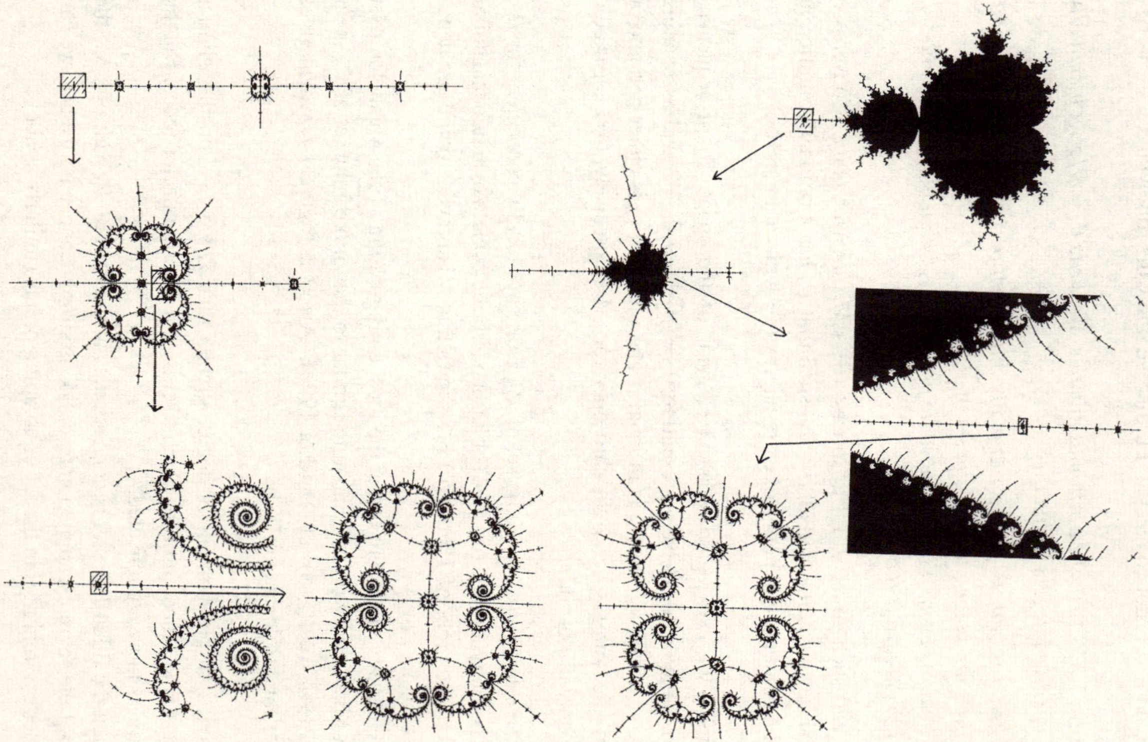

Fig. 1.7. Local similarity

1.4 Hausdorff dimension and capacity

1.4.1 Hausdorff measure

There are several ways to measure a fractal set quantitatively. One of them is the Hausdorff measure.

Definition Let a closed subset E of $\mathbb{C}$ be given. Fix a positive $d \in \mathbb{R}$. Then the d-dimensional *Hausdorff measure* $H_d(E)$ of E is defined as follows. For every covering $\mathcal{B} = \{U_j\}$ of E by countably many sets, we call the supremum of diameters $\operatorname{diam}(U_j)$ of all U_j the *size* of $\mathcal{B}$, and denote it by $\delta(\mathcal{B})$. Then set

$$H_d(E) = \lim_{\epsilon \to 0} \inf_{\{\mathcal{B}=\{U_j\} | \delta(\mathcal{B}) \leq \epsilon\}} \sum_{j=1}^{\infty} \operatorname{diam}(U_j)^d.$$

Next set

$$\dim(E) = \inf\{d > 0 \mid H_d(E) = 0\}$$

and call it the *Hausdorff dimension* of E.

Remark When $d > \dim(E)$, then $H_d(E) = 0$ by definition. And it is easy to see that, when $d < \dim(E)$, then $H_d(E) = +\infty$. Also note that, when $d_0 = \dim(E)$, $H_{d_0}(E)$ can take any non-negative value including $+\infty$.

For every monotonically increasing function $h(x)$ with $h(0) = 0$ on $[0, 1]$, we can define the Hausdorff h-measure similarly as above.

It is difficult to determine the Hausdorff dimension of a given set. A successful example is the canonical Cantor set defined below.

First, we give the definition of a general Cantor set.

Definition We call a topological space X a *Cantor set* if X is a perfect, totally disconnected compact Hausdorff space.

This definition is equivalent to that in §1.1.

Example 1.4.1 (The generalized ternary sets) *Take a sequence* $\{p_j\}_{j=1}^{\infty}$ *of numbers greater than* 1. *Set* $E(p_0) = [0, 1]$, *and for every positive n, we define* $E(p_0, \dots, p_n)$ *inductively as the union of all sets*

$$I_v - \left(x_v - \frac{p_n - 1}{2p_n} m_1(I_v), x_v + \frac{p_n - 1}{2p_n} m_1(I_v) \right),$$

where I_v *is any component of* $E(p_0, \dots, p_{n-1})$, x_v *is the midpoint of the interval* I_v *and* m_1 *denotes the one-dimensional Lebesgue measure.*

Then clearly, $E(p_0, \ldots, p_n)$ consists of 2^n components and the total length measured by the one-dimensional Lebesgue measure is $\prod_{j=1}^{n}(1/p_j)$. Set

$$E(\{p_j\}) = \bigcap_{n=1}^{\infty} E(p_0, \ldots, p_n),$$

and we have a totally disconnected perfect set which is a Cantor set in the sense of §1.1, with total length $\prod_{j=1}^{\infty}(1/p_j)$. Equipping it with the relative topology, we have a Cantor set in the above sense. We call $E(\{p_j\})$ the generalized ternary set of Cantor *with respect to $\{p_j\}$.*

When $p_j = 3/2$ for every j, we call the resulting Cantor set simply the ternary set of Cantor.

When all p_j are the same, say

$$p_j = 2^{(1-d)/d}$$

with $d \in (0, 1)$, then we see that the d-dimensional Hausdorff measure of the resulting Cantor set E_d is finite and positive, which implies that

$$\dim(E_d) = d.$$

In particular, the Hausdorff dimension of the ternary set of Cantor is $\log 2/\log 3$.

Remark For every d with $0 \leq d \leq 2$, we can construct a Cantor set with Hausdorff dimension d, which has positive area with respect to the Hausdorff measure H_d. See, for instance, Sario and Nakai (1970).

Also note that every Cantor set is homeomorphic to the ternary set of Cantor.

Next we gather some of the basic facts on the Hausdorff dimension of the Julia set of a polynomial.

Proposition 1.4.1 *Let P be a polynomial of degree not less than 2. Then the Hausdorff dimension of the Julia set J_P is positive.*

Theorem 1.4.2 (Hamilton) *Let P be as above. If J_P is a simple closed curve, but neither a circle nor a line, then the Hausdorff dimension of J_P is greater than 1 and hence J_P is a fractal set.*

Remark The above two results are valid for rational functions. Actually, Proposition 1.4.1 is a special case of Theorem 5.4.3.

Theorem 1.4.3 (Shishikura) *There is a generic set E in the boundary $\partial\mathcal{M}$ of the Mandelbrot set such that* $\dim J_{P_c} = 2$ *for every $c \in E$.*

Remark On the other hand, we will show as the corollary to Theorem 5.4.9 that for every P_c with c in a component of the interior of $\mathcal{M}$ containing a point of E in Theorem 1.3.5,

$$\dim(J_{P_c}) < 2.$$

In the case of entire functions discussed in Chapter 3, the Julia sets with positive area are not exceptional. But in the case of quadratic polynomials, for instance, we may pose the following.

JA2 conjecture The Julia set of any quadratic polynomial has vanishing area.

Finally, Shishikura also showed the following.

Theorem 1.4.4 (Shishikura) *The Hausdorff dimension of the boundary of the Mandelbrot set is* 2.

Here, we cite the following.

MA conjecture The area of the boundary of the Mandelbrot set is zero:

$$H_2(\partial\mathcal{M}) = 0.$$

1.4.2 Hyperbolicity

The last remark in the previous subsection indicates a quadratic polynomial corresponding to a point in a component of the interior of $\mathcal{M}$ containing a point of E has some tame character. The general case of rational functions and polynomial automorphisms of $\mathbb{C}^n$ $(n \geq 2)$ will be discussed in §4.4 and §9.3.

Definition Let $P(z)$ be a polynomial of degree not less than 2. We say that $P(z)$ is *hyperbolic* if

$$J_P \cap \overline{C^+(P)} = \emptyset,$$

where

$$C^+(P) = \bigcup_{n=1}^{\infty} P^n(C_P),$$

and $PC_P = \overline{C^+(P)}$ is called the *postcritical set* of P.

Every P_c with c in a component of the interior of $\mathcal{M}$ containing a point of E is a typical example of a hyperbolic polynomial. Since $\dim(J_{P_c}) < 2$ for such a P_c, $m_2(J_{P_c}) = 0$. This fact can be shown rather easily by using some of the basic facts in Chapter 2. For the sake of convenience, we include the proof.

Lemma 1.4.5 *Let $P(z)$ be a hyperbolic polynomial. Then*

$$\lim_{n\to\infty}\left(\min_{z\in J_P}|(P^n)'(z)|\right)=+\infty.$$

Proof If not, we have a subsequence $\{n(m)\}_{m=1}^{\infty}$ and a sequence $\{z_m\}_{m=1}^{\infty}$ of points in J_P such that the $(P^{n(m)})'(z_m)$ are bounded. Further, we may assume that $w = \lim_{m\to\infty} P^{n(m)}(z_m)$ exists. Consider a simply connected domain U containing w and disjoint from $C^+(P)$. Then every branch of any $(P^n)^{-1}$ is single valued and holomorphic on U. We denote by $\mathcal{F}$ the set of all such branches. Then by Lemma 2.4.3, $\mathcal{F}$ is a normal family, and since $w \in J_P$, every limit function of a sequence in $\mathcal{F}$ is a constant by Lemma 2.4.4.

Now, let $s_m(z) \in \mathcal{F}$ be the branch of $(P^{n(m)})^{-1}$ such that

$$s_m(P^{n(m)}(z_m)) = z_m,$$

and $s(z)$ be a limit function of a subsequence of $\{s_m(z)\}$, which should be a constant. But the assumption implies that

$$|s'(w)| = \lim_{m\to\infty}\frac{1}{|(P^{n(m)})'(z_m)|} \neq 0,$$

which is a contradiction. □

Theorem 1.4.6 *For every hyperbolic polynomial $P(z)$, the Julia set J_P has vanishing area.*

Proof Considering some power P^N instead of P, we may assume by the preceding lemma that there is a constant $A > 1$ satisfying

$$|P'(z)| > A \quad (z \in J_P).$$

Here we use the fact that $J_{P^N} = J_P$. In the case of polynomials, this is clear because every orbit $\{P^n(z)\}$ of a point z is either bounded or escaping.

Then, since $P^{-1}(J_P) = J_P$ and P^{-1} is (locally well defined and) uniformly contracting on J_P, we can construct an open set V containing J_P such that $P^{-1}(V) \subset V$, $V \cap C^+(P) = \emptyset$ and

$$|P'(z)| \geq A \quad (z \in V).$$

Hence, on any sufficiently small disk B in V, every single valued branch of $(P^n)^{-1}$ satisfies

$$|(P^n)^{-1}(z) - (P^n)^{-1}(w)| \leq \frac{1}{A^n}|z - w|$$

for every $z, w \in B$. Since $1/P'$ is bounded on V, the Cauchy estimate implies that

$$\frac{((P^n)^{-1})'(w)}{((P^n)^{-1})'(z)} = \prod_{j=1}^{n}(1 + A^{-j}O(|z-w|)) = 1 + O(|z-w|),$$

where the estimates are independent of n and locally uniform on V.

Now, fix an ϵ sufficiently small, and set $B(\alpha) = \{|z-\alpha| \le \epsilon\}$ for every $\alpha \in J_P$. Then since J_P has no interior points, we can find a constant δ less than 1 such that $H_2(B(\alpha) \cap J_P) \le \delta\pi\epsilon^2$ for every $\alpha \in J_P$. On the other hand, fix a $z_0 \in J_P$ arbitrarily, and set $\alpha_n = P^n(z_0)$ for every n. Then the two inequalities above give that the diameter $2r_n$ of the connected component B_n of each $(P^n)^{-1}(B(\alpha_n))$ containing z_0 tends to 0 as $n \to \infty$, and $\{(z-z_0)/r_n \mid z \in B_n\}$ are approximately the unit disk for a sufficiently small ϵ.

Thus z_0 cannot be a density point of J_P. Since z_0 is arbitrary, we conclude that J_P has no density points, which implies that J_P has vanishing area. □

Hamilton's result (Theorem 1.4.2) implies that, if the Julia set is a simple closed curve, then it is a fractal set except for rare cases. But it may be helpful to include here the following classical result due to Fatou, for the proof is again rather easy.

Theorem 1.4.7 (Fatou) *Let P be a hyperbolic polynomial, and suppose that J_P is connected, and neither a circle/line nor a subarc of a circle/line. Then J_P has a tangent line at no point of J_P, and hence is not locally rectifiable at any point.*

Proof Take any $z_0 \in J_P$ and choose a sufficiently small simply connected neighborhood U of z_0 so that $U \cap C^+(P) = \emptyset$. We consider the family $\mathcal{F}$ consisting of all single valued branches of $(P^n)^{-1}$ on U. Then for any $z_0 \in J_P$, Lemma 2.4.4 and Proposition 2.4.5 give a sequence $\{s_m\}_{m=1}^{\infty}$ of elements in $\mathcal{F}$ such that s_m converges to a constant z_0 locally uniformly on U. Suppose that s_m is a branch of $(P^{n(m)})^{-1}$ and set $z_m = P^{n(m)}(z_0)$. Here we may further assume without loss of generality that $z_\infty = \lim_{m\to\infty} z_m$ exists and belongs to U.

Set

$$f_m(z) = \frac{s_m(z) - z_0}{s'_m(z_m)}.$$

Then every f_m is univalent and holomorphic on U and satisfies $f_m(z_m) = 0$ and $f'_m(z_m) = 1$. Hence the Koebe distortion theorem below implies that the family $\{s_m\}$ is a normal family. Hence taking a subsequence if necessary, we may assume that $f = \lim_{m\to\infty} f_m$ exists on U. Then the Hurwitz theorem gives that $f(z)$ is again a univalent function on U. Moreover, $f(z_\infty) = 0$, and when $z \in J_P \cap U - \{z_\infty\}$, we have

$$\lim_{m\to\infty} \arg(s_m(z) - z_0) = \lim_{m\to\infty} \arg s'_m(z_m) + \arg f(z) \pmod{\pi}.$$

Now if J_P has a tangent line at z_0, then since the left hand side is a constant on U, $f(U \cap J_P)$ should be a segment. Hence $U \cap J_P$ should be an analytic arc. Thus by the self-similarity (Proposition 2.4.5), we conclude that J_P is analytic. Then using the complete invariance of J_P, we can show that J_P should be one of the excluded cases. □

Theorem 1.4.8 (Koebe distortion theorem) *Let $f(z)$ be a univalent function on the unit disk Δ such that $f(0) = 0$ and $f'(0) = 1$. Then*

$$\frac{1 - |z|}{(1 + |z|)^3} \leq |f'(z)| \leq \frac{1 + |z|}{(1 - |z|)^3},$$

$$\frac{1 - |z|}{(1 + |z|)^2} \leq |f(z)| \leq \frac{1 + |z|}{(1 - |z|)^2}.$$

Remark Finally, we state the following conjecture, which is one of the central conjectures about complex dynamics.

HD conjecture In the space of all polynomial of the same degree (≥ 2), hyperbolic ones are dense.

Remark In the case of quadratic polynomials, the HD2 conjecture is equivalent to the HD conjecture.

Also, in the space of real quadratic polynomials, hyperbolic ones are dense (see, for instance, McMullen 1994a, and Graczyk and Swiatek 1998).

1.4.3 Capacity

As one of the other ways to know the size of a set, some kinds of capacity of the set are used. Here we use the logarithmic capacity. Also see §8.3.

Definition For a compact set E in $\mathbb{C}$, let D be the unbounded connected component of $\mathbb{C} - E$. The *Green function* $g(z, \infty)$ on D with a logarithmic pole at ∞ is by definition the infimum of the Perron family $\mathcal{G}$ consisting of all non-negative superharmonic functions $s(z)$ on D such that $s(z) - \log|z|$ is bounded near ∞, i.e.,

$$g(z, \infty) = \inf_{s \in \mathcal{G}} s(z).$$

Here, when $\mathcal{G} = \varnothing$, we set $g(z, \infty) = +\infty$.

If the Green function exists, namely if $g(z, \infty) \neq +\infty$, then the Green function $g(z, \infty)$ is positive harmonic on D, and

$$\gamma_\infty = \lim_{z \to \infty} \{g(z, \infty) - \log|z|\}$$

is a finite value, which we call the *Robin constant* of D at ∞. Also set

$$\operatorname{cap}(E) = e^{-\gamma_\infty},$$

and call it the *logarithmic capacity* of E with respect to ∞.

If the Green function does not exist, we consider that $\gamma_\infty = +\infty$ and $\operatorname{cap}(E) = 0$.

Remark When the boundary of D is regular with respect to the Dirichlet boundary value problem, or equivalently, each boundary point of D has a barrier, we denote by $u(z)$ the solution of the Dirichlet boundary value problem for the boundary value $\log|z|$. Then

$$g(z, \infty) = \log|z| - u(z),$$

where we assume that the closure of D does not contain 0.

Example 1.4.2 *Let D be a simply connected domain in $\widehat{\mathbb{C}}$ containing ∞, and $f(z)$ a Riemann map of D with $f(\infty) = 0$. Then*

$$g(z, \infty) = -\log|f(z)|.$$

The following result is known. See, for instance, Sario and Nakai (1970).

Proposition 1.4.9 *For a generalized ternary set $E = E(\{p_j\})$ of Cantor with p_j bounded away from* 1, $\operatorname{cap}(E) = 0$ *if and only if*

$$\sum_{j=1}^{\infty} \frac{\log p_j}{2^j} = +\infty.$$

Now, the following remarkable fact holds.

Theorem 1.4.10 *Let*

$$P(z) = z^k + a_1 z^{k-1} + \cdots + a_{k-1} z + a_k$$

be a monic polynomial of degree $k \geq 2$, and $g(z,\infty)$ be the Green function on I_P with the logarithmic pole ∞. Then we have

(i)

$$g(P(z),\infty) = kg(z,\infty),$$

(ii)

$$\lim_{z\to\partial I_P} g(z,\infty) = 0,$$

(iii)

$$\mathrm{cap}(J_P) = \mathrm{cap}(K_P) = 1.$$

Proof The second assertion has been shown as Proposition 1.2.2. The first assertion is nothing but the logarithm of equation (1.2). Since $\gamma_\infty = k\gamma_\infty$ by (i), $\gamma_\infty = 0$, which implies (iii). □

Thus the logarithmic capacity of the Julia set is always 1 for every monic polynomial. Similarly, we see the following.

Theorem 1.4.11 (Douady–Hubbard) *The logarithmic capacity of the Mandelbrot set equals* 1, *i.e.,*

$$\mathrm{cap}(\mathcal{M}) = \mathrm{cap}(\partial\mathcal{M}) = 1.$$

Proof The map

$$\Phi(c) = c\prod_{n=0}^{\infty}\left(1+\frac{c}{P_c^n(c)^2}\right)^{2^{-n-1}}$$

of $\widehat{\mathbb{C}} - \mathcal{M}$ to $\widehat{\mathbb{C}} - \bar{\Delta}$ in the proof of Theorem 1.3.2 is $\Phi(c) = c + O(1)$ as $c \to \infty$. Since $1/\Phi$ is a Riemann map of $\widehat{\mathbb{C}} - \mathcal{M}$, we conclude that the Robin constant at ∞ is 0. □

Finally, the Robin constant of the Julia set can be considered as the logarithmic energy of the equilibrium distribution. For a monic polynomial $P(z)$ of degree $k \geq 2$, fix a point $z_0 \in I_P - C^+(P)$, and define a measure μ_n as the sum of the point masses with weight k^{-n} supported at k^n points of $(f^n)^{-1}(z_0)$, which we count including multiplicity. Then these μ_n converge weakly to a measure $\mu = \mu_P$. Compare with the Lyubich measure in §5.4. Further see §8.5.

Theorem 1.4.12 (Brolin) *Let $P(z)$ be a monic polynomial of degree not less than 2, and $\mu = \mu_P$ be constructed as above. Then*

$$\mu(\mathbb{C} - J_P) = 0, \quad \mu(J_P) = 1,$$

and the energy of μ equals γ_∞, i.e.,

$$I(\mu) = \iint_{\mathbb{C}^2} \log \frac{1}{|z - \zeta|} d\mu(z) d\mu(\zeta) = \gamma_\infty.$$

Moreover, μ is P-invariant and strongly mixing, i.e.,

$$\mu(P^{-1}(E)) = \mu(E) \quad (E \subset J_P)$$

and

$$\lim_{n \to \infty} \int_{J_P} f \circ P^n \cdot g d\mu = \int_{J_P} f d\mu \cdot \int_{J_P} g d\mu$$

for every $f, g \in L^2(d\mu)$.

1.5 Polynomial-like mappings

In the Julia set of a polynomial of higher degree, we sometimes find a part of it similar to the Julia set of a polynomial of lower degree. Such a phenomenon indicates that a polynomial may act locally like a polynomial of lower degree. Here it is important to describe such local behavior explicitly. The local behavior near periodic points has been discussed in §1.2, and here we discuss a more synthetic kind of local behavior. The following notion due to Douady and Hubbard is crucial for this purpose.

Definition We call the triple (f, U, V) of bounded simply connected domains U and V such that $\overline{U} \subset V$ and a holomorphic proper map $f : U \to V$ a *polynomial-like map*.

The number of points in $f^{-1}(z_0)$ including multiplicity is called the *degree* of this polynomial-like map (f, U, V).

A polynomial-like map of degree 2 is called a *quadratic-like map*.

For any polynomial $P(z)$, taking a sufficiently small neighborhood W of ∞, we have a polynomial-like map

$$(P(z), \mathbb{C} - P^{-1}(\overline{W}), \mathbb{C} - \overline{W})$$

of the same degree. Every polynomial can be regarded as a polynomial-like map in this way.

Definition For any polynomial-like map (f, U, V) of degree not less than 2, we define the *filled Julia set* K_f of (f, U, V) by setting

$$K_f = \{z \in U \mid f^n(z) \in U \text{ for every positive integer } n\}.$$

Example 1.5.1 *Let $P(z)$ be a monic polynomial of degree 3, which has two distinct critical points z_1 and z_2. Suppose that*

$$a = g(z_1, \infty) > g(z_2, \infty),$$

where $g(z, \infty)$ is the Green function on I_P with a logarithmic pole at ∞.

Let U be the connected component of $\mathbb{C} - \{z \mid g(z, \infty) \geq a\}$ containing z_2, and set

$$V = \mathbb{C} - \{z \mid g(z, \infty) \geq 3a\},$$

which is connected and simply connected. Then the triple (f, U, V) is a quadratic-like map.

Example 1.5.2 *For the entire function*

$$f(z) = 2\cos z - 2,$$

set

$$U = \left\{z \in \mathbb{C} \mid |\mathrm{Re}\, z| < 2, |\mathrm{Im}\, z| < 3\right\}.$$

Then $(f, U, f(U))$ is a quadratic-like map.

Example 1.5.3 *For the entire function*

$$f(z) = \frac{1}{2} z^2 e^{2-z}$$

(cf. Example 3.4.6), let W be the connected component of $f^{-1}(\Delta)$ containing 0, where Δ is the unit disk. Then $\overline{W} \subset \Delta$ and (f, W, Δ) is a quadratic-like map.

A remarkable fact is that a polynomial-like map globally looks like a polynomial. To explain this, we begin with the definition of conformal-like maps.

Definition We call an orientation-preserving homeomorphism f of a domain D onto another, D', a *quasiconformal map* if f is ACL on every rectangle $R = \{z = x + iy \mid a \leq x \leq b, c \leq y \leq d\}$, i.e., $f(x + iy)$ is absolutely continuous on $[a, b]$ and on $[c, d]$, respectively, with respect to x and y for almost every fixed y and x, and there is a constant $k < 1$ such that

$$|f_{\bar{z}}| \leq k|f_z|$$

almost everywhere on D.

Here, set $K = (1+k)/(1-k)$ and we call f also a *K-qc* map. Further, the infimum of such K is called the *maximal dilatation* of f.

When we can take 0 as k, namely when f is 1-qc, then a classical lemma of Weyl means that f is actually a conformal map.

In general, f_z is non-vanishing almost everywhere on D, and hence we can consider the quotient

$$\mu(f) = \frac{f_{\bar{z}}}{f_z},$$

which is called the *complex dilatation* of f. Note that, if f is K-qc, then the essential supremum of $\mu(f)$ satisfies

$$\|\mu(f)\|_\infty = \text{ess.sup}_D|\mu(f)| = k < 1,$$

and the maximal dilatation of f equals $(1+k)/(1-k)$.

Theorem 1.5.1 (Straightening theorem) *Every polynomial-like map* (f, U, V) *of degree* $k \geq 2$ *is hybrid-equivalent to a polynomial* P *of degree* k*, i.e., we can find a quasiconformal map* h *of a neighborhood* W *of the filled Julia set* K_f *in* U *into* $\mathbb{C}$ *which satisfies*

(i) $h(K_f) = K_P$,
(ii) $\mu(h) = 0$ *almost everywhere on* K_f,
(iii) $h \circ f = P \circ h$ *on* $W \cap f^{-1}(W)$.

Such a polynomial P *is determined up to affine conjugacy.*

Definition The image of the circle by a quasiconformal map of $\mathbb{C}$ is called a *quasicircle.*

Corollary *For every c in the main cardioid of the Mandelbrot set, the Julia set* J_{P_c} *is a quasicircle.*

Proof By Theorem 2.4.1, $P^n(0)$ tends to the unique attracting fixed point in the interior of K_P. Hence we can find a connected compact set E in the interior of K_P such that $P(E)$ is contained in the interior of E and contains 0. Then $(P, \widehat{\mathbb{C}} - E, \widehat{\mathbb{C}} - P(E))$ considered as a function of $1/z$ is a quadratic-like map. Hence by the straightening theorem, it is hybrid-equivalent to a quadratic polynomial P^*. Since P^* has 0 as a fixed point, $P^*(z) = az^2$. Hence the Julia set of P^* is a circle. □

Remark Note that $P_{1/4}$ is a Jordan curve but not a quasicircle.

To prove Theorem 1.5.1, the following fact is crucial. For the proof, see Ahlfors (1966) or Imayoshi and Taniguchi (1992).

Proposition 1.5.2 (Measurable Riemann mapping theorem) *Let $\mu(z)$ be a measurable function on $\mathbb{C}$ satisfying $\|\mu\|_\infty < 1$. Then there exists a quasiconformal map $f : \mathbb{C} \to \mathbb{C}$ with $\mu(f) = \mu$ almost everywhere on $\mathbb{C}$. Such an f is unique under the conditions that $f(0) = 0$ and $f(1) = 1$.*

Proof of Theorem 1.5.1 For the sake of simplicity, we consider only the case that K_f is connected (cf. Lemma 4.3.4). Fix a simply connected domain W in U containing K_f, and assume that $A = f(\overline{W}) - W$ is diffeomorphic to a closed annulus $A' = \{(1 <) \, r \leq |z| \leq r^k\}$. Fix a conformal map ϕ of $\mathbb{C} - f(\overline{W})$ onto $\{r^k < |z| < \infty\}$ and extend it to a diffeomorphism of A onto A' such that

$$\phi \circ f = \phi^k$$

on the boundary of W. Then we can extend f by setting $f(z) = \phi^{-1}(\phi(z)^k)$ on $\mathbb{C} - W$.

Since ϕ is a quasiconformal map of the interior of A, we have a measurable function $\mu(\phi)$ on the interior of A. Pulling it back as far as possible by f, we have a measurable function on $f(W) - K_f$, which gives a measurable function μ on the whole of $\mathbb{C}$, by setting $\mu = 0$ on $\mathbb{C} - f(W)$ and on K_f. This μ clearly satisfies that $\|\mu\|_\infty < 1$ and is invariant under the action of f, that is the pullback of μ by f is equal to μ almost everywhere. By Proposition 1.5.2, we have a quasiconformal map $h : \mathbb{C} \to \mathbb{C}$ with $\mu(h) = \mu$. And the invariance implies that $F = h \circ f \circ h^{-1}$ is holomorphic on the whole of $\mathbb{C}$, which implies that F is a polynomial of degree k. □

The next result gives the reason why we find small copies of the Mandelbrot set in the connected locus for polynomials of higher degree.

Theorem 1.5.3 (McMullen 1997) *Let $\{f_t(z) \mid t \in \mathbb{C}\}$ be a family of polynomials (or more generally, rational functions) whose coefficients are rational functions of t. Then (except for some rare cases, the family has bifurcations: the set $B(f)$ of t where the Julia set of f_t does not move continuously is not empty, and) arbitrarily similar copies of the Mandelbrot set are dense in $B(f)$.*

Here, an arbitrarily similar copy of $\mathcal{M}$ is a set E such that there is a quasi-conformal map f of a neighborhood of $\mathcal{M}$ with maximal dilatation arbitrarily near to 1 *and satisfying $f(\mathcal{M}) = E$.*

Remark A function f_t corresponding to a point t where the Julia set of f_t moves continuously is called J-stable, and will be discussed in §5.3.

Finally, the Feigenbaum universality for unimodal maps of an interval is related to a fixed point theorem for the renormalization operator on the space of unimodal maps. Such renormalization can be formulated by using quadratic-like maps.

Definition Let m be a positive integer, and $P(z) = P_c(z)$. We say that P^m is *renormalizable* if there are simply connected domains U and V such that $0 \in U$, $(f = P^m|_U, U, V)$ is a quadratic-like map, and the filled Julia set K_f of this quadratic-like map is connected. Also we call (f, U, V) a *renormalization* of P^m.

If P^m is renormalizable for infinitely many m, then we say that P is *infinitely renormalizable.*

Theorem 1.5.4 (Yoccoz) *Suppose that $c \in \mathcal{M}$ and P_c is not infinitely renormalizable. Then $\mathcal{M}$ is locally connected at c.*

Moreover, if every periodic point of P_c is repelling (i.e., neither attracting nor indifferent), then the filled Julia set of P_c is locally connected.

Remark Shishikura announced that the subset of $c \in \partial\mathcal{M}$ such that P_c is not infinitely renormalizable has vanishing area.

A sufficient condition for renormalizability is the divisibility.

Definition A continuous degree 1 map ϕ is a *quotient map* of $(\widehat{\mathbb{C}}, A)$ to $(\widehat{\mathbb{C}}, B)$ if A and B are closed, $\phi(A) = B$, $\phi^{-1}(z)$ is connected for every z, and $\phi^{-1}(z)$ consists of a single point if z does not belong to B.

We say that P_1 is a *quotient* of P_0 if there are quotient maps ϕ_0 and ϕ_1 of $(\widehat{\mathbb{C}}, PC_{P_0})$ to $(\widehat{\mathbb{C}}, PC_{P_1})$ such that $\phi_0 = \phi_1$ on PC_{P_0}, there is a homotopy $\{\phi_t \mid t \in [0, 1]\}$ of quotient maps relative to PC_{P_0}, and

$$P_1 \circ \phi_0 = \phi_1 \circ P_0.$$

We further say that the quotient is *proper* if quotient maps are not injective.

Theorem 1.5.5 (McMullen 1994a) *If P_c has a proper quotient $P_{c'}$ with $c' \neq 0$ and $PC_{P'_c}$ is finite, then P_c is renormalizable.*

For the proof, see McMullen (1994a)

Remark For the relation between the Feigenbaum universality and the renormalization of quadratic-like maps, see Melo and van Strien (1993) and McMullen (1994a).

2

The Fatou set and the Julia set

When we discuss the stability of an orbit under the iteration of an endomorphism of a planar domain D, the concept of the normal family is particularly useful. In this chapter, we divide the domain D considered into two parts, the stable part and the chaotic part, and call them the Fatou set and the Julia set. We also explain some of the fundamental properties of them.

First, in §2.1, we introduce hyperbolic domains, and give a proof of the Denjoy–Wolff theorem, which is one of the keys for investigation of complex dynamics. We define the Fatou set and the Julia set by the normal family property, and give the fundamental classification theorem of the Fatou set as an application of the Denjoy–Wolff theorem

Next we need several covering properties of holomorphic endomorphisms considered as branched and not unlimited coverings of the domain onto itself. We explain those in §2.2 as a brief introduction to the value distribution theory on meromorphic functions.

In §2.3 we explain basic properties of the Julia set, including the first fundamental theorem, which states that repelling periodic points are dense in the Julia set.

Finally, in §2.4, we include interrelation between components of the Fatou set and the singular values. First we explain basic such facts on periodic components, and next, after defining the Teichmüller spaces of complex dynamics, we give a proof of the weak second fundamental theorem, which states that an endomorphism with a finite number of singular values has no simply connected wandering domains.

2.1 Holomorphic endomorphisms of planar domains

2.1.1 Hyperbolic domains

We begin with the following.

Definition We call a domain in the Riemann sphere $\widehat{\mathbb{C}}$ or more generally a planar Riemann surface D a *hyperbolic domain* if D is conformally equivalent to none of $\widehat{\mathbb{C}}$, $\mathbb{C}$ and $\mathbb{C}^* = \mathbb{C} - \{0\}$.

Then the famous uniformization theorem implies that the universal covering surface of a hyperbolic domain is always conformally equivalent to the unit disk Δ, and the corresponding covering transformation group consists of Möbius transformations preserving Δ. Since such a Möbius transformation T preserves the Poincaré metric on Δ, i.e.,

$$ds_\Delta^2 = \frac{4|dz|^2}{(1-|z|^2)^2} = \frac{4|dw|^2}{(1-|w|^2)^2}$$

with $w = T(z)$, we can project the Poincaré metric on Δ to a metric $ds_D^2 = \lambda_D^2(\zeta)|d\zeta|^2$ $(\zeta \in D)$ by the relation

$$\lambda_D(\pi(z))|\pi'(z)| = \frac{2}{1-|z|^2},$$

where $z \in \Delta$ and $\pi : \Delta \to D$ is the universal covering projection. It is clear that ds_D^2 is well defined by this relation. We call this ds_D^2 the *hyperbolic metric* on D.

Example 2.1.1 *Set* $D = \{0 < |z| < 1\}$*; then*

$$\lambda_D(z) = \frac{1}{|z| \log |\frac{1}{z}|} \quad (z \in D).$$

Definition For every rectifiable path L in a hyperbolic domain D, the *hyperbolic length* $\ell(L)$ of L is defined by

$$\ell(L) = \int_L ds_D.$$

The *hyperbolic distance* $d_D(z, w)$ between two points z and w is the infimum of the hyperbolic lengths of all rectifiable paths connecting z and w.

The hyperbolic distance is complete on D, i.e., every closed ball with a finite hyperbolic radius is compact. In this setting, the classical lemma of Pick, which says that for every holomorphic function f on the unit disk Δ into itself,

$$\frac{|f'(z)|}{1-|f(z)|^2} \leq \frac{1}{1-|z|^2},$$

can be reformulated as the following *principle of contraction.*

Theorem 2.1.1 (Contraction principle) *For every holomorphic endomorphism $f : D \to D$ of a hyperbolic domain D,*

$$d_D(f(z), f(w)) \leq d_D(z, w).$$

If the equality holds for some distinct points z and w, then f is a self-covering map of D.

Remark When a planar domain D is neither $\mathbb{C}^*$ nor $\Delta - \{0\}$, then a self-covering map of D is actually a conformal automorphism of D.

For a general planar domain D, we denote by $\mathrm{Aut}(D)$ the group of all conformal automorphisms of D.

Example 2.1.2 $\mathrm{Aut}(\widehat{\mathbb{C}})$ *is equal to the group* $\mathrm{M\ddot{o}b}(\widehat{\mathbb{C}})$ *of all Möbius transformations.* $\mathrm{Aut}(\mathbb{C})$ *is the group of all holomorphic linear maps.*

In addition to these two cases, $\mathrm{Aut}(\Delta)$, $\mathrm{Aut}(\mathbb{C}^*)$, $\mathrm{Aut}(\Delta - \{0\})$ *and* $\mathrm{Aut}(\{0 < r < |z| < R < \infty\})$ *contain an uncountable subgroup* $S^1 = \{T(z) = e^{i\theta}z \mid \theta \in \mathbb{R}\}$.

For every D conformally equivalent to none of the above domains, $\mathrm{Aut}(D)$ *is a countable group.*

For a general hyperbolic domain D in $\mathbb{C}$, it is very hard to determine the density $\lambda_D(z)$ of the hyperbolic metric. Hence it is useful to obtain estimates of λ_D by other quantities such as the Euclidean distance $d(z, \partial D)$ between $z \in D$ and the boundary ∂D of D.

Theorem 2.1.2 *For every hyperbolic domain D,*

$$\lambda_D(z) \leq \frac{2}{d(z, \partial D)} \quad (z \in D).$$

Moreover, if D is simply connected, then we have

$$\lambda_D(z) \geq \frac{1}{2d(z, \partial D)} \quad (z \in D).$$

Proof First, set $D_1 = \{\zeta \in \mathbb{C} \,|\, |\zeta - z| < d(z, \partial D)\}$, and we have $D_1 \subset D$ and hence $\lambda_D(z) \le \lambda_{D_1}(z) = 2/d(z, \partial D)$.

Conversely, take the inverse $f : \Delta \to D$ of a Riemann map such that $f(0) = z$ and $f'(0) > 0$. Then

$$\zeta = g(w) = \frac{1}{2}\lambda_D(z)(f(w) - z)$$

is a univalent function on Δ satisfying $g(0) = 0$ and $g'(0) = 1$. Hence by Koebe's $\frac{1}{4}$-theorem below, the image $g(\Delta)$ contains $\{|\zeta| < 1/4\}$, which implies that

$$\frac{1}{2}\lambda_D(z)d(z, \partial D) \ge \frac{1}{4}.$$

□

Theorem 2.1.3 (Koebe's $\frac{1}{4}$-theorem) *For every univalent function $\zeta = g(z)$ on Δ with $g'(0) = 1$, the image $g(\Delta)$ contains the disk $\{|\zeta - g(0)| < 1/4\}$.*

Proof We may assume that $g(0) = 0$. Fix a point c not contained in the image $g(\Delta)$, and set

$$G(z) = \frac{cg(z)}{c - g(z)} = z + \left(a_2 + \frac{1}{c}\right)z^2 + \cdots,$$

where we set $g(z) = z + a_2 z^2 + \cdots$. Then $G(z)$ also is univalent on Δ, and $|a_2| \le 2$.

Indeed, $h(z) = 1/\sqrt{g(z^2)}$ has a single valued branch on Δ, and we may assume that

$$h(z) = \frac{1}{z} - \frac{a_2}{2}z + \cdots.$$

It is easy to see that $h(z)$ is injective on Δ. Hence, letting $\Delta_r = \{|z| < r < 1\}$, the area of the complement $D_r = \widehat{\mathbb{C}} - h_r(\Delta_r)$ is

$$\iint_{D_r} dxdy = -\frac{1}{2i}\int_{\partial\Delta_r} \bar{h}dh = \pi\left\{\frac{1}{r^2} - \left|\frac{a_2}{2}\right|^2 r^2 - \cdots\right\}.$$

When r tends to 1, we have $|a_2| \le 2$ as desired.

Thus we conclude that

$$\frac{1}{|c|} \le |a_2| + \left|a_2 + \frac{1}{c}\right| \le 2 + 2,$$

which shows the assertion. □

Remark Actually, for every univalent function $g(z) = z + \sum_{k=2}^{\infty} a_k z^k$ on Δ, we have $|a_k| \leq k$ for every k. This had been called the Bieberbach conjecture, and was proved by de Branges. (See, for instance, Wernstein 1991.)

For a general hyperbolic domain D, we know

$$\lambda_D(z) \geq \frac{a}{d(z, \partial D) \log(2 + 1/d(z, \partial D))},$$

with a positive constant a. And as a condition equivalent to boundedness of $1/(\lambda_D(z) d(z, \partial D))$, the following concept is known.

Definition We say that a compact subset E in $\widehat{\mathbb{C}}$ is *uniformly perfect* if E contains more than two points and the moduli of doubly connected domains in $\widehat{\mathbb{C}} - E$ separating points of E are bounded.

Here, we say that a doubly connected domain R in $\widehat{\mathbb{C}} - E$ separates E if each component of $\widehat{\mathbb{C}} - R$ contains a point of E.

Remark Actually, Beardon and Pommerenke (1978) showed that, for every hyperbolic domain D in $\mathbb{C}$, $1/(\lambda_D(z) d(z, \partial D))$ is bounded if and only if $\mathbb{C} - D$ is uniformly perfect. For more quantitative treatment, see Sugawa (1998).

2.1.2 Normal families

Now, we introduce the concept of a normal family.

Definition We say that a family $\mathcal{F}$ consisting of holomorphic maps of a planar domain D to $\widehat{\mathbb{C}}$ is a *normal family* if any sequence of elements of $\mathcal{F}$ contains a subsequence converging uniformly on compact sets of D. Here we assume that $\widehat{\mathbb{C}}$ is equipped with the spherical metric $d\chi^2$.

Since $\widehat{\mathbb{C}}$ is compact, we can restate the Arzelà–Ascoli theorem as follows.

Theorem 2.1.4 *A family $\mathcal{F}$ of holomorphic maps of a planar domain D to $\widehat{\mathbb{C}}$ is a normal family if and only if $\mathcal{F}$ is locally equicontinuous on D, i.e., for every point $z \in D$ and every $\epsilon > 0$, we can find $\delta > 0$ such that the spherical distance $\chi(f(z_1), f(z_2)) < \epsilon$ for every z_1, z_2 in D with $\chi(z_1, z_2) < \delta$.*

The boundedness condition for normality due to Montel can be generalized to the three-point condition.

Theorem 2.1.5 (Three-point condition) *Let $\mathcal{F}$ be a family of holomorphic maps of a planar domain D to $\widehat{\mathbb{C}}$. If there exist three distinct points $\alpha_1, \alpha_2, \alpha_3$*

such that the image $f(D)$ contains none of them for every $f \in \mathcal{F}$, then $\mathcal{F}$ is a normal family.

Proof Fix a point $z_0 \in D$ arbitrarily, and take disks U and V with the center z_0 such that $\overline{V} \subset U \subset D$. Consider each $f \in \mathcal{F}$ as a holomorphic map of U to another hyperbolic domain $\Omega = \widehat{\mathbb{C}} - \{\alpha_1, \alpha_2, \alpha_3\}$, and then the contraction principle implies that the hyperbolic diameters of the image $f(\overline{V})$ are bounded.

In particular, we can choose a subsequence from any given sequence in $\mathcal{F}$ such that the images of $\overline{V}$ are bounded away from one of these $\alpha_1, \alpha_2, \alpha_3$ with respect to the spherical distance. Then the classical theorem of Montel implies that this subsequence is a normal family on V. Thus the given sequence contains a subsequence converging uniformly on a compact neighborhood of z_0. Since z_0 is arbitrary, we have the assertion. □

Corollary 1 *The family of all holomorphic maps of a planar domain to a hyperbolic domain is always a normal family.*

For a general planar domain D, we denote by $\mathrm{End}(D)$ the family of all holomorphic endomorphisms of D.

Corollary 2 *For every hyperbolic domain D,* $\mathrm{End}(D)$ *is a normal family. In particular, for every $f \in \mathrm{End}(D)$, the sequence $\{f^n\}_{n=1}^{\infty}$ of iterates of f contains a subsequence converging uniformly on compact sets of D.*

Hence, after preparing the general Denjoy–Wolff theorem on the sequence as in Corollary 2 above, in Chapters 3 and 4, we will discuss the iterations of an element in $\mathrm{End}(D)$ with non-hyperbolic planar domains D.

Recall that there are three such domains: $\widehat{\mathbb{C}}$, $\mathbb{C}$ and $\mathbb{C}^*$. $\mathrm{End}(\widehat{\mathbb{C}})$ is the family consisting of all rational functions, while $\mathrm{End}(\mathbb{C})$ is that of all entire functions. The former will be discussed in Chapter 4 and the latter in Chapter 3. In this book, we omit the systematic study of the family $\mathrm{End}(\mathbb{C}^*)$, but see §3.4.2.

Example 2.1.3 *Every element f of* $\mathrm{End}(\mathbb{C}^*)$ *has a form*

$$f(z) = z^m \exp\left(g(z) + h(1/z)\right)$$

with an integer m and entire functions g and h. A typical example is

$$F_{a,b}(z) = z e^{ia} e^{b(z-1/z)/2}$$

with real a, b. When we consider these functions as a map between arguments, we have the standard family

$$f_{a,b}(\theta) = \theta + a + b \sin\theta \pmod{2\pi}.$$

Also note that $f_{a,b}$ *is induced from a logarithmic lift, defined in §3.4.2, of* $F_{a,b}$.

The subset of $\mathrm{End}(\widehat{\mathbb{C}})$ consisting of elements whose restriction to $\mathbb{C}$ belongs to $\mathrm{End}(\mathbb{C})$ is the set Poly of all polynomials

$$\mathrm{Poly} = \mathrm{End}(\widehat{\mathbb{C}}) \cap \mathrm{End}(\mathbb{C}).$$

Also, $\mathrm{Möb}(\widehat{\mathbb{C}})$ is a subset of $\mathrm{End}(\widehat{\mathbb{C}})$, the behavior of whose elements is rather simple and well known. See §5.1.

Definition We denote by Ent and Rat, respectively, the set of all transcendental entire functions and the set of all rational functions of degree not less than 2:

$$\begin{aligned} \mathrm{Ent} &= \mathrm{End}(\mathbb{C}) - \mathrm{Poly}, \\ \mathrm{Rat} &= \mathrm{End}(\widehat{\mathbb{C}}) - \mathrm{Möb}(\widehat{\mathbb{C}}). \end{aligned}$$

2.1.3 The Julia set and the Fatou set

In this subsection we assume that D is either $\mathbb{C}$ or $\widehat{\mathbb{C}}$.

Definition For every $f \in \mathrm{End}(D)$, the *Fatou set* of f is the set of all the points z in D such that there is a neighborhood of z where $\{f^n\}_{n=1}^{\infty}$ is a normal family, and we denote it by $F(f)$. A connected component of the Fatou set is called a *Fatou component*.

We call the complement of the Fatou set of f in D the *Julia set* of f and denote it by $J(f)$. Here, when f belongs to Poly, we take $\mathbb{C} - F(f)$ as $J(f)$.

From the definition, it is clear that the Fatou set is open, and hence the Julia set is closed. Further, the theorems below follow directly from the definitions.

Theorem 2.1.6 *Let* f *be an element of* Ent $\cup$ Rat. *Then the Julia set and the Fatou set of* f *are completely invariant. More precisely,*

$$\begin{aligned} f(J(f)) \subset J(f), \quad f^{-1}(J(f)) = J(f), \\ f(F(f)) \subset F(f), \quad f^{-1}(F(f)) = F(f). \end{aligned}$$

Moreover, when f *belongs to* Rat, $f(J(f)) = J(f)$ *and* $f(F(f)) = F(f)$.

Proof It is clear that $f^{-1}(F(f)) \subset F(f)$. Indeed, if z_0 belongs to $F(f)$, then $\{f^n\}$ is equicontinuous near (in a neighborhood of) z_0, and since f is continuous, so is $\{f^n\}$ near every point of $f^{-1}(z_0)$. Next, suppose again that $z_0 \in F(f)$. Then $\{f^n\}$ is equicontinuous near z_0, and hence so is it also near $f(z_0)$, for f is an open map. Thus we conclude that $f(z_0) \in F(f)$.

Thus we have $f(F(f)) \subset F(f)$ and $f^{-1}(F(f)) \subset F(f)$. The former implies that $F(f) \subset f^{-1}(F(f))$, and hence $f^{-1}(F(f)) = F(f)$. Moreover, when $f \in \mathrm{Rat}$, the latter implies $F(f) \subset f(F(f))$, and hence $f(F(f)) = F(f)$. (If $f \in \mathrm{Ent}$, $F(f) - f(f^{-1}(F(f)))$ may be non-empty and consist of Picard exceptional values.)

The assertions for the Julia set follow also from the above. □

Theorem 2.1.7 *Let f be an element of* Ent ∪ Rat. *Then, for every positive integer N,*

$$J(f) = J(f^N), \quad F(f) = F(f^N).$$

Proof It is clear that $F(f) \subset F(f^N)$. Next fix a point $z_0 \in F(f^N)$. Then $\{(f^N)^n\}_{n=1}^{\infty}$ is equicontinuous near z_0, and as before so are $\{(f^N)^n \circ f\}, \ldots, \{(f^N)^n \circ f^{N-1}\}$. Hence $\{f^n\}$ itself is equicontinuous near z_0, which implies that $z_0 \in F(f)$. □

Next the three-point condition can be restated as follows.

Theorem 2.1.8 *Let f be an element of* Ent ∪ Rat. *Then, for every point z_0 in $J(f)$ and every neighborhood U of z_0,*

$$\widehat{\mathbb{C}} - \bigcup_{n=1}^{\infty} f^n(U)$$

consists of at most two points.

Here note that it is very hard, not to say impossible, in general to determine whether a given point belongs to the Fatou set or not. (See §4.1.) A rather easy case is that of periodic points.

Definition Let $f : D \to \widehat{\mathbb{C}}$ be a meromorphic function on D, and z_0 be a periodic point of f of period N, i.e., $f^N(z_0) = z_0$ and $f^j(z_0) \neq z_0$ $(1 \leq j \leq N-1)$. We define the *multiplier* λ at z_0 of f by

$$\lambda = (f^N)'(z_0).$$

Here, when $z_0 = \infty$, we consider that $(f^N)'(z_0)$ is the value of

$$\left(\frac{1}{f^N(1/z)}\right)'$$

at 0.

(i) When $|\lambda| > 1$, we call z_0 a *repelling periodic* point.

(ii) When $|\lambda| < 1$, we call z_0 an *attracting periodic* point. In particular, if $\lambda = 0$, then we call z_0 a *superattracting periodic* point.

(iii) When $|\lambda| = 1$, we call z_0 an *indifferent periodic* point. When there exists a positive integer m such that $\lambda^m = 1$, we call z_0 a *rationally indifferent periodic* point, or *parabolic periodic* point. Otherwise, we call z_0 an *irrationally indifferent periodic* point.

Finally, we call $\{z_0, f(z_0), \ldots, f^{N-1}(z_0)\}$ the *cycle* of z_0.

Remark The multiplier at z_0 is the same as that at any other point in the cycle of z_0.

Theorem 2.1.9 *Let f be an element of* Ent ∪ Rat. *Then, every repelling periodic point or parabolic periodic point of f belongs to the Julia set. Every (super)attracting periodic point belongs to the Fatou set.*

An irrationally indifferent periodic point z_0 belongs to the Fatou set if and only if f^N is linearizable at the fixed point z_0, that is, if the Schröder equation for f^N at z_0 has a solution.

Proof The first two of the assertions can be seen from Theorems 1.2.3, 1.2.4, 1.2.6 and 2.1.7. The last assertion follows from the fact that the Schröder equation for f^N at z_0 has a solution if and only if $\{(f^N)^n\}$ are uniformly bounded near z_0. Indeed, if $\{(f^N)^n\}$ are uniformly bounded near z_0, assuming that $N = 1$ and $z_0 = 0$,

$$\phi_m(z) = \frac{1}{m} \sum_{k=1}^{m} \frac{f^k(z)}{\lambda^k}$$

forms a normal family, and hence a subsequence converges to a solution of the Schröder equation. □

2.1.4 The classification theorem

In this subsection, we derive the classification theorem for Fatou components by using the Denjoy–Wolff theorem, which is a crucial fact on the behavior of iterations of any holomorphic endomorphism of a hyperbolic domain.

Definition Let f be an element of Ent ∪ Rat. Fix a Fatou component U of $F(f)$. For every non-negative n, let U_n be the component of $F(f)$ containing $f^n(U)$. In particular, $U_0 = U$.

When $\{U_n\}_{n=0}^{\infty}$ are disjoint domains, i.e., $U_i \cap U_j = \varnothing$ for every distinct i and j, then we say that U is a *wandering domain* of f.

If not, there exist distinct non-negative integers M and N ($> M$) such that $U_M = U_N$. Here we may further assume that $U_j \neq U_M$ for every j with $M < j < N$. Then $\mathcal{U} = \{U_n\}_{n=M}^{\infty} = \{U_M, U_{M+1}, \ldots, U_{N-1}\}$.

If U itself belongs to this set $\mathcal{U}$, then we call U a *periodic component* of $F(f)$ and $\mathcal{U}$ the *cycle* of the periodic component U. The *period* of U is by definition $N - M$. If not, we call U a *preperiodic* component of $F(f)$.

Finally, if the period of a periodic component U is 1, then we call U an *invariant* component. Further, a *completely invariant* component is an invariant component U such that $f^{-1}(U) \subset U$.

It is remarkable that the behavior of iterates on periodic components is completely classified as follows.

Theorem 2.1.10 (Classification theorem of periodic components) *Let f be an element of* Ent $\cup$ Rat, *and U be a periodic component of f with period k. Then one and only one of the following conditions is satisfied.*

(i) *U contains a (super)attracting periodic point z_0.*

(ii) *There exists a parabolic periodic point z_0 on the boundary ∂U of U, and for every $z \in U$,*

$$\lim_{n\to\infty} f^{nk}(z) = z_0.$$

(iii) *U contains an irrationally indifferent periodic point.*

(iv) *U is conformally equivalent to an annulus, and f^k restricted to U is an element of* Aut(U) *of infinite order.*

(v) *$f \in$* Ent, *U is unbounded in $\mathbb{C}$, and for every $z \in U$,*

$$\lim_{n\to\infty} f^{nk}(z) = \infty.$$

Definition When a periodic component U as above satisfies condition (iii), or (iv), or (v), then we call U a *Siegel disk*, or a *Herman ring*, or a *Baker domain*, respectively.

When U satisfies condition (i) or (ii) with the (super)attracting or parabolic periodic point z_0, we denote by $A(z_0)$ the set of all z such that $\lim_{n\to\infty} f^{kn+j}(z) = z_0$ with some j. This $A(z_0)$ is called the *basin* of z_0. According as (i) or (ii) is satisfied, we call $A(z_0)$ a *(super)attracting basin* or a *parabolic basin*, respectively. The union of components of $A(z_0)$ containing a point in the cycle of z_0 is called the *immediate basin* of the cycle of z_0.

Remark When condition (i) or (ii) is satisfied, U is a component of the immediate basin of the cycle of z_0.

If we regard a polynomial P as an element of $\mathrm{End}(\widehat{\mathbb{C}})$, then the set I_P of escaping points equals the immediate superattracting basin of the fixed point ∞.

Now to prove Theorem 2.1.10, we begin with the following generalization of the Denjoy–Wolff theorem.

Theorem 2.1.11 (Denjoy–Wolff theorem) *For every holomorphic endomorphism g of a hyperbolic domain Ω, one of the following conditions holds.*

(i) *For every $z \in \Omega$, $g^n(z)$ tends to the boundary $\partial\Omega$ as $n \to \infty$, i.e., for every compact set K of Ω, $\{m \mid g^m(z) \in K\}$ is a finite set.*

(ii) *There exists a fixed point z_0 of g in Ω, and*

$$\lim_{n\to\infty} g^n(z) = z_0$$

for every $z \in \Omega$.

(iii) *There exists a conformal map h of Ω onto one of Δ, $\Delta^* = \Delta - \{0\}$, and annuli which satisfies*

$$h \circ g \circ h^{-1}(w) = e^{i\theta} w \quad (\theta \in \mathbb{R} - \mathbb{Q}).$$

(iv) *The endomorphism g belongs to* $\mathrm{Aut}(\Omega)$ *and has a finite order.*

Proof Assume that condition (i) does not hold. Then there is a point $z_1 \in \Omega$ such that $g^n(z_1)$ accumulate to some point in Ω. So fix a subsequence $\{g^{n_k}(z_1)\}_{k=1}^{\infty}$ which converges to a point, say z_0, in Ω.

First suppose that g is a strict contraction with respect to the hyperbolic metric at every point in Ω. Take a geodesic segment I between z_1 and $g(z_1)$. Then all $g^{n_k}(I)$ are contained in a compact subset K of Ω. Since there is a constant $c < 1$ such that

$$d_\Omega(g(z), g(w)) \le c \cdot d_\Omega(z, w) \quad (z, w \in K),$$

the hyperbolic length of $g^n(I)$ tends to 0 as $n \to \infty$, and we have

$$g(z_0) = g\left(\lim_{n\to\infty} g^n(z_1)\right) = \lim_{n\to\infty} g^n(z_1) = z_0.$$

Thus g has a fixed point z_0 and the same argument shows the second assertion in condition (ii).

Next, if g is a local isometry with respect to the hyperbolic metric on Ω, then by Theorem 2.1.1 g is a self-covering map of Ω. When g does not belong to $\mathrm{Aut}(\Omega)$, Ω should be conformally equivalent to Δ^* and hence $\lim_{n\to\infty} g^n(z)$

is the puncture 0 for every $z \in \Omega$, which contradicts the assumption. Hence we may assume that $g \in \mathrm{Aut}(\Omega)$. If the order of g is infinite, Ω should be conformally equivalent to one of Δ, $\Delta^* = \Delta - \{0\}$, and annuli, and hence condition (iii) holds.

Indeed, if Ω is not conformally equivalent to one of Δ, $\Delta^* = \Delta - \{0\}$, and annuli, then the fundamental group $\pi_1(\Omega)$ of Ω is non-abelian. Here it is well known that there is a fixed-point free, discrete subgroup Γ of $\mathrm{M\ddot{o}b}(\widehat{\mathbb{C}})$ isomorphic to $\pi_1(\Omega)$, such that Δ/Γ is conformally equivalent to Ω, which is non-elementary in the sense of §5.1. Such a Γ is called a Fuchsian model of Ω. By the assumption that $\lim_{n\to\infty} g^{n_k}(z_1) = z_0$, we can choose a lift γ_k of g^{n_k} on Δ for every k such that $\gamma_k(\zeta_1)$ converges to ζ_0, where ζ_j is a lift of z_j for each j. Hence we may assume that γ_k and γ_k^{-1} converge to Möbius transformations. Then $\delta_k = \gamma_k^{-1} \circ \delta \circ \gamma_k \in \Gamma$ converge as $k \to \infty$ for a given element δ of Γ. Since Γ is discrete, δ_k is the same for every sufficiently large $k \geq k_0$. Then such a $\gamma_k \circ \gamma_{k_0}^{-1}$, which is a lift of $g^{n_k - n_{k_0}}$, is commutative with δ and hence fixes the set of the fixed points of δ. Thus by considering non-commutative elements of Γ, we see that some iterate of g should be the identity (cf. §5.1), or equivalently, condition (iv) holds. □

To use Theorem 2.1.11 in the proof of Theorem 2.1.10, we need to show that each component of the Fatou set is hyperbolic. Actually, we prove, independently of these theorems, in §2.3 that the Julia set of $f \in \mathrm{Ent} \cup \mathrm{Rat}$ is a non-empty perfect set, and hence contains uncountably many points. Assuming Theorem 2.3.6, we give a proof of Theorem 2.1.10.

Proof of Theorem 2.1.10 As noted above, U is a hyperbolic domain. Considering f^k, we may assume that $f(U) \subset U$. Then apply Theorem 2.1.11 with $\Omega = U$ and $g = f$.

If condition (ii) of Theorem 2.1.11 holds, then the point z_0 is a (super)attracting fixed point and U is the immediate attracting basin of z_0. If condition (iii) holds, then U is conformally equivalent to one of Δ, $\Delta^* = \Delta - \{0\}$, and annuli. In the first case, U is a Siegel disk, and in the last case, U is a Herman ring. In the second case, the Julia set should have an isolated point, which contradicts Theorem 2.3.6.

If condition (iv) holds, then the identity theorem implies that f is of finite order, which cannot occur.

Finally, if condition (i) holds, then we need to show that U is either a parabolic basin or a Baker domain. For this purpose, fix a point $z \in U$ and let I be the geodesic segment between z and $f(z)$. Then by Theorem 2.1.1, the hyperbolic lengths of $I_n = f^n(I)$ are bounded. Since I_n tends to the boundary and for every z

$$\lambda_U(z) \geq \frac{a}{d(z, \partial U)\log(2 + 1/d(z, \partial U))},$$

the spherical diameters of I_n tend to 0 as $n \to \infty$. Let L be the curve obtained from all I_n connected naturally. Then the set $\overline{L} - L$ of accumulation points of L in $\widehat{\mathbb{C}}$ is connected, and the same argument as in the proof of Theorem 2.1.11 shows that each point of $(\overline{L} - L)$, possibly except for ∞, is a fixed point of f. Since f is not the identity, $(\overline{L} - L)$ should consist of a single point, say z_0. Then as before we see that

$$\lim_{n\to\infty} f^n(z) = z_0 \quad (z \in U).$$

Here, if $f \in \mathrm{Ent}$ and $z_0 = \infty$, then U is a Baker domain. If not, we can see by the following snail lemma that U is a parabolic immediate basin. (When $f \in \mathrm{Rat}$ and $z_0 = \infty$, we may consider that $z_0 \in \mathbb{C}$ by taking a conjugation.) □

Proposition 2.1.12 (Snail lemma of Fatou) *Let U be an invariant component of $F(f)$ for some $f \in$ Ent ∪ Rat. Suppose that*

$$\lim_{n\to\infty} f^n(z) = z_0$$

for every point $z \in U$ with $z_0 \in \partial U \cap \mathbb{C}$. Then z_0 is a parabolic fixed point of f.

Proof By the assumption z_0 is an indifferent fixed point. So we set $f'(z_0) = e^{2\pi i\theta}$ $(\theta \in \mathbb{R})$. If $\theta \in \mathbb{Q}$, then the petal theorem (Theorem 1.2.6) implies the assertion. Hence assume that $\theta \in \mathbb{R} - \mathbb{Q}$.

Fix a point $w \in U$ and a relatively compact simply connected subdomain V of U containing w and $f(w)$. Since f is injective in a neighborhood of z_0, we may assume that every f^n is injective on V. Normalize f^n as

$$\phi_n(z) = \frac{f^n(z) - z_0}{f^n(w) - z_0} \quad (z \in V).$$

Let $\psi : \Delta \to V$ be a conformal map with $\psi(0) = w$, and set

$$h_n(\zeta) = \phi_n \circ \psi(\zeta) - 1.$$

Then every h_n is a univalent function on Δ such that $h_n(0) = 0$ and the image $h_n(\Delta)$ does not contain -1. Hence, Koebe's $\frac{1}{4}$-theorem (Theorem 2.1.3) implies that $|h_n'(0)| \leq 4$. And the distortion theorem of Koebe (Theorem 1.4.8) gives that the h_n are locally uniformly bounded, and hence form a normal family. Thus we conclude that $\{\phi_n\}$ is a normal family on V.

Take a converging subsequence $\{\phi_{n_k}\}$ such that $\lim_{k\to\infty} e^{n_k 2\pi i\theta} = 1$, and let $\phi(z)$ be the limit of it. Then ϕ is either the constant 1 or a univalent function on V by Hurwitz's theorem.

On the other hand, the assumption implies that

$$\phi(f(z)) = \lim_{k\to\infty} \frac{f(f^{n_k}(z)) - f(z_0)}{f^{n_k}(z) - z_0} \phi(z) = e^{2\pi i\theta}\phi(z) \quad (z \in V).$$

Hence, if ϕ is the constant 1, this is impossible. If not, $f^{n_k}(z) = \phi^{-1}(e^{n_k 2\pi i\theta}\phi(z))$ should tend to z, which contradicts the assumption. Thus we have a contradiction. □

Remark The reason why we call this lemma the snail lemma is as follows. In the above proof, if θ were irrational, then the orbit $f^n(V)$ of V would tend to z_0. Since $f^{n+1}(V) \cap f^n(V)$ is non-empty, we can show that this orbit would tend to z_0 so slowly that the union of it would fill a neighborhood of z_0, just like a trace of a snail.

2.2 Value distribution theory

Each element of End(D) (with D being either $\mathbb{C}$ or $\widehat{\mathbb{C}}$) can be considered as a holomorphic covering, possibly branched and incomplete, of D into itself. For instance, every element of Rat gives a branched complete covering of $\widehat{\mathbb{C}}$ onto itself. Since $\widehat{\mathbb{C}}$ is compact, such a topological viewpoint gives several useful tools, including the Riemann–Hurwitz relation.

On the other hand, the situation for an element of Ent is not so easy to handle. But a powerful device has been developed as value distribution theory. In this section, we summarize some of the fundamental results in this field. For the details, see, for instance, Nevanlinna (1970) and Hayman (1964).

2.2.1 The first main theorem and the characteristic function

In this subsection, let f always be a non-constant meromorphic function on the whole plane. Then the following, Jensen's formula, which states that value distributions of zeros and poles are essentially the same for such a function, is the starting point of the general theory.

Theorem 2.2.1 (Jensen's formula) *Let f be a non-constant meromorphic function on $\mathbb{C}$. Denote by $\{a_j\}_{j=1}^J$ and $\{b_k\}_{k=1}^K$ zeros and poles, respectively, of f, where J and K may be infinite. Here, zeros and poles are counted including multiplicity and ordered so that the absolute values are non-decreasing.*

Assume that $a_1 \neq 0$, $b_1 \neq 0$, *and there are no zeros and poles on* $\{|z| = R\}$. *Then*

$$\log|f(0)| = \frac{1}{2\pi}\int_0^{2\pi} \log|f(Re^{i\theta})|d\theta + \log\left\{\prod_{|a_j|<R} \frac{|a_j|}{R}\right\} - \log\left\{\prod_{|b_k|<R} \frac{|b_k|}{R}\right\}.$$

Proof Set

$$g(z) = f(z) \prod_{|b_k|<R} \frac{R(z-b_k)}{R^2 - \overline{b_k}z} \prod_{|a_j|<R} \frac{R^2 - \overline{a_j}z}{R(z-a_j)},$$

and we get an entire function g with no zeros in $\{|z| < R\}$ such that $|g(z)| = |f(z)|$ on $\{|z| = R\}$. In particular, $\log|g(z)|$ is harmonic in a neighborhood of $\{|z| \leq R\}$ and the assertion follows from the mean value theorem of Gauss. □

Here note that we can represent, in general, the third term of the right hand side as

$$\int_0^R \{n(t, f) - n(0, f)\}\frac{dt}{t} + n(0, f)\log R,$$

where $n(t, f)$ is the number of poles in $\{|z| \leq t\}$ counting multiplicity. Hence, the last two terms indicate how often the zeros and the poles are taken.

On the other hand, the first term indicates how much $f(\{|z| = R\})$ comes near to infinity or to the origin. So we divide it as follows.

Definition Let f be a non-constant meromorphic function on $\mathbb{C}$, and r be any positive number. Then we set

$$m(r, f) = \frac{1}{2\pi}\int_0^{2\pi} \log^+|f(re^{i\theta})|d\theta$$

and call it the *proximity function* of f. Here recall that $\log^+ x = \max\{\log x, 0\}$.

Next, we set

$$N(r, f) = \int_0^r \{n(t, f) - n(0, f)\}\frac{dt}{t} + n(0, f)\log r$$

and call it the *counting function* (for the poles) of f. Further, we set

$$T(r, f) = m(r, f) + N(r, f)$$

and call it the *characteristic function* of f in the sense of Nevanlinna.

Then the first term in the right hand side of Jensen's formula can be rewritten as $m(r, f) - m(r, 1/f)$, and the formula itself gives the following.

Theorem 2.2.2 (The first main theorem of Nevanlinna) *For every non-constant meromorphic function f on the whole plane and every constant $a \in \mathbb{C}$,*

$$T\left(r, \frac{1}{f-a}\right) = T(r, f) + O(1) \quad (r \to +\infty).$$

Here it may be more natural to take the *chordal distance* on $\widehat{\mathbb{C}}$ in the definition of the proximity function. More precisely, let

$$[a, b] = \begin{cases} \frac{|a-b|}{\sqrt{1+|a|^2}\sqrt{1+|b|^2}}, & a, b \in \mathbb{C}, \\ \frac{1}{\sqrt{1+|b|^2}}, & a = \infty,\ b \in \mathbb{C}. \end{cases}$$

And define the *chordal proximity function* of f by

$$\hat{m}(r, a, f) = \hat{m}(r, a) = \frac{1}{2\pi} \int_0^{2\pi} \log \frac{1}{[f(re^{i\theta}), a]} d\theta.$$

Further, modify the counting function so that

$$\widehat{N}(r, a, f) = \widehat{N}(r, a) = N\left(r, \frac{1}{f-a}\right) - c$$

with $c = \lim_{r \to 0}(\hat{m}(r, a) + N(r, 1/(f-a)))$. Then we have the *characteristic function*

$$\widehat{T}(r, f) = \hat{m}(r, a) + \widehat{N}(r, a)$$

of Ahlfors. Here note that this quantity is independent of a.

The characteristic function of Ahlfors has a simple geometric meaning. To state this, consider how many times a holomorphic map of the closed disk $\{|z| \leq r\}$ covers $\widehat{\mathbb{C}}$ for every positive r. This is represented by

$$S(r) = \frac{1}{\pi} \iint_{\{|z| \leq r\}} \frac{|f'(z)|^2}{(1 + |f(z)|^2)^2} dx dy,$$

which is called the *mean number of sheets*.

Theorem 2.2.3 (Ahlfors–Shimizu) *Let f be a non-constant meromorphic function on the whole plane. Then*

$$\widehat{T}(r, f) = \int_0^r \frac{S(t)}{t} dt + O(1) \quad (r \to \infty).$$

Proof Integrate both sides of

$$\widehat{T}(r, f) = \hat{m}(r, a) + \widehat{N}(r, a)$$

with respect to the spherical volume form $d\omega_z = (1 + |z|^2)^{-2} dx dy$ on $\widehat{\mathbb{C}}$, and we have

$$\begin{aligned}
\pi \widehat{T}(r, f) &= \iint_{\widehat{\mathbb{C}}} \widehat{N}(r, a) d\omega_a + \frac{1}{2\pi} \int_0^{2\pi} \left\{ \iint_{\widehat{\mathbb{C}}} \log \frac{1}{[f(re^{i\theta}), a]} d\omega_a \right\} d\theta \\
&= \int_0^r \frac{1}{t} \left\{ \iint_{\widehat{\mathbb{C}}} n(t, 1/(f - a)) d\omega_a \right\} dt + C \\
&= \int_0^r \frac{\pi S(t)}{t} dt + C
\end{aligned}$$

with a suitable constant C. □

Remark When f is a rational function of degree k, then we can easily see that

$$\widehat{T}(r, f) = (k + o(1)) \log r.$$

And for transcendental functions, the characteristic function plays a role similar to that of the degree for rational functions.

Theorem 2.2.4 *An element f in* End($\mathbb{C}$) *is a rational function if and only if*

$$\lim_{r \to \infty} \frac{\widehat{T}(r, f)}{\log r} < +\infty.$$

For two transcendental entire functions f and g,

$$\lim_{r \to \infty} \frac{\widehat{T}(r, f)}{\widehat{T}(r, g \circ f)} = \lim_{r \to \infty} \frac{\widehat{T}(r, g)}{\widehat{T}(r, g \circ f)} = 0.$$

Definition We call

$$\limsup_{r \to \infty} \frac{\log^+ T(r, f)}{\log r}$$

the *order* of f for every meromorphic function f on $\mathbb{C}$.

2.2.2 The second main theorem and direct consequences

On the other hand, the second main theorem of Nevanlinna states that the characteristic function can be estimated by counting functions for three different values.

Theorem 2.2.5 (The second main theorem) *Fix values $\{\alpha_\nu\}_{\nu=1}^q$ in $\widehat{\mathbb{C}}$ with $q \geq 3$. Then for every r outside an open set E_f of finite total length, i.e., finite one-dimensional Hausdorff measure,*

$$(q-2)\widehat{T}(r, f) \leq \sum_{\nu=1}^{q} \widehat{N}(r, \alpha_\nu) - N_1(r) + O(\log^+ r + \log^+ \widehat{T}(r, f)),$$

where, letting $n_1(r, f)$ be the number of all critical points including multiplicity in $\{|z| \leq t\}$, we set

$$N_1(r) = \int_0^r \frac{n_1(t, f) - n_1(0, f)}{t} dt + n_1(0, f) \log r.$$

Remark The number $n_1(r, f)$ of all critical points (including multiplicity) is just the number of branched points (including multiplicity) of the covering $f : \{|z| \leq t\} \to \widehat{\mathbb{C}}$, and $N_1(r)$ is called the *ramification function.*

Recall that there are sharper forms of the second main theorem. But for our purpose, the above form is enough.

Proof Consider a density on $\widehat{\mathbb{C}}$ with a finite total mass and concentrating near the given $\{\alpha_\nu\}_{\nu=1}^q$. For instance, we take some modification

$$\rho(z) = \prod_{\nu=1}^{q} \frac{1}{[z, \alpha_\nu]^2} \left(\sum_{\nu=1}^{q} \log \frac{1}{[z, \alpha_\nu]} \right)^{-2}$$

of the hyperbolic density as the density and set

$$d\mu_z = c\rho(z) d\omega_z,$$

where $c = 1/\int_{\widehat{\mathbb{C}}} \rho(z) d\omega_z$.

First, by the convexity of the logarithm, we have

$$2\sum_{\nu=1}^{q} \hat{m}(r, \alpha_\nu) - 2\log\left(\sum_{\nu=1}^{q} \hat{m}(r, \alpha_\nu) \right) \leq \frac{1}{2\pi} \int_0^{2\pi} \log c\rho(f(re^{i\theta})) d\theta + O(1).$$

On the other hand, since $\widehat{T}(r, f) \geq \widehat{N}(r, \alpha)$,

$$\begin{aligned} \widehat{T}(r, f) &\geq \iint_{\widehat{\mathbb{C}}} \widehat{N}(r, \alpha) d\mu_\alpha \\ &= \int_0^r \frac{1}{t} \left\{ \iint_{\{|z| \leq t\}} \frac{|f'(z)|^2}{(1 + |f(z)|^2)^2} c\rho(f(z)) dx dy \right\} dt + O(1). \end{aligned}$$

Here set $z = Re^{i\theta}$ and

$$\lambda(R) = \int_0^{2\pi} \frac{|f'(Re^{i\theta})|^2}{(1 + |f(Re^{i\theta})|^2)^2} c\rho(f(Re^{i\theta})) d\theta.$$

Then since

$$\widehat{T}(r, f) \geq \int_0^r \frac{1}{t}\left\{\int_0^t \lambda(u)udu\right\} dt + O(1),$$

we see that

$$\log \lambda(r) = O(\log^+ r + \log^+ \widehat{T}(r, f))$$

outside an open set E_f with finite total length.

Moreover, the convexity of the logarithm implies that

$$\begin{aligned}\log \frac{\lambda(R)}{2\pi} &\geq \frac{1}{2\pi}\int_0^{2\pi} \log \frac{|f'(Re^{i\theta})|^2}{(1+|f(Re^{i\theta})|^2)^2} d\theta \\ &+ \frac{1}{2\pi}\int_0^{2\pi} \log\{c\rho(f(Re^{i\theta}))\} d\theta.\end{aligned}$$

Since a direct counting gives that

$$n_1(r, f) = n(r, 1/f') - n(r, f') + 2n(r, f),$$

Jensen's formula and the first main theorem show that

$$\frac{1}{4\pi}\int_0^{2\pi} \log \frac{|f'(Re^{i\theta})|^2}{(1+|f(Re^{i\theta})|^2)^2} d\theta = N_1(R) - 2\widehat{T}(R, f) + O(1).$$

Summing up these inequalities, we conclude that

$$2\sum_{\nu=1}^{q} \hat{m}(r, \alpha_\nu) \leq 2\{2\widehat{T}(r, f) - N_1(r) + O(\log^+ \lambda(r) + \log^+ \widehat{T}(r, f))\}$$

holds when r is outside E_f and tends to $+\infty$, which implies the assertion. □

When we consider f as a holomorphic covering, we measure how strongly smoothness breaks over each value α by the following quantity.

Definition For a non-constant meromorphic function f on $\mathbb{C}$ and any $\alpha \in \widehat{\mathbb{C}}$, we denote by $\bar{n}(t, \alpha)$ the number of zeros of $f - \alpha$ in $\{|z| \leq t\}$ ignoring multiplicity. Further we set

$$\bar{N}(r, \alpha) = \int_0^r \{\bar{n}(t, \alpha) - \bar{n}(0, \alpha)\}\frac{dt}{t} + \bar{n}(0, \alpha)\log r$$

and

$$\Theta(\alpha, f) = \Theta(\alpha) = 1 - \limsup_{r\to\infty} \frac{\bar{N}(r, \alpha)}{\widehat{T}(r, f)},$$

which we call the *deficiency index* of f at α.

Then the following consequence of the second main theorem can be considered as an alternative to the Riemann–Hurwitz formula.

Theorem 2.2.6 (Nevanlinna inequality) *For every transcendental meromorphic function f on $\mathbb{C}$,*

$$\sum_{\alpha\in\mathbb{C}} \Theta(\alpha) \leq 2.$$

Proof For every finite subset $\{\alpha_\nu\}_{\nu=1}^{q}$ of $\widehat{\mathbb{C}}$ consisting of more than two points, the second main theorem implies that

$$\begin{aligned}(q-2)\widehat{T}(r,f) &\leq \sum_{\nu=1}^{q} \widehat{N}(r,\alpha_\nu) - N_1(r) + o(\widehat{T}(r,f)) \\ &\leq \sum_{\nu=1}^{q} \bar{N}(r,\alpha_\nu) + o(\widehat{T}(r,f))\end{aligned}$$

as r, outside an open set with finite total length, tends to $+\infty$. Hence we have

$$\sum_{\nu=1}^{q} \Theta(\alpha_\nu) \leq 2.$$

Since q is arbitrary, we have the desired inequality. □

Remark In the case of a rational function f of degree k, the Riemann–Hurwitz relation says that

$$2k - V = 2,$$

where V is the total number of critical points including multiplicity.

Hence let $\bar{n}(\alpha)$ be the number of preimages of α and set $\Theta(\alpha) = 1-\bar{n}(\alpha)/k$. Then $k\sum_{\alpha\in\widehat{\mathbb{C}}} \Theta(\alpha) = V$, or equivalently,

$$\sum_{\alpha\in\widehat{\mathbb{C}}} \Theta(\alpha) = \frac{V}{k} = \frac{2k-2}{k} < 2.$$

The Nevanlinna inequality can be considered the case that $k = \infty$.

In the case of a transcendental meromorphic function f on $\mathbb{C}$, we call α a *Picard exceptional value* of f if the preimage $f^{-1}(\alpha)$ of α is finite in number. If α is a Picard exceptional value, then the definition implies that $\Theta(\alpha) = 1$, and hence the Nevanlinna inequality means that there are at most two Picard exceptional values. This is the Picard little theorem.

On the other hand, if the preimages $f^{-1}(\alpha)$ are branched points except for a finite number of points, then we say that α is a *completely branched value*. Since $\Theta(\alpha) \geq 1/2$, the Nevanlinna inequality gives the following.

Corollary *Every transcendental meromorphic function on* $\mathbb{C}$ *has at most four completely branched values. In particular, every element of* Ent *has at most two completely branched values.*

As another direct application of the second main theorem, we note the following theorem of Rosenbloom.

Theorem 2.2.7 *Every element of* Ent *has infinitely many fixed points or infinitely many periodic points of order* 2.

Proof Suppose that an $f \in$ Ent has only a finite number of fixed points and periodic points of order 2. Then consider

$$h(z) = \frac{f(f(z)) - z}{f(z) - z}.$$

Then the assumption implies that $\widehat{N}(r, h)$ and $\widehat{N}(r, 1/h)$ are $O(\log r)$, and that $\widehat{N}(r, 1/(h-1)) = O(\widehat{T}(r, f))$. Hence the second main theorem gives that

$$\widehat{T}(r, h) = O(\widehat{T}(r, f))$$

as r, outside an open set with finite total length, tends to $+\infty$.

On the other hand, we can see that $\widehat{T}(r, h)$ should have the same growth as $\widehat{T}(r, f^2)$ by Theorem 2.2.4. This is a contradiction. □

Remark Actually, for every $f \in$ Ent and every integer N (≥ 2), there are infinitely many periodic points of f with period N. This is a theorem due to Baker and Bergweiler. See Bergweiler (1993a).

In the case of rational functions, we can easily show by a direct calculation the following.

Proposition 2.2.8 *For every* $f \in$ Rat, *there are either two fixed points or a cycle of periodic points of order* 2.

A rational function which does not have periodic points of a given period is very unusual. See Theorem 4.2.5.

2.2.3 More on entire functions

In the case of entire functions, there are several possibilities for the characteristic functions. We explain two of them. One is the maximum value function, i.e., for every $f \in \mathrm{Ent}$, set

$$M(r, f) = \max_{|z|=r} |f(z)|.$$

Theorem 2.2.9 *For every* $f \in \mathrm{Ent}$ *and every* $r > 0$,

$$T(r, f) \leq \log^+ M(r, f) \leq 3T(2r, f).$$

Another is the value of the maximal term, which is defined as follows.

Definition Let $f \in \mathrm{Ent}$ have the Taylor expansion $\sum_{j=0}^{\infty} a_j z^j$. For every $r > 0$, we call the term $|a_{j_0}|r^{j_0}$ the *maximal term* if

$$|a_{j_0}|r^{j_0} = \max_j |a_j|r^j,$$

and write the value $|a_{j_0}|r^{j_0}$ as $\mu(r, f)$. Since $|a_j|r^j$ tends to 0 as $j \to \infty$, the maximal index among j_0 as above exists. We call this maximal index the *central index* and denote it by $\nu(r, f)$.

It is clear that both of $\mu(r, f)$ and $\nu(r, f)$ are monotonically increasing and tend to $+\infty$ as $r \to +\infty$. Furthermore, we have the following.

Theorem 2.2.10 *For every* $f \in \mathrm{Ent}$ *and every* $R > r > 0$,

$$\mu(r, f) \leq M(r, f) \leq \mu(r, f) \left\{ \nu(R, f) + \frac{R}{R - r} \right\}.$$

Moreover, when the order of f *is finite,*

$$\lim_{r \to \infty} \frac{\log^+ M(r, f)}{\log^+ \mu(r, f)} = 1.$$

Proof The first inequality is the Cauchy estimate of the coefficients. Next, since

$$M(r, f) \leq \left\{ \sum_{j=0}^{\nu(R,f)-1} + \sum_{j=\nu(R,f)}^{\infty} \right\} |a_j|r^j$$

and since

$$\sum_{j=\nu(R,f)}^{\infty} |a_j|r^j \leq \mu(r, f) \sum_{j=\nu(R,f)}^{\infty} \frac{|a_j|R^j}{\mu(R, f)} \left(\frac{r}{R}\right)^{j-\nu(R,f)},$$

we reach the second inequality.

Further suppose that the order of f is finite. The second inequality in the theorem implies that there exists a positive integer N such that

$$M(r, f) \leq \mu(r, f)r^N + C.$$

Since f is transcendental,

$$\lim_{r\to\infty} \frac{\log \mu(r, f)}{\log r} = +\infty.$$

Hence the second assertion follows. □

Finally, we recall (a somewhat weaker form of) a fundamental theorem of Wiman and Valiron, which gives a polynomial approximation of a transcendental entire function f near the point where f equals its maximum value.

Theorem 2.2.11 (Wiman–Valiron) *For every $f \in \mathrm{Ent}$ and every $r > 0$, we can represent each $f^{(\ell)}$ with non-negative integer ℓ as*

$$f^{(\ell)}(z) = \nu(r, f)^\ell \left(\frac{z}{z_0}\right)^{\nu(r,f)} \frac{f(z_0)}{z_0^\ell}(1 + \epsilon_\ell(r)),$$

where z_0 satisfies

$$|z_0| = r, \quad |f(z_0)| = M(r, f),$$

z satisfies

$$|z - z_0| < \frac{r}{\nu(r, f)^{(1/2)+\eta}} \quad (\eta > 0),$$

and ϵ_ℓ tends to 0 as r, outside an open set E_ℓ such that $\int_{E_\ell}(dt/t)$ is finite, tends to $+\infty$.

Proof For the sake of simplicity, we give an outline of the proof for $\ell = 0$. For the details, see, for instance, Hayman (1974).

Set $\mu = \mu(r, f)$ and $\nu = \nu(r, f)$, and assume that the Landau symbols below are valid outside an open set E_ℓ as in the theorem. Then first, since $|a_\nu|r^\nu$ is not smaller than all other terms by definition, we can find an N such that $N = o(\nu^{(1/2)+\eta})$ and that

$$\left| f(z) - \sum_{|\nu - j|\leq N} a_j z^j \right| = O(\mu |z/z_0|^\nu)$$

for every z with $|z - z_0| < \frac{r}{\nu(r,f)^{(1/2)+\eta}}$. Here we write $\sum_{|\nu-j|\leq N} a_j z^j$ as $z^q P(z)$ with $q = \nu - N$ and a polynomial $P(z)$ of degree at most $2N$. Then

we have

$$|z_0^{-q} f(z_0) - P(z_0)| \leq r^{-q} O(\mu) = O(r^{-q} M(r, f)).$$

On the other hand, for the polynomial P, we see by direct calculations that

$$|P(z) - P(z_0)| = O(r^{-q} M(r, f)).$$

Hence we conclude that

$$\left| f(z) - \left(\frac{z}{z_0}\right)^q f(z_0) \right| \leq O(M(r, f)(|z/z_0|^\nu + |z/z_0|^q)).$$

Here since $q = \nu(1 + o(1))$, the assertion follows. □

2.3 Basic properties of the Julia set

2.3.1 Other characterizations of the Julia set

We have defined the Julia set as the complement of the Fatou set. But there are equivalent direct definitions. We explain some of them. We begin by rewriting the three-point condition.

Theorem 2.3.1 *Let f be an element of* Rat ∪ Ent. *Then either the Julia set $J(f)$ of f contains no interior points, or the Fatou set $F(f)$ is empty.*

Proof Assume that $J(f)$ contains an interior point z_0, and let U be a neighborhood of z_0 contained in $J(f)$. Then the three-point condition implies that $\widehat{\mathbb{C}} - \bigcup_{n=1}^{\infty} f^n(U)$ contains at most two points. Since $J(f)$ is completely invariant and closed, we conclude that the Fatou set is empty. □

Actually, functions whose Fatou set are empty do exist. In the case of transcendental entire functions, we will give an example in Chapter 3. Here we give two typical examples in the case of rational functions. For this purpose, we recall the $\wp$-function of Weierstrass, which we use to construct the desired examples.

Definition Fix a pair $\{\omega_1, \omega_2\}$ of non-zero complex numbers such that $\mathrm{Im}\{\omega_2/\omega_1\} > 0$. Set

$$\wp(z) = \wp(z \mid \omega_1, \omega_2) = \frac{1}{z^2} + \sum_{(m,n)\in\mathbb{Z}^2-(0,0)} \left\{ \frac{1}{(z - \omega_{m,n})^2} - \frac{1}{\omega_{m,n}^2} \right\},$$

where $\omega_{m,n} = m\omega_1 + n\omega_2$, and call it the $\wp$-function of Weierstrass.

It is well known that $\wp(z)$ is a doubly periodic meromorphic function on $\mathbb{C}$ having ω_1 and ω_2 as the basis of periods and satisfying the differential equation

$$\wp'(z)^2 = 4(\wp(z) - e_1)(\wp(z) - e_2)(\wp(z) - e_3),$$

where

$$e_1 = \wp(\omega_1/2), \quad e_2 = \wp(\omega_2/2), \quad e_3 = \wp((\omega_1 + \omega_2)/2).$$

When we consider the torus $T = T_{\omega_1,\omega_2}$ by identifying each pair of opposite sides of the parallelogram with the vertices $0, \omega_1, \omega_2, \omega_1 + \omega_2$, we can regard $\wp(z)$ as a two-sheeted covering of T onto $\widehat{\mathbb{C}}$ branched over e_1, e_2, e_3 and ∞, and as a meromorphic function, these four values are just the completely branched value of $\wp(z)$.

The important fact here is the following formulas.

Proposition 2.3.2 (i)

$$\wp(z + w) = -\wp(z) - \wp(w) + \frac{1}{4}\left\{\frac{\wp'(z) - \wp'(w)}{\wp(z) - \wp(w)}\right\}^2.$$

(ii)

$$\wp(2z) = -2\wp(z) + \frac{1}{4}\left(\frac{\wp''(z)}{\wp'(z)}\right)^2.$$

Proof First consider the canonical product of Weierstrass

$$\sigma(z) = z \prod_{(m,n)\in\mathbb{Z}^2-(0,0)} \left\{1 - \frac{z}{\omega_{m,n}}\right\} \exp\left(\frac{z}{\omega_{m,n}} + \frac{1}{2}\left(\frac{z}{\omega_{m,n}}\right)^2\right).$$

Then $\sigma(z)$ is an odd entire function and its logarithmic derivative

$$\zeta(z) = \frac{\sigma'(z)}{\sigma(z)} = \frac{1}{z} + \sum_{(m,n)\in\mathbb{Z}^2-(0,0)} \left\{\frac{1}{z - \omega_{m,n}} + \frac{1}{\omega_{m,n}} + \frac{z}{\omega_{m,n}^2}\right\}$$

is again an odd function and satisfies

$$\zeta'(z) = -\wp(z).$$

Next it is easy to see that $\zeta(z + \omega_j) - \zeta(z)$ is a constant, which we denote by η_j, for each j. Hence by integration, we have

$$\sigma(z + \omega_j) = C_j \sigma(z) e^{\eta_j z}$$

for each j. Since $\sigma(z)$ is odd, by setting $z = -\omega_j/2$, we see that $C_j = -e^{\eta_j\omega_j/2}$, which in turn implies that

$$-\frac{\sigma(z-w)\sigma(z+w)}{\sigma(z)^2\sigma(w)^2}$$

is a doubly periodic meromorphic function having ω_1 and ω_2 as periods. Moreover, comparing the coefficients of the Laurent expansion at the origin, we can show that this function equals $\wp(z) - \wp(w)$. Hence taking the logarithmic derivatives with respect to z, we have

$$\frac{\wp'(z)}{\wp(z)-\wp(w)} = \zeta(z-w) + \zeta(z+w) - 2\zeta(z),$$

or by symmetrizing it, we have

$$\frac{\wp'(z)-\wp'(w)}{\wp(z)-\wp(w)} = 2\zeta(z+w) - 2\zeta(z) - 2\zeta(w).$$

Differentiation of both sides with respect to z gives

$$\frac{\wp''(z)}{\wp(z)-\wp(w)} - \frac{\wp'(z)(\wp'(z)-\wp'(w))}{(\wp(z)-\wp(w))^2} = -2\wp(z+w) + 2\wp(z).$$

Here write the differential equation following the definition of the $\wp$-function as $\wp'(z)^2 = 4\wp(z)^3 - g_2\wp(z) - g_3$, and we have $\wp''(z) = 6\wp(z)^2 - g_2/2$. Canceling $\wp''(z)$ from the above equation, and symmetrizing the result, we obtain equation (i).

Next, letting w tend to z, we get equation (ii). □

Here the right hand side of equation (ii) is a rational function of $\wp(z)$, which is the key to the following examples.

Example 2.3.1 *Let*

$$R(z) = \frac{(z^2+1)^2}{4z(z^2-1)}.$$

Then $J(R) = \widehat{\mathbb{C}}$.

Verification Assume that $e_1 = -1$, $e_2 = 1$ and $e_3 = 0$. Then $\wp'^2 = 4\wp(\wp^2 - 1)$ and hence the above equation (ii) gives that

$$\wp(2z) = -2\wp(z) + \frac{(6\wp(z)^2-2)^2}{16\wp(z)(\wp(z)^2-1)} = \frac{(\wp(z)^2+1)^2}{4\wp(z)(\wp(z)^2-1)}.$$

Hence the map $z \to 2z$, which gives a four-sheeted smooth self-covering map of the corresponding torus, is semiconjugate to R, i.e., $R \circ \wp(z) = \wp(2z)$.

Now set

$$E = \{z \in \mathbb{C} \mid \text{there are integers } \ell, m, n \text{ such that } 2^{\ell} z = \omega_{m,n}\}.$$

Then for each point a in the image $\wp(E)$, some point in the orbit of a by R is a fixed point of R, which is clearly repelling. Since $\wp(E)$ is dense in $\widehat{\mathbb{C}}$, we obtain the claim.

Example 2.3.2 *Let*

$$R(z) = -\frac{i}{2}\frac{z^2 - 1}{z}.$$

Then $J(R) = \widehat{\mathbb{C}}$.

Verification Again assume that $e_1 = -1$, $e_2 = 1$ and $e_3 = 0$. In this case, symmetry (or the Schwarz–Christoffel formula) implies that $\omega_2/\omega_1 = i$. Hence the map $z \to (1+i)z$ also induces a two-sheeted smooth self-covering F of the corresponding torus.

Equation (i) gives that

$$\wp((1+i)z) = \frac{1}{4}\left\{\frac{1-i}{2}\frac{\wp'(z)}{\wp(z)}\right\}^2 = -\frac{i}{2}\frac{\wp(z)^2 - 1}{\wp(z)}.$$

Hence R is the rational function corresponding to F. And a similar argument as in the above example shows that $J(R) = \widehat{\mathbb{C}}$.

Now to give another characterization of the Julia set, we should exclude points whose backward orbit consists of a finite number of points.

Definition Let f be in Rat $\cup$ Ent. For every value $z \in \widehat{\mathbb{C}}$, we denote by $O^-(z)$ the backward orbit of z by f, i.e.,

$$O^-(z) = \{w \mid f^n(w) = z \text{ for some } n \in \mathbb{N}\}.$$

In general, $O^-(z)$ contains infinitely many points. But for some z, it may consist of finitely many points. The point ∞ for any polynomial is a typical example. Anyway, such phenomena are exceptional, and we call every z_0 such that $O^-(z_0)$ is a finite set an *exceptional point* for f.

Theorem 2.3.3 (i) *For every* $f \in$ Rat $\cup$ Ent, *there are at most two exceptional points for* f.

(ii) *Every* $f \in \mathrm{End}(\mathbb{C})$ *has* ∞ *as an exceptional point.*

(iii) *When* $f \in$ Ent *has two exceptional points* ∞ *and* 0, *it has a form as*

$$f(z) = z^m e^{g(z)}$$

with a non-negative integer m and $g \in \mathrm{End}(\mathbb{C})$.

(iv) *When $f \in \mathrm{Rat}$ has two exceptional points ∞ and 0, it has a form as*

$$f(z) = cz^m$$

with an integer m with $|m| \geq 2$ and $c \in \mathbb{C}^*$. In particular, both of the exceptional points belong to the Fatou set of f.

Proof First, suppose that $f \in \mathrm{Ent}$. If there are three exceptional points, then every point is by definition a Picard exceptional value. Hence we have a contradiction. And if there is one exceptional point z_0 other than ∞, then $f^{-1}(z_0)$ should be contained in $\{z_0\}$, for every other point has infinitely many preimages. Hence it is easy to see that f has a form as in (iii).

Next, suppose that $f \in \mathrm{Rat}$, and fix an exceptional point α. Then since $O^-(\alpha)$ is finite, we can find a point $\beta \in O^-(\alpha)$ and two distinct positive integers n and m such that $f^n(\beta) = f^m(\beta) = \alpha$, which implies that α and all its forward orbit also belong to $O^-(\alpha)$. Thus a finite set $O^-(\alpha)$ is completely invariant, and f gives a permutation of $O^-(\alpha)$. In particular, every element of $O^-(\alpha)$ has only one preimage by f, which means that all of them are critical points of order $k - 1$, where k is the degree of f. By the Riemann–Hurwitz formula, at most two such points can exist, and hence $O^-(\alpha)$ consists of at most two points. The same argument shows that there can exist at most two exceptional points. If there are two, it clearly has a form as in (iv). □

Now we have the following backward orbit characterization of the Julia set.

Theorem 2.3.4 *Let f be an element of* Rat ∪ Ent. *For every non-exceptional point z_0 for f in $J(f)$,*

$$J(f) = \overline{O^-(z_0)}.$$

Proof Take a $z \in J(f)$, and fix a neighborhood U of z. The assumption means that $O^-(z_0)$ contains infinitely many points, the three-point condition gives that $\bigcup_{n=1}^{\infty} f^n(U)$ intersects $O^-(z_0)$, and hence $U \cap O^-(z_0)$ is non-empty. Since z and U are arbitrary, we have the assertion. □

By the same argument, we also reach the following conclusion.

Theorem 2.3.5 *Let E be a completely invariant set with respect to f, containing more than two points. Then the closure of E contains $J(f)$.*

Corollary 1 *The Julia set $J(f)$ is the smallest set among all completely invariant closed sets with respect to f containing more than two points.*

Corollary 2 *Let Ω be a union of Fatou components which is completely invariant with respect to f. Then the Julia set $J(f)$ is coincident with the boundary of Ω. In particular, for any completely invariant Fatou component D, the boundary of D is coincident with $J(f)$.*

Corollary 3 *When f is a polynomial, the Julia set defined in Chapter 1 is coincident with the Julia set $J(f)$ defined in this chapter.*

Here as another application of Theorem 2.3.3, we show the following theorem, which we have used in §2.1.

Theorem 2.3.6 *For every $f \in \text{Rat} \cup \text{Ent}$, the Julia set $J(f)$ is a non-empty closed perfect set. In particular, $J(f)$ is uncountable.*

Proof When $f \in \text{Rat}$, then by Proposition 2.2.8, there are two distinct points z_1 and z_2 such that $f^2(z_j) = z_j$ for each j. When $f \in \text{Ent}$, Rosenbloom's theorem gives infinitely many such z_j.

If one of such z_j is non-exceptional, and belong to the Julia set, then Theorem 2.3.4 implies that $J(f)$ contains infinitely many points. If two z_j belong to the Fatou set, then the classification theorem (Theorem 2.1.10) means that these two points should be contained in different periodic components of $F(f)$, and hence $J(f)$ contains again infinitely many points. Moreover, since there are at most two exceptional points and, when there are two and $f \in \text{Rat}$, both belong to $F(f)$ by Theorem 2.3.3, we see that one of the above two cases always occurs. Thus we have shown that $J(f)$ contains infinitely many points.

Next, fix a point $z \in J(f)$ and a neighborhood U of z. Then there are infinitely many points of $J(f)$ not contained in the forward orbit $\{f^n(z)\}_{n=0}^{\infty}$. Indeed, this is clear if $f(z) = z$. And if $f(z) \neq z$, then at least one of these is not exceptional, by Theorem 2.3.3. Hence either $O^-(z)$ or $O^-(f(z))$ contains infinitely many points belonging to $J(f) - \{f^n(z)\}_{n=0}^{\infty}$.

Hence we can take three distinct points in $J(f) - \{f^n(z)\}_{n=0}^{\infty}$. Then the three-point condition implies that $\bigcup_{n=1}^{\infty} f^n(U)$ contains one of these three points, which implies that $(U - \{z\}) \cap J(f)$ is non-empty. Thus no $z \in J(f)$ are isolated. □

2.3.2 *The first fundamental theorem*

Now it is not difficult to show that periodic points are dense in the Julia set.

Theorem 2.3.7 (Fatou) *For every* $f \in \mathrm{Rat} \cup \mathrm{Ent}$, *periodic points are dense in the Julia set* $J(f)$.

Proof Suppose that there were a point $z_0 \in J(f)$, a neighborhood U of which contained no periodic points of f. We may also assume that U is simply connected and that z_0 has at least two distinct, and non-critical, preimages, say ζ_1 and ζ_2 different from z_0. Then there are disjoint neighborhoods V_j of ζ_j such that $U' = f(V_1) = f(V_2)$, which is a neighborhood of z_0 contained in U and disjoint from $V_1 \cup V_2$. Moreover, we have two conformal maps, branches of f^{-1}, h_1 and h_2 of U' onto V_1 and V_2, respectively.

Consider the family of functions

$$\phi_n(z) = \left(\frac{f^n(z) - h_1(z)}{f^n(z) - h_2(z)}\right) \Big/ \left(\frac{z - h_1(z)}{z - h_2(z)}\right)$$

on U'. Then since $f^n(z) \neq h_j(z)$ on U' for every n and j, every ϕ_n has three omitted values $0, 1, \infty$. Hence the three-point condition implies that $\{\phi_n\}$ is a normal family, which implies that U' should be contained in $F(f)$, which is a contradiction. □

A deeper result is the following.

Theorem 2.3.8 (The first fundamental theorem) *For every* $f \in \mathrm{Rat} \cup \mathrm{Ent}$, *repelling periodic points are dense in the Julia set* $J(f)$.

In the case of rational functions, we can use Theorem 2.3.7 to prove Theorem 2.3.8, which is due to Fatou, and we will explain such a proof in §4.2. Also note that the first proof in the case of transcendental entire functions was given by Baker by using Ahlfors theory on covering surfaces. Here we give a proof due to Schwick (1997).

For this purpose, first we recall that equicontinuity on $\widehat{\mathbb{C}}$ can be rewritten as follows.

Theorem 2.3.9 (Marty) *A family* $\mathcal{F}$ *of meromorphic functions is a normal family if and only if the spherical derivatives*

$$\chi[f](z) = \frac{2|f'(z)|}{1 + |f(z)|^2} \quad (f \in \mathcal{F})$$

are locally uniformly bounded.

Using this criterion, we obtain the following convergence theorem, which is interesting in itself.

Theorem 2.3.10 (Rescaling lemma of Zalcman) *A family $\mathcal{F}$ of meromorphic functions on the unit disk Δ is not a normal family if and only if there are sequences $\{f_n\}_{n=1}^{\infty}$ of elements of $\mathcal{F}$, $\{z_n\}_{n=1}^{\infty}$ of points in a compact set of Δ, and $\{\rho_n\}_{n=1}^{\infty}$ of positive numbers converging to 0 such that*

$$h_n(\zeta) = f_n(z_n + \rho_n \zeta)$$

converges to a non-constant meromorphic function $h(\zeta)$ uniformly on compact sets of $\mathbb{C}$.

Moreover, when $\mathcal{F}$ consists only of holomorphic functions, then also $h(\zeta)$ is holomorphic.

Proof Suppose that $\mathcal{F}$ is not a normal family, then by Marty's theorem, we can find a sequence $\{f_n\}$, a compact set $K_r = \{|z| \le r < 1\}$ of the unit disk, and a sequence $\{\zeta_n\}$ in K_r such that

$$\chi[f_n](\zeta_n) \to +\infty \quad (n \to +\infty).$$

Fix another r^* with $r < r^* < 1$, and set

$$M_n = \max_{z \in K_{r^*}} \left(1 - \frac{|z|^2}{r^{*2}}\right) \chi[f_n](z).$$

Then this maximum is attained by a point, say z_n, in the interior of K_{r^*}. Set

$$\rho_n = \frac{1}{M_n}\left(1 - \frac{|z_n|^2}{r^{*2}}\right),$$

and we have

$$\lim_{n \to +\infty} \frac{\rho_n}{r^* - |z_n|} = 0.$$

Moreover, if we set

$$h_n(\zeta) = f_n(z_n + \rho_n \zeta),$$

then h_n is defined at least on $K_{(r^* - |z_n|)/\rho_n}$ and satisfies

$$\chi[h_n](0) = \rho_n \chi[f_n](z_n) = 1.$$

Also a similar argument shows that $\{\chi[h_n]\}$ are locally uniformly bounded on $\mathbb{C}$, which implies that $\{h_n\}$ contains a subsequence which converges uniformly on compact sets of $\mathbb{C}$. Hence we may assume that the h_n converge to a meromorphic function h uniformly on compact sets of $\mathbb{C}$. Then since

$$\chi[h](0) = 1,$$

h should be non-constant. Here it is clear that, when all h_n are holomorphic, so is h.

Conversely, suppose that $\mathcal{F}$ is a normal family. Then for all sequences $\{f_n\}$, $\{z_n\}$ relative compact in Δ, and $\{\rho_n\}$ of positive numbers such that $\lim_{n\to\infty} \rho_n = 0$, Marty's theorem implies that

$$\chi[h_n](\zeta) = \rho_n \chi[f_n](z_n + \rho_n \zeta)$$

converges to 0 uniformly on compact sets of $\mathbb{C}$. Thus any limit should be a constant. □

Proof of Theorem 2.3.8 First assume that $f \in$ Ent. Let A be the set of all completely branched values of f. Then the corollary to Theorem 2.2.6 gives that A contains at most two points.

Now, fix a point w in $J(f) - A$, and a neighborhood U of w contained in $(\mathbb{C} - A)$. Then Zalcman's theorem gives sequences $\{f^{n_k}\}$, $\{z_k\}$ converging to a point w_0 in U, and $\{\rho_k\}$ of positive numbers such that $\lim_{k\to\infty} \rho_k = 0$ and that

$$h_k(\zeta) = f^{n_k}(z_k + \rho_k \zeta)$$

converges to a non-constant holomorphic function $h(\zeta)$ uniformly on compact sets of $\mathbb{C}$. Note that

$$f^{n_k+1}(z_k + \rho_k \zeta) \to (f \circ h)(\zeta).$$

Since w_0 does not belong to A, there are infinitely many distinct solutions $\{\zeta_j\}_{j=1}^{\infty}$ of $f(\zeta) - w_0 = 0$ satisfying $f'(\zeta_j) \neq 0$. If the limit h is transcendental and entire, similarly we can find some ζ_j such that a solution α of the equation $h(\zeta) - \zeta_j = 0$ satisfies $h'(\alpha) \neq 0$. Existence of such an α is clear when h is a polynomial.

On the other hand, from the construction,

$$f^{n_k+1}(z_k + \rho_k \zeta) - (z_k + \rho_k \zeta) \to (f \circ h)(\zeta) - w_0$$

uniformly on compact sets of $\mathbb{C}$, and since α is a solution of $f \circ h(\zeta) - w_0 = 0$, Hurwitz's theorem gives that there is a solution α_k of

$$f^{n_k+1}(z_k + \rho_k \zeta) - (z_k + \rho_k \zeta) = 0$$

converging to α. Here

$$\tilde{z}_k = z_k + \rho_k \alpha_k$$

is a fixed point of f^{n_k+1}, that is a periodic point of f, and since

$$\rho_k \left. \frac{df^{n_k+1}}{dz} \right|_{z=\tilde{z}_k} = \left. \frac{df^{n_k+1}}{d\zeta}(z_k + \rho_k \zeta) \right|_{\zeta=\alpha_k} \to (f \circ h)'(\alpha) \neq 0,$$

$(f^{n_k+1})'(\widetilde{z_k})$ tends to $+\infty$. Hence $\widetilde{z_k}$ is a repelling periodic point of f, and converges to w_0, which show the assertion when $f \in \mathrm{Ent}$.

The case that $f \in \mathrm{Rat}$ is easier. In this case, let A consist of ∞, $f(\infty)$ and all images of critical points, and fix a point w and a neighborhood U of w as before. Recall that there are at most four values such that every preimage is critical in this case, as is seen from the Riemann–Hurwitz formula.

Now Zalcman's theorem again gives sequences $\{f^{n_k}\}$, $\{z_k\}$ converging to a point w_0 in U, and $\{\rho_k\}$ of positive numbers converging to 0 such that

$$h_k(\zeta) = f^{n_k}(z_k + \rho_k \zeta)$$

tend to a non-constant meromorphic function $h(\zeta)$ uniformly on compact sets of $\mathbb{C}$.

Considering f^3 instead of f if necessary, we may assume that the degree of f is greater than 5. Then there are more than five simple solutions of $f(z) = w_0$, and we can find such α as before. Hence the same argument shows the assertion also in this case. □

2.4 Singular values and Fatou components

Definition For every element f of $\mathrm{Rat} \cup \mathrm{Ent}$, we call $\alpha \in \widehat{\mathbb{C}}$ a *singular value* if f is not a smooth covering map over any neighborhood of α. We denote the set of all singular values by $\mathrm{sing}(f^{-1})$.

If α is a non-singular value of f, then there exists a neighborhood V of α where every branch of f^{-1} in V is well defined and is a conformal map of V.

Remark In the case of a rational function f, $\mathrm{sing}(f^{-1})$ is nothing but the set of all critical values, that is images of critical points.

This set $\mathrm{sing}(f^{-1})$ plays very important roles in the investigations of complex dynamics. In this section, we explain how the singular values govern the structure of Fatou components.

2.4.1 Singular values and periodic components

In this subsection, we show some fundamental facts concerning the relationship between singular values and periodic Fatou components.

First, it is clear by definition that the immediate basin of any cycle of superattracting periodic points contains a singular value in the cycle. Such an assertion is still valid for attracting periodic points.

Theorem 2.4.1 *For every $f \in \mathrm{Rat} \cup \mathrm{Ent}$, the immediate basin of any cycle of attracting periodic points contains a singular value of f.*

Proof It is enough to consider the case of the immediate basin U of an attracting but not superattracting fixed point z_0 of f, where we may assume that $z_0 \in \mathbb{C}$.

Let ϕ be a solution of the Schröder equation

$$\phi \circ f(z) = \lambda \phi(z) \quad (|\lambda| < 1)$$

in a simply connected neighborhood V of z_0 in U, where $\lambda = f'(z_0)$. Then $\psi = \phi^{-1}$ is a conformal map of a neighborhood of the origin onto V with $\psi(0) = z_0$. If there are no singular values of f in U, then we can continue ψ analytically to a holomorphic map of $\mathbb{C}$ into U by using the equation

$$f(\psi(w)) = \psi(\lambda w).$$

This contradicts Picard's theorem, for the image is contained in a hyperbolic domain U. □

Theorem 2.4.2 *For every $f \in \mathrm{Rat} \cup \mathrm{Ent}$, the immediate basin of any cycle of parabolic periodic points contains a singular value of f.*

Proof Again, it is enough to consider the immediate basin U of a parabolic fixed point $z_0 \in \mathbb{C}$ with $f'(z_0) = 1$. Fix an attracting petal U_0 in U, and we have a conformal map ϕ, which we call the Fatou function in §1.2, of U_0 into $\mathbb{C}$ such that

$$\phi \circ f(z) = \phi(z) + 1$$

and that $\phi(U_0)$ contains the right half plane.

If $U \cap \mathrm{sing}(f^{-1})$ is empty, then we can continue the conformal map $\psi = \phi^{-1}$ of the right half plane to a holomorphic map of $\mathbb{C}$ into U by using the equation

$$f(\psi(w)) = \psi(w + 1).$$

This contradicts Picard's theorem again, and hence we have the assertion. □

To discuss the case of other periodic components, we first note the following.

Lemma 2.4.3 *Fix a domain Ω, and consider a family $\mathcal{F}$ consisting of all holomorphic branches of some $(f^m)^{-1}$ on Ω. Then $\mathcal{F}$ is a normal family.*

Proof It suffices to consider the case that $\Omega \cap J(f)$ is non-empty. Since periodic points are dense in the Julia set, there are finite sets A_1 and A_2 of periodic points of f such that A_j contains more than two points, $f(A_j) = A_j$ for each j, and $A_1 \cap A_2$ is empty. Then since $h(\Omega - A_j) \cap A_j = \varnothing$ for every $h \in \mathcal{F}$, $\mathcal{F}$ is a normal family on $\Omega - A_j$ for each j, which implies the assertion. □

Lemma 2.4.4 *Let Ω and $\mathcal{F}$ be as in the previous lemma. If $\Omega \cap J(f)$ is not empty, then every limit of any convergent sequence in $\mathcal{F}$ is a constant.*

Proof We may assume that Ω is sufficiently small. Let z_0 be a point in $\Omega \cap J(f)$, and suppose that a sequence $\{h_j\}$ of elements of $\mathcal{F}$ converges to a non-constant h uniformly on compact sets of Ω. Then $h(z_0) \in J(f)$, and Hurwitz's theorem means that h is univalent. Hence there is a sufficiently small neighborhood V of $h(z_0)$ such that $h_j(\Omega)$ contains V for every sufficiently large j. Letting h_j be a branch of $(f^{n_j})^{-1}$, we have $n_j \to \infty$ and

$$f^{n_j}(V) \subset f^{n_j} \circ h_j(\Omega) = \Omega.$$

But this contradicts the following proposition. □

Proposition 2.4.5 *Let $f \in$ Rat $\cup$ Ent, $\alpha \in J(f)$, and V an arbitrary neighborhood of α. Then for every compact set K containing no exceptional points, $f^n(V) \supset K$ for any sufficiently large n.*

Proof Let E be the set of all exceptional points. Then for every open set W containing points of $J(f)$,

$$\bigcup_{n=1}^{\infty} f^n(W) \supset \widehat{\mathbb{C}} - E,$$

for E is completely invariant.

Take the given V as W, and fix a repelling periodic point z_0 in V. Let V_0 be a sufficiently small neighborhood of z_0 contained in V. Then letting q be the period of z_0, we have

$$V_0 \subset f^q(V_0) \subset f^{2q}(V_0) \subset \cdots.$$

Also we have

$$\bigcup_{n=1}^{\infty} f^n(V_j) \supset \widehat{\mathbb{C}} - E \supset K$$

for every $V_0, V_1 = f(V_0), \ldots, V_{q-1} = f^{q-1}(V_0)$. Since K is compact, it is easy to see that there is an N such that $f^N(V_j) \supset K$ for every j, which implies that $f^n(V_0)$ contains K for every $n \geq N$. Since $V_0 \subset V$, the assertion follows. □

Now we denote the union of all forward orbits of singular values by $S^+(f)$:

$$S^+(f) = \bigcup_{n=0}^{\infty} f^n(\mathrm{sing}(f^{-1})).$$

Theorem 2.4.6 *Let $\{U_1, \ldots, U_q\}$ be a cycle of a Siegel disk or a Herman ring of $f \in \mathrm{Rat} \cup \mathrm{Ent}$. Then*

$$\overline{S^+(f)} \supset \bigcup_{j=1}^{q} \partial U_j.$$

Proof It suffices to consider an invariant Siegel disk or an invariant Herman ring U. Suppose that there is a point $\zeta \in \partial U$ having a neighborhood V such that $V \cap \overline{S^+(f)} = \varnothing$.

Since $f : U \to U$ is a conformal automorphism, we can consider the inverse $h : U \to U$ of f. Then the assumption implies that h^n can be extended to a holomorphic map of V. By Lemma 2.4.3, $\{h^n\}$ is a normal family on V and by Lemma 2.4.4, any limit function of a convergent subsequence should be a constant. This is impossible, for there is a subsequence converging to the identity on $U \cap V$. □

Remark For a polynomial P, the critical orbit $C^+(P)$ is nothing but the singular orbit $S^+(P)$. Furthermore, if P is hyperbolic, then any accumulating point of $C^+(P)$ is a (super)attracting periodic point by Theorems 2.4.1, 2.4.2, the classification theorem (Theorem 2.1.10), and the second fundamental theorem below. See §4.4 for the case of general rational functions.

Definition We call an irrationally indifferent periodic point in $J(f)$ of $f \in \mathrm{Rat} \cup \mathrm{Ent}$ a *Cremer point* of f.

Theorem 2.4.7 *Every Cremer point is contained in $\overline{S^+(f)}$.*

Proof Again it suffices to consider an irrationally indifferent fixed point z_0 in $J(f)$. Suppose that there is a neighborhood V of z_0 such that $S^+(f) \cap (V - \{z_0\}) = \varnothing$. Then there exists a single valued holomorphic branch h_n of $(f^n)^{-1}$ such that $h_n(z_0) = z_0$ for every n. Then again by Lemmas 2.4.3 and 2.4.4, we have a subsequence of $\{h_n\}$ converging to a constant, which contradicts the fact that $|(h^n)'(z_0)| = 1$. □

2.4.2 The Teichmüller space of complex dynamics

To see that singular values govern the stable family of complex dynamics, we first recall the definition of the Teichmüller spaces of complex dynamics, following McMullen and Sullivan (preprint).

Let f be an element of Ent ∪ Rat. Then f induces an equivalence relation on the domain D $(= \mathbb{C}$ or $\widehat{\mathbb{C}})$.

Definition We say that z and w are *grand orbit equivalent* if there are positive integers n and m such that $f^n(z) = f^m(w)$. For every $z \in D$, we denote the equivalence class of z by $[z]$, and call it the *grand orbit* of z.

Also, for any subset $U \subset D$, we set

$$[U] = \{w \mid w \in [z], \text{ for some } z \in U\}$$

and call it the *grand orbit* of U.

Next we define the stable family of the given f (also see §5.3).

Definition We say that two quasiconformal automorphisms ϕ_1 and ϕ_2 of D are *equivalent* if $\phi_1 = \psi \circ \phi_2$ with some conformal automorphism ψ of D.

The deformation space Def(D, f) of f is the set of all equivalence classes of quasiconformal automorphisms ϕ of D which satisfy

$$\phi \circ f = g \circ \phi$$

with other holomorphic maps $g = g_\phi$ of D. We say that such a g is *quasiconformally conjugate* to f.

The measurable mapping theorem in Chapter 1 implies that this space Def(D, f) is identified with the unit ball $M_1(D, f)$ of the space $M(D, f)$ of all invariant Beltrami differentials for f. Here, an *invariant Beltrami differential* μ for f on D is by definition a bounded measurable function on the whole plane which satisfies

$$f^*\mu(z) = \mu(f(z))\overline{f'(z)}/f'(z) = \mu(z)$$

almost everywhere.

Now, it is natural to identify those Beltrami differentials which induce the same map.

Definition Let $\mathrm{QC}_0(D, f)$ be the set of all quasiconformal automorphisms $F(z)$ of D admitting a *uniform quasiconformal isotopy to the identity compatible with* f: there are a K and a homotopy $F_t(z)$ with $0 \le t \le 1$ such that

(i) $F_0(z) = z, \quad F_1(z) = F(z)$,
(ii) for a fixed t, $F_t(z)$ is a K-quasiconformal self-map of D satisfying $F_t \circ f = f \circ F_t$.

The *Teichmüller space* $\mathrm{Teich}(D, f)$ of f is the quotient space of $\mathrm{Def}(D, f)$ by $\mathrm{QC}_0(D, f)$: two quasiconformal self-maps ϕ_1 and ϕ_2 of D determine the same point of $\mathrm{Teich}(D, f)$, or equivalently, are *Teichmüller equivalent* on D if there are an $F \in \mathrm{QC}_0(D, f)$ and a conformal automorphism ψ of D such that $\phi_2 \circ F = \psi \circ \phi_1$.

The Teichmüller space is equipped with the usual Teichmüller topology and has the natural complex structure induced from that of $M_1(D, f)$.

Now the structure of the Teichmüller space heavily depends on the set of singular values. In the rest of this section, we discuss the case that *the singular values of f are finite in number*. Recall that any $f \in \mathrm{Rat}$ always satisfies this assumption. As to Ent, see §3.1.2.

Definition We denote by $\widehat{J}(f)$ the closure of the set which consists of the grand orbit of $\mathrm{sing}(f^{-1})$ and the grand orbits of all periodic points of f. We denote the complement of $\widehat{J}(f)$ by $\widehat{F}(f)$. The subset $F^{dis}(f)$ of $\widehat{F}(f)$ consists of points whose grand orbits are discrete in $\widehat{F}(f)$, and is called the *discrete part*. Finally, we set $F^{fol}(f) = \widehat{F}(f) - F^{dis}(f)$, and call it the *foliated part*.

Since repelling periodic points of f are dense in $J(f)$ by the first fundamental theorem, $\widehat{J}(f) \supset J(f)$.

Next, since $\widehat{J}(f)$, $F^{dis}(f)$, and $F^{fol}(f)$ are completely invariant, every $\mu \in M(D, f)$ restricted to any one of them is again an invariant Beltrami differential for f. Hence $M(D, f)$ is the product of three spaces:

$$M(\widehat{J}(f), f), \quad M(F^{dis}(f), f), \quad M(F^{fol}(f), f).$$

Here $M(X, f)$ is the set of all invariant Beltrami differentials whose supports are contained in the completely invariant set X. The unit balls of $M(F^{dis}(f), f)$ and $M(F^{fol}(f), f)$ are considered as the set of equivalence classes of quasiconformal maps of $F^{dis}(f)$ and $F^{fol}(f)$, respectively, which we denote by $\mathrm{Def}(F^{dis}(f), f)$ and $\mathrm{Def}(F^{fol}(f), f)$.

Then we can define the Teichmüller spaces

$$\mathrm{Teich}(F^{dis}(f), f), \quad \text{and} \quad \mathrm{Teich}(F^{fol}(f), f),$$

as the quotient spaces of $\mathrm{Def}(F^{dis}(f), f)$ and $\mathrm{Def}(F^{fol}(f), f)$ by the sets of all quasiconformal self-homeomorphisms $F(z)$ of $F^{dis}(f)$ and $F^{fol}(f)$, respectively, admitting a uniform quasiconformal isotopy to the identity relative to the boundary and compatible with f.

Theorem 2.4.8 (McMullen–Sullivan) *Suppose that* $\mathrm{sing}(f^{-1})$ *is a finite set. Then the Teichmüller space* $\mathrm{Teich}(D, f)$ *can be identified with the product*

$$M_1(\widehat{J}(f), f) \times \mathrm{Teich}(F^{fol}(f), f) \times \mathrm{Teich}(F^{dis}(f), f).$$

Proof Since every quasiconformal homeomorphism $F(z)$ in $\mathrm{QC}_0(D)$ fixes periodic points of f pointwise, $F(z)$ is equal to the identity on $\widehat{J}(f)$, and hence can be considered as a quasiconformal homeomorphism of $F^{dis}(f)$ and of $F^{fol}(f)$ which fixes the boundary pointwise. Hence Teichmüller equivalence on D induces those on $F^{dis}(f)$ and on $F^{fol}(f)$, which implies that the above decomposition of $\mu \in M_1(D, f)$ induces a holomorphic map of $\mathrm{Teich}(D, f)$ onto the desired product.

Since the converse is clear, we conclude with the assertion. □

The important fact is that the dimension of the Teichmüller space can be bounded by the number of singular values.

Theorem 2.4.9 *Let q be the number of points in* $\mathrm{sing}(f^{-1})$*; then*

$$\dim_{\mathbb{C}} \mathrm{Teich}(D, f) \leq q.$$

Proof To prove this theorem, it suffices to show that the set $\mathrm{Rep}(f)$ of all holomorphic self-maps h of D quasiconformally conjugate to f has a dimension at most $q + N$, where $N = 2$ or 3 according as $D = \mathbb{C}$ or $\widehat{\mathbb{C}}$.

More generally, we consider the set

$$T_f = \{\text{holomorphic self-maps } h \text{ of } D \text{ topologically equivalent to } f\}.$$

Here we say that h is *topologically equivalent* to f if there are homeomorphisms φ and ϕ of D such that

$$\phi \circ f = h \circ \varphi.$$

Fix an N-tuple $\{\beta_j\}_{j=1}^N$ of mutually distinct points in D which satisfies $f(\beta_j) \notin \mathrm{sing}(f^{-1})$ for every j. Let $T_f(\{\beta_j\})$ be the set of all $h \in T_f$ such that there is a φ, as in the definition of topological equivalence, satisfying $\varphi(\beta_j) = \beta_j$ for every j. Then, T_f is covered by $\{T_f(\{\beta_j\})\}$, where $\{\beta_j\}$ moves all N-tuples which are pairwise disjoint and $f(\{\beta_j\}) \cap \mathrm{sing}(f^{-1}) = \varnothing$. Fixing such a $\{\beta_j\}$, we set $\mathrm{sing}(f^{-1}) = \{a_1, \ldots, a_q\}$ and $a_{q+j} = f(\beta_j)$. Then we have a point $\mathcal{T}(h) = (\phi(a_1), \phi(a_2), \ldots, \phi(a_{q+N}))$ in D^{q+N}, where $\mathrm{sing}(h^{-1}) = \{\phi(a_1), \ldots, \phi(a_q)\}$ and $\phi(a_{q+j}) = h(\beta_j)$, and hence $\mathcal{T}(h)$ depends only on h.

Since every element h of $T_f(\{\beta_j\})$ is *quasiconformally equivalent* to f, i.e., it can be represented as $\phi \circ f \circ \varphi^{-1}$ with quasiconformal maps ϕ and φ of D, we can define a topology on $T_f(\{\beta_j\})$ compatible with that of $\mathrm{Rep}(f)$. Hence Theorem 2.4.9 follows from the lemma below. □

Lemma 2.4.10 *The map $\mathcal{T}$ of $T_f(\{\beta_j\})$ into D^{q+N} is injective.*

Proof It suffices to show that, if $X = \mathcal{T}(h_0) = \mathcal{T}(h_1)$ for two elements $h_0 = \phi_0 \circ f \circ \varphi_0^{-1}$ and $h_1 = \phi_1 \circ f \circ \varphi_1^{-1}$ of $T_f(\{\beta_j\})$, then $h_0 = h_1$. Here the assumption implies

$$h_0 \circ (\varphi_0 \circ \varphi_1^{-1}) = (\phi_0 \circ \phi_1^{-1}) \circ h_1.$$

And $\phi_0 \circ \phi_1^{-1}$ admits a uniformly quasiconformal isotopy to the identity which fixes X pointwise. Since each $h_j : D - h_j^{-1}(X) \to D - X$ is a covering map, this isotopy can be lifted by the above equation to an isotopy of $\varphi_0 \circ \varphi_1^{-1}$ to a conformal automorphism of D, which fixes three points of $\widehat{\mathbb{C}}$, and hence should be the identity. Thus we conclude with the assertion. □

Remark It is known that the space $\mathrm{Teich}(F^{fol}, f)$ is the product of several unit disks and $\mathrm{Teich}(F^{dis}, f)$ is just the Teichmüller space of a union of Riemann surfaces.

2.4.3 The second fundamental theorem

The second fundamental theorem asserts that finiteness of the number of singular values implies non-existence of wandering domains. In this subsection, we show the following.

Theorem 2.4.11 (Weak second fundamental theorem) *Every $f \in \mathrm{Rat} \cup \mathrm{Ent}$ with a finite number of singular values has no simply connected wandering domains.*

To show this theorem, we start with the following.

Definition For a hyperbolic domain Ω, we denote the hyperbolic metric on Ω by $\lambda_\Omega(z)^2|dz|^2$. We say that a Beltrami differential μ is *harmonic* if, with some holomorphic function φ on Ω,

$$\mu(z) = \lambda_\Omega(z)^{-2}\overline{\varphi(z)}.$$

We denote by $H(\Omega)$ the space of all harmonic Beltrami differentials on Ω.

A Beltrami differential μ on the invariant open set $[\Omega]$, the grand orbit of a wandering domain Ω, is said to be *eventually harmonic* for f if the following two conditions hold:

(i) μ is invariant, i.e., $f^*\mu(z) = \mu(z)$ for almost all $z \in [\Omega]$;
(ii) $\mu|_{f^n(\Omega)}$ is harmonic for any sufficiently large n.

We denote the space of all eventually harmonic (invariant) Beltrami differentials on $[\Omega]$ for f by $EH([\Omega])$.

For every μ in the unit ball of $EH([\Omega])$, there is a quasiconformal mapping φ^μ of $\mathbb{C}$ with the Beltrami coefficient μ on $[\Omega]$, which fixes 0 and 1 and is conformal in the interior of $\mathbb{C} - [\Omega]$. And

$$\theta(\mu) = \varphi^\mu \circ f \circ (\varphi^\mu)^{-1}$$

defines a holomorphic map $\theta : EH([\Omega]) \to \mathrm{Rep}(f)$.

Lemma 2.4.12 *The dimension of $EH([\Omega])$ is at most $q + N$.*

Proof Assume that $EH([\Omega])$ contains $q+N+1$ linearly independent elements, say $\mu_1, \mu_2, \ldots, \mu_{q+N+1}$. We choose a sufficiently small neighborhood W of the origin in $\mathbb{C}^{q+N+1}$ such that, for every $(c_1, \ldots, c_{q+N+1}) \in W$,

$$\mu = c_1\mu_1 + c_2\mu_2 + \cdots + c_{q+N+1}\mu_{q+N+1}$$

satisfies $\|\mu\|_\infty < 1$. Consider the map

$$(c_1, c_2, \ldots, c_{q+N+1}) \mapsto \theta(\mu) = f^\mu;$$

then it is a holomorphic map of W into $\mathrm{Rep}(f)$ and hence can be considered one into D^{q+N}. Hence the dimension of the inverse image of some point in $\mathrm{Rep}(f)$ is at least 1 and thus it contains a non-degenerate submanifold V. This implies that there is an injective holomorphic map of $\{\zeta \mid |\zeta| < 2\}$ into V such that

$$\nu_\zeta(z) = \sum_{j=1}^{q+N+1} h_j(\zeta)\mu_j(z)$$

is eventually harmonic for f. Set

$$\Phi_\zeta(z) = \varphi^{\nu_\zeta}(z),$$

and fix $\zeta \neq 0$. The fact that V is contained in the inverse image of the same point implies that

$$\Phi_\zeta \circ f \circ \Phi_\zeta^{-1} = \Phi_0 \circ f \circ \Phi_0^{-1}.$$

Hence $\Phi_0^{-1} \circ \Phi_\zeta$ is commutative with f. Since $\Phi_0^{-1} \circ \Phi_\zeta$ is continuous with respect to ζ and is the identity for $\zeta = 0$, it fixes every repelling periodic point. The first fundamental theorem implies that

$$\Phi_0^{-1} \circ \Phi_\zeta = \mathrm{id}$$

on $J(f)$ and thus $\Phi_0^{-1} \circ \Phi_\zeta$ fixes each component Ω' of $[\Omega]$.

Since ν_ζ is an eventually harmonic Beltrami differential for f, $\nu_0|_{f^n(\Omega)}$ and $\nu_\zeta|_{f^n(\Omega)}$ are harmonic for sufficiently large n. Moreover, $\Phi_0^{-1} \circ \Phi_\zeta$ is isotopic relative to the boundary to the identity on each component of $F(f)$. Hence Φ_0 and Φ_ζ determine the same point of the Teichmüller space of $f^n(\Omega)$. Since the harmonic Beltrami differential is uniquely determined from the Teichmüller equivalence class, they should coincide on $f^n(\Omega)$ for a sufficiently large n, which implies that $\nu_0 = \nu_\zeta$. This is a contradiction. □

Now the weak second fundamental theorem can be restated as follows.

Proposition 2.4.13 *If there is an eventually singular-value free, simply connected wandering domain, then* Teich(F^{dis}, f) *is of infinite dimension.*

Proof Assume that f has an eventually singular-value free, simply connected wandering domain Ω. Then we may assume that

$$\left(\bigcup_{n \geq 0} f^n(\Omega)\right) \cap \mathrm{sing}(f^{-1}) = \varnothing.$$

Hence $f^n|_\Omega$ is a conformal mapping to $f^n(\Omega)$. Set $h_n = (f^n|_\Omega)^{-1}$. Since $H(\Omega)$ can be identified with some space of holomorphic functions on Ω which contains bounded ones, it is an infinite-dimensional space. Let $\mu_1, \mu_2, \ldots$ be linearly independent elements of $H(\Omega)$ and define

$$\hat{\mu}_i(z) = (h_n)^* \mu_i(z)$$

for $z \in f^n(\Omega)$ $(n \geq 0)$, and then set

$$\hat{\mu}_i(z) = (f^n)^* \hat{\mu}_i(z)$$

for z satisfying $f^n(z) \in f^m(\Omega)$ $(n \geq 0, m \geq 0)$. Then $\hat{\mu}_i \in EH([\Omega])$ and $\hat{\mu}_1, \hat{\mu}_2, \ldots$ are linearly independent, which implies the assertion. □

In §3.1 and §4.2, we will show that every $f \in$ Rat and $f \in$ Ent with a finite number of singular values has no multiply connected wandering domains. Assuming these results, we obtain the following.

Theorem 2.4.14 (The second fundamental theorem) *Every* $f \in \mathrm{Rat} \cup \mathrm{Ent}$ *with a finite number of singular values has no wandering domains.*

As an application, we have the precise dimension of the Teichmüller spaces in this case. To state this, we say that z and z' are *foliated equivalent* if the closures of $[z]$ and $[z']$ are coincident. Let N_{AC} be the number of the foliated equivalence classes of acyclic (i.e. neither periodic nor preperiodic) singular values in the Fatou set. Also we denote by $N(P)$ and $N(HR)$ the numbers of cycles of parabolic periodic points and of Herman rings, respectively.

Theorem 2.4.15 (McMullen–Sullivan) *For every* $f \in \mathrm{Rat} \cup \mathrm{Ent}$ *with a finite number of singular values,*

$$\dim_{\mathbb{C}} \mathrm{Teich}(D, f) = \dim_{\mathbb{C}} M(J(f), f) + N_{AC} + N(HR) - N(P).$$

Here the second fundamental theorem states that there are no wandering domains. Further, when $f \in$ Ent, the assumption implies that there are no Baker domains, as will be shown in §3.1. Also, we have the following.

Lemma 2.4.16 *The set* $\widehat{J}(f) \cap F(f)$ *consists of a countable number of points and analytic curves. In particular,*

$$M(\widehat{J}(f), f) = M(J(f), f).$$

Proof A periodic point in $F(f)$ is either (super)attracting or in a Siegel disk. Hence the closure of the grand orbit is a countable set.

As to a singular value z, if z is in $F^{dis}(f)$, then the closure of $[z]$ in $F^{dis}(f)$ is countable in number. If z is in $F^{fol}(f)$, then z is in the grand orbit of either a Siegel disk, or a Herman ring, or a superattracting basin. Then the closure of $[z]$ consists of a countable number of analytic curves. □

Proof of Theorem 2.4.15 Let W be a periodic component of $F^{fol}(f)$, and k the number of the foliated equivalence classes of acyclic singular values in W. When W is an invariant Herman ring, the closure of $[\mathrm{sing}(f^{-1})]$ divides W into $k+1$ ring domains. When W is an invariant Siegel disk, it divides W

into k ring domains and a once-punctured disk. And when W is an invariant superattracting component, it divides W into domains, which contain k ring domains whose grand orbits are disjoint and cover all others.

The crucial fact here is that each of such ring domains as above contributes one complex dimension to the Teichmüller space $\text{Teich}(F^{fol}(f), f)$, while each once-punctured disk does nothing.

Next, every periodic component W of $F^{dis}(f)$ is contained in either a parabolic basin or an attracting basin. Let k be as before, then the space of all grand orbit equivalence classes in $[W]$ is a Riemann surface of $(0, k+2)$-type or of $(1, k)$-type, respectively. Hence the Teichmüller space is $(k-1)$- or k-dimensional.

Thus summing up, we conclude that

$$\dim_{\mathbb{C}} \text{Teich}(F^{fol}, f) + \dim_{\mathbb{C}} \text{Teich}(F^{dis}, f) = N_{AC} + N(HR) - N(P),$$

and the assertion follows by Theorem 2.4.8 and Lemma 2.4.16. □

Remark In the case of rational functions of degree k (≥ 2), we will show in §4.3 that $N_{AC} + N(HR) \leq 2k - 2$.

3
Dynamics of entire functions

The point at infinity is the essential singularity of a transcendental entire function. The Picard theorem shows that, in a neighborhood of the point at infinity, the action of such a function is strongly 'explosive'. Thus, in general, iteration of transcendental entire functions is much more complicated than that of polynomials.

In §3.1, first, we explain several fundamental properties of the Julia and the Fatou sets of a general entire function. Next, we consider some special classes of transcendental entire functions, including one which satisfies the assumption of the second fundamental theorem. We also direct our attention to the set of escaping points. The relationship between the Julia set and the set of escaping points illuminates the other side of the character of the Julia set of a transcendental entire function.

In §3.2, we study the exponential family. Dynamical properties of exponential functions are rather better known than those of any other transcendental functions. The Julia set of a certain kind of exponential function is characterized as the Cantor bouquet. We also discuss the size of it. Finally, we briefly look at the representation space of the exponential family.

In §3.3, we show the topological completeness of the decorated exponential families, which is crucial to discussing the representation spaces of them.

In §3.4 we exhibit various examples of transcendental entire functions whose Fatou sets contain wandering domains or Baker domains, which never appear in the case of polynomials and rational functions.

3.1 Basic dynamical properties of entire functions

First, we consider topological properties of the Fatou sets and the Julia sets, especially by looking at multiply connected Fatou components. The existence of such components will be postponed to §3.4. Next, we consider some special classes and show the second fundamental theorem for some transcendental entire functions. We also study escaping points which give a new character to the Julia sets of transcendental entire functions. For example, they play an important role in estimating the sizes of the Julia sets of exponential maps in §3.2.

3.1.1 Topological properties of the Fatou set and the Julia set

For an arbitrary polynomial, the point at infinity is always a superattracting fixed point. Hence the Julia set of a polynomial is compact in $\mathbb{C}$. On the other hand, we have the following.

Proposition 3.1.1 *The Julia set of a transcendental entire function is unbounded in* $\mathbb{C}$.

Proof Let f be a transcendental entire function. Choose a point ζ in $J(f)$ which is not an exceptional value in the sense of Picard. Let U be any neighborhood of the point at infinity. The Picard theorem shows that there is a point η in U satisfying $f(\eta) = \zeta$. Since the Julia set is backward invariant, η is in $J(f)$. Thus $J(f)$ is unbounded. □

Again, recall that, in the case of a polynomial, the immediate basin of the point at infinity is completely invariant. Hence its boundary is the Julia set of the polynomial. It follows that all Fatou components except for this basin are simply connected. Further, it is infinitely connected, if it contains some finite critical points. The situation changes drastically in the case of transcendental entire functions.

First of all, we study basic properties of multiply connected Fatou components of transcendental entire functions.

Proposition 3.1.2 *The Fatou set of an entire function contains no cycles of Herman rings.*

Proof Suppose that there were a Herman ring H. Since $\{f^n\}_{n=1}^{\infty}$ is uniformly bounded on H, the maximum principle shows that $\{f^n\}_{n=1}^{\infty}$ is uniformly bounded also in the bounded component U of $\mathbb{C} - H$. This is a contradiction, for U contains points of $J(f)$. □

Remark This proposition does not imply non-existence of multiply connected components of the Fatou set for transcendental entire functions. We will give an example of a multiply connected Fatou component due to Baker (Theorem 3.4.1).

In what follows, we denote by $\operatorname{ind}_a \gamma$ the winding number of a closed curve γ with respect to $a \in \mathbb{C}$. The following lemma is crucial.

Lemma 3.1.3 *Let f be in* Ent *and D a multiply connected Fatou component. Then $\{f^n\}_{n=1}^{\infty}$ converges uniformly to ∞ on compact sets of D. Moreover, for any Jordan curve γ in D which is not contractible in D,* $\operatorname{ind}_0 f^n(\gamma) \neq 0$ *for all sufficiently large n.*

Proof Assume that there were a subsequence $\{f^{n_k}\}_{k=1}^{\infty}$ which is uniformly bounded on γ. Then as in the proof of Proposition 3.1.2, we have a contradiction.

Next, we assume that there were a subsequence $\{f^{n_k}\}_{k=1}^{\infty}$, each of whose elements satisfies $\operatorname{ind}_0 f^{n_k}(\gamma) = 0$. This means that f^{n_k} does not have zeros in the domain U surrounded by γ. Hence the minimum value of $|f^{n_k}|$ in $\overline{U}$ is obtained on γ. Thus $\{f^{n_k}\}$ converges to ∞ in U. On the other hand, since U contains points of $J(f)$, which is a perfect set, where repelling periodic points are dense. This is a contradiction. □

Remark In general, an attracting or parabolic periodic component is either simply or infinitely connected. But, in the case of transcendental entire functions, it is always simply connected.

Proposition 3.1.4 *If $f \in$* Ent *is bounded on a curve Γ which tends to ∞, then every component of $F(f)$ is simply connected.*

Proof Assume that there were a Fatou component D which is not simply connected. We take a Jordan curve γ in D which is not contractible in D. Lemma 3.1.3 shows that $f^n(\gamma) \cap \Gamma \neq \varnothing$ for all sufficiently large n. Take a point $z_n \in f^n(\gamma) \cap \Gamma$, and we have $f(z_n) \to \infty$. This is a contradiction. □

Theorem 3.1.5 *Every unbounded Fatou component of a transcendental entire function is simply connected.*

Proof Let f be in Ent, and assume that there is an unbounded, multiply connected Fatou component D. Let γ_0 be a Jordan curve in D which is not contractible in D. From Lemma 3.1.3, $\gamma_n = f^n(\gamma_0)$ intersects D for all sufficiently large n. Hence D is forward invariant. Let V be a bounded domain which contains γ_0 and γ_1 and whose closure is contained in D. Using Lemma 3.1.3 again, we see that $h_n(z) = \log|f^n(z)|$ is a positive harmonic function on V for all sufficiently large n. Since $\gamma_0 \cup \gamma_1$ is compact, by the Harnack inequality, there is a constant c independent of n such that $h_n(z_1) \leq ch_n(z_0)$ for all $z_j \in \gamma_j$ $(j = 0, 1)$ and for all sufficiently large n. Hence we have $|f^n(z_1)| \leq |f^n(z_0)|^c$.

On the other hand, since f is transcendental, there is a point ζ_n on γ_n which satisfies $|f(\zeta_n)| > |\zeta_n|^c$ for all sufficiently large n. Choose such an n, take $z_0 \in \gamma_0$ satisfying $\zeta_n = f^n(z_0)$ and set $z_1 = f(z_0) \in \gamma_1$; then we have $|f^n(z_1)| > |f^n(z_0)|^c$. This is a contradiction. □

The theorem above shows the following difference between the Julia sets of transcendental entire functions and those of polynomials.

Corollary *The Julia set of a transcendental entire function is never totally disconnected.*

Theorem 3.1.6 *Let f be in* Ent *and assume that $F(f)$ contains an unbounded component D. Then every component of $F(f)$ is simply connected.*

Proof If there is a multiply connected component U, then, by Theorem 3.1.5, U is bounded and so is $f^n(U)$. Thus $f^n(U)$ does not intersect D for all n. This contradicts Lemma 3.1.3. □

We give a remarkable theorem on multiply connected components of the Fatou sets of transcendental entire functions.

Theorem 3.1.7 *For $f \in$* Ent, *a multiply connected component of $F(f)$ is a wandering domain.*

Proof By Theorem 3.1.5, a multiply connected component D of $F(f)$ is bounded. Hence $f^n(D)$ cannot be contained in any Baker domain for every n. Unless D is a wandering component, Theorem 2.1.10 shows that $f^n(D)$ is eventually contained in either an immediate attracting basin, an immediate parabolic basin, or a cycle of Siegel disks. Then $\{f^n\}_{n=1}^{\infty}$ is uniformly bounded in any compact set of D. This contradicts Lemma 3.1.3. □

3.1.2 Special classes of entire functions

In this section, we consider the following three classes of entire functions.

$$\mathcal{B} = \{f \in \text{Ent} \mid \text{sing}(f^{-1}) \text{ is a bounded set}\},$$

$$\mathcal{S} = \{f \in \text{Ent} \mid \text{sing}(f^{-1}) \text{ is a finite set}\},$$
$$\mathcal{C} = \{f \in \text{Ent} \mid d(S^+(f), J(f)) > 0\}.$$

Here $S^+(f) = \bigcup_{n \geq 0} f^n(\text{sing}(f^{-1}))$ and recall that $d(A, B)$ means the Euclidean distance between sets A and B in $\mathbb{C}$.

We call a transcendental entire function of *bounded singular type* if it belongs to $\mathcal{B}$. A transcendental entire function is said to be of *finite singular type* or to belong to the *Speiser class* if it belongs to $\mathcal{S}$. We will see that the properties of functions belonging to $\mathcal{S}$ resemble those of rational functions in the sense that their Fatou sets contain no wandering domain and no Baker domain.

We now explain some properties of these families. First, we take an $f \in \mathcal{B}$. In what follows, we denote the open disk centered at z of radius $r > 0$ by $D(z, r)$. Choose $R > 0$ so that $\text{sing}(f^{-1}) \subset D(0, R/2)$ and set $A = \mathbb{C} - \overline{D(0, R)}$ and $G = f^{-1}(A)$. Each component V of G is simply connected and the boundary consists of only one analytic curve which tends to infinity along both directions. Thus $f : V \to A$ is a universal covering map. Further, we may assume $|f(0)| < R$, that is, $0 \notin G$. Hence, for any component W of $U = \log(G)$, $\exp : W \to G$ is a conformal map. It follows that there is a holomorphic function φ on U which satisfies

$$f \circ \exp = \exp \circ \varphi$$

and maps each component of U conformally onto $H = \log(A) = \{\zeta \mid \text{Re}\,\zeta > \log R\}$.

Theorem 3.1.8 *Let f be in $\mathcal{B}$. Then, for all $z \in F(f)$, $\{f^n(z)\}_{n=0}^{\infty}$ does not converge to ∞.*

Proof Assume that, for some point $z_0 \in F(f)$, $\{f^n(z_0)\}_{n=0}^{\infty}$ converges to ∞. Then $\{f^n\}_{n=0}^{\infty}$ converges uniformly on some $B_0 = D(z_0, r)$. If necessary, we change z_0 and we may assume that $B_n = f^n(B_0) \subset G$ for all n. We denote a component of $\log(B_0)$ by C_0 and set $C_n = \varphi^n(C_0)$. Hence, $\exp(C_n) = B_n$ and $C_n \subset U$. Note that the vertical width of a component of U is no longer than 2π. Hence, considering the inverse function of φ on the disk $D(\varphi(z), \text{Re}\,\varphi(z) - \log R)$ for each $z \in U$, or more precisely, considering

$$\phi(\zeta) = \frac{\varphi^{-1}(\varphi(z) + (\text{Re}\,\varphi(z) - \log R)\zeta) - z}{(\varphi^{-1})'(\varphi(z))(\text{Re}\,\varphi(z) - \log R)}$$

on the unit disk, and applying the Koebe $\frac{1}{4}$-theorem, we have

$$|\varphi'(z)| \geq \frac{1}{4\pi}(\operatorname{Re}\varphi(z) - \log R).$$

Next we take $\zeta_0 \in C_0$ and set $\zeta_n = \varphi^n(\zeta_0) \in C_n$. Besides, by d_n we denote the supremum of the radius of the disk centered at ζ_n which is contained in C_n. Using again the Koebe $\frac{1}{4}$-theorem, we obtain

$$d_{n+1} \geq \frac{1}{4} d_n |\varphi'(\zeta_n)|.$$

Since $\operatorname{Re}\zeta_n$ converges uniformly to ∞ on C_0 by the definition, we have $|\varphi'(\zeta_n)| \to \infty$, which implies $d_n \to \infty$. This is a contradiction. □

Corollary 1 *If $f \in \mathcal{B}$, then $F(f)$ contains no Baker domains.*

By Lemma 3.1.3, we also obtain the following.

Corollary 2 *If $f \in \mathcal{B}$, then every component of $F(f)$ is simply connected.*

Thus we have the following fundamental theorem.

Theorem 3.1.9 (The second fundamental theorem for the class $\mathcal{S}$) *If f is in $\mathcal{S}$, then $F(f)$ contains no wandering domains.*

Remark Fatou asked whether there might be wandering domains in the iteration of rational or entire functions. And about sixty years later, Sullivan answered that, in the case of rational functions, there are no wandering domains. This was proved by using quasiconformal maps, which play crucial roles in the modern theory of complex dynamics (see §2.4.3 and §4.2).

Proof Since $f \in \mathcal{S}$, the weak second fundamental theorem (Theorem 2.4.11) implies that there are no simply connected wandering domains. On the other hand, Corollary 2 to Theorem 3.1.8 implies that there are no multiply connected ones. □

Next, we show some consequences of this theorem. Because of $\mathcal{S} \subset \mathcal{B}$, using Corollary 1 to Theorem 3.1.8 and Theorem 3.1.9, we can classify the orbit of a point in $F(f)$ for $f \in \mathcal{S}$ as follows.

Theorem 3.1.10 *Assume that $f \in \mathcal{S}$ and $F(f) \neq \varnothing$. Then, for any $z \in F(f)$, the orbit of z is eventually contained in either an immediate attracting basin, an immediate parabolic basin, or a cycle of Siegel disks.*

Combining Theorems 2.4.1, 2.4.2, 2.4.6 and 3.1.9, we have the following.

Theorem 3.1.11 *Assume that $f \in \mathcal{S}$. If the orbit of each point in* $\operatorname{sing}(f^{-1})$ *either is preperiodic or converges to ∞, then $F(f) = \varnothing$.*

Corollary *The Fatou set of e^z is empty.*

Remark Whether $F(e^z)$ is empty or not is also a problem posed by Fatou. The first answer to it was given by Misiurewicz (1981), whose proof was to study the action of e^z minutely.

We will see by Example 3.1.1 that $\mathcal{B} - \mathcal{S} \neq \varnothing$. It is an unsolved problem whether the Fatou set of a function of singular bounded type contains wandering domains. But we obtain the following variant of the second fundamental theorem.

Theorem 3.1.12 *If $f \in \mathcal{B} \cap \mathcal{C}$, then $F(f)$ contains no wandering domains.*

To show this, we note the following two results on the limit functions of wandering domains.

Proposition 3.1.13 *On a wandering domain of $f \in$* Ent, *each limit function of $\{f^n\}_{n=1}^{\infty}$ is a constant (including ∞). In particular, if D is multiply connected, then it is always ∞. Moreover if the limit constant is finite, then it is in the Julia set.*

Proof Assume a subsequence $\{f^{n_i}\}_{i=1}^{\infty}$ converges uniformly to a non-constant limit function φ on compact sets of D. This induces a contradiction because $\varphi(D)$ and $f^{n_i}(D)$ are open sets and $f^{n_i}(D)$ do not intersect each other. Let $\zeta \neq \infty$ be a limit function on D. Then each $f^{n_i}(D)$ is a component of $F(f)$. Hence we have $\zeta \in J(f)$.

If D is multiply connected, then the claim is obvious due to Lemma 3.1.3. □

Remark In §3.4, we give examples of simply connected wandering domains where the limit function is ∞ only. Also we will give the example constructed by Eremenko and Lyubich (Theorem 3.4.3) of a transcendental entire function whose Fatou set contains a wandering domain on which there are infinitely many limit functions.

Moreover we know the following theorem due to Bergweiler, Haruta, Kriete, Meier and Terglane (1993).

Theorem 3.1.14 *Let D be a wandering domain of $f \in \text{Ent}$ and X the derived set of $S^+(f)$. Then every limit function of $\{f^n\}_{n=0}^{\infty}$ on D is a constant contained in $X \cup \{\infty\}$.*

Proof If D is multiply connected, then the claim follows by Lemma 3.1.3.

Suppose that D is a simply connected wandering domain and that $\alpha \in \mathbb{C} - X$ is the limit of some subsequence $\{f^{n_k}\}$ on D. Then we may assume that all $f^{-n_k} : f^{n_k}(D) \to D$ are well defined and univalent. In particular, the $f^{n_k}|_D$ are univalent.

Assume that $\alpha = 0$. Choose $r > 0$ so that

$$D(0, r) \cap (S^+(f) - \{0\}) = \emptyset.$$

Take $z_0 \in \mathbb{C}$ and $R > 0$ such that $\overline{D(z_0, R)} \subset D$, and we may assume that $f^{n_k}(D(z_0, R)) \subset D(0, r) - \{0\}$. Then the Koebe $\frac{1}{4}$-theorem implies that

$$|(f^{n_k})'(z_0)| \leq 4|f^{n_k}(z_0)|/R.$$

Define a function $g_k : D(z_0, R) \to H = \{z \mid \text{Re}\, z < \log r\}$ by setting $g_k(z) = \log f^{n_k}(z)$ with some branch of the logarithm. Then the inverse functions $h_k : H \to \mathbb{C}$ of g_k are well defined.

Suppose that every h_k is univalent. Then since $h_k(H)$ omits every cycle of periodic points of f outside of $D(z_0, r)$, an argument similar to that in the proof of Theorem 3.1.8 shows that $(h_k)'(g_k(z_0))$ tends to 0. This contradicts the fact that $(h_k)'(g_k(z))(g_k)'(z) = 1$, for $|(g_k)'(z_0)| \leq 4/R$ by the above inequality. So we may assume that the h_k are not univalent in H. Then h_k is periodic with period $2\pi l_k i$ and univalent in $\{z \mid \text{Re}\, z < \log r,\ c < \text{Im}\, z < c + 2\pi l_k\}$ for all k. Further, if $l_k \to \infty$, then we have a contradiction similarly as above. Hence we may assume that all l_k are the same l.

Now, set $G_k = \exp(g_k/l)$ and $H_k(z) = h_k(l \log z)$. Then we have $H_k(G_k(z)) = z$ for $z \in D(z_0, R)$. Further, by the Koebe $\frac{1}{4}$-theorem, we have $|G_k'(z_0)| \to 0$. On the other hand, H_k can be continued univalently on $D(0, r^{1/l})$, and again by the Koebe $\frac{1}{4}$-theorem, we have

$$|H_k'(z_k)| \leq \frac{4M}{r^{1/l} - |z_k|}$$

with $z_k = G_k(z_0)$. Since $|z_k| \to 0$, this contradicts the fact that $H_k'(z_k)G_k'(z_0) = 1$. □

Proof of Theorem 3.1.12 If $F(f)$ contains a wandering domain D, then Proposition 3.1.13 and Theorem 3.1.8 show that the limit a of $\{f^n\}_{n=0}^{\infty}$ on D is in

Fig. 3.1. The Julia set of $(\sin z)/z$

$J(f)$. Denoting the derived set of $S^+(f)$ by X, from Theorem 3.1.14 we have $a \in X$. This contradicts the fact that $X \cap J(f) = \varnothing$ because of $f \in \mathcal{C}$. □

Example 3.1.1 *Let $f(z) = (\sin z)/z$; then $f(z) \in \mathcal{B}-\mathcal{S}$. Moreover $F(f) \neq \varnothing$ and $F(f)$ does not contain wandering components (see Figure 3.1).*

Verification The function f has infinitely many singular values, which are all in the closed interval $[-1, 1]$. Hence $f \in \mathcal{B} - \mathcal{S}$. Further, f has an attracting fixed point on $[0, 1]$, to which the orbit of every singular value converges. This implies that $f \in \mathcal{C}$ and Theorem 3.1.12 shows that $F(f)$ contains no wandering domains.

3.1.3 Escaping points

The first fundamental theorem shows that the set of repelling periodic points is dense in the Julia set. Hence the set of the points whose orbits are bounded is dense in the Julia set. Here, we discuss the relation between the Julia set $J(f)$ and the set of escaping points

$$I_f = \left\{ z \in \mathbb{C} \,\middle|\, \lim_{n\to\infty} f^n(z) = \infty \right\},$$

which is defined in Chapter 1. If f is a polynomial, it is clear that $I_f \neq \varnothing$, for I_f is the immediate basin of the superattracting fixed point ∞. However, this is not clear in the case that $f \in \text{Ent}$.

Theorem 3.1.15 *For every f in* Ent,

$$I_f \neq \emptyset.$$

Proof We denote the central index of f for $r > 0$ by $\nu(r) = \nu(r, f)$ and choose $w(r)$ so that

$$|w(r)| = r, \quad |f(w(r))| = M(r),$$

where $M(r)$ is the maximum value function $\max_{|z|=r} |f(z)|$.

By the Wiman–Valiron Theorem (Theorem 2.2.11), if

$$|z - w(r)| < r(\nu(r))^{-\alpha}$$

for $\alpha > 1/2$, then we have

$$\begin{aligned} f(z) &= \left(\frac{z}{w(r)}\right)^{\nu(r)} f(w(r))(1+\epsilon_1), \\ f'(z) &= \nu(r)\left(\frac{z}{w(r)}\right)^{\nu(r)} \frac{f(w(r))}{w(r)}(1+\epsilon_2), \end{aligned}$$

where $\epsilon_i = \epsilon_i(r, z)$ tends to 0 uniformly with respect to z as $r \to \infty$ outside an exceptional set L whose logarithmic measure is finite. Choosing a sufficiently large $r_1 \notin L$ so that $\nu(r_1) > 10^4$,

$$M(r) > 4r \quad \text{and} \quad |\log(1+\epsilon_i)| < 1$$

for all $r \geq r_1$ and $r \notin L$. Set $w_1 = w(r_1)$ and

$$S_1' = \left\{ z \;\middle|\; \left|\log\left|\frac{z}{w_1}\right|\right| < \frac{5}{\nu(r_1)}, \left|\arg\frac{z}{w_1}\right| < \frac{5}{\nu(r_1)} \right\}.$$

Then by the mapping

$$z \mapsto \nu(r_1)(\log z - \log w_1) + \log f(w_1)$$

S_1' is mapped univalently onto the square

$$\left\{\zeta = \xi + i\eta \;\middle|\; |\xi - \log|f(w_1)|| < 5, \left|\eta - \arg f(w_1)\right| < 5\right\}.$$

From Theorem 2.2.11 and Rouché's theorem, there is a subdomain S_1 of S_1' which is mapped onto the square

$$Q_1 = \left\{\zeta \;\middle|\; |\xi - \log|f(w_1)|| < 4, \left|\eta - \arg f(w_1)\right| < 4\right\}$$

by the map $\log f(z)$. Then $f(S_1)$ contains

$$A_1 = \left\{z \mid e^{-4}M(r_1) < |z| < e^4 M(r_1)\right\}.$$

Next we choose $r_2 \notin L$ so that

$$\frac{1}{2}M(r_1) < r_2 < 2M(r_1).$$

Then

$$S_2' = \left\{ z \,\middle|\, \left|\log\left|\frac{z}{w_2}\right|\right| < \frac{5}{\nu(r_2)}, \left|\arg\frac{z}{w_2}\right| < \frac{5}{\nu(r_2)} \right\}$$

is contained in A_1 for $w_2 = w(r_2)$.

Repeating this construction, we have a sequence of domains S_j such that $S_j \to \infty$, $S_{j+1} \subset f(S_j)$ and that f^{-1} is defined on S_{j+1}. Since $\overline{B_{j+1}} \subset B_j$ for some component B_j of $f^{-j}(S_{j+1})$, we have $\bigcap_{j=1}^{\infty} B_j \neq \varnothing$. Any point in this set belongs to I_f. □

Theorem 3.1.16 *For every element* f *in* Ent,

$$J(f) = \partial I_f.$$

Proof It is clear that $J(f) \supset \partial I_f$.

Choose $z \in J(f)$. Since I_f is an infinite set, we see

$$\left(\bigcup_{n \geq 0} f^n(V)\right) \cap I_f \neq \varnothing$$

for any neighborhood V of z. The complete invariance of I_f means

$$V \cap I_f \neq \varnothing.$$

Since V is arbitrary, we have $z \in \overline{I_f}$. Since interior points of I_f are contained in $F(f)$, we obtain $z \in \partial I_f$. □

In the case of a polynomial f, $J(f) = \partial I_f$ and further $I_f \subset F(f)$. On the other hand, we have the following theorem in the case of transcendental entire functions.

Theorem 3.1.17 *For every element* f *in* Ent,

$$I_f \cap J(f) \neq \varnothing.$$

Proof If $F(f)$ contains a multiply connected component D, which is bounded by Theorem 3.1.5, then Lemma 3.1.3 implies that $f^n(D) \to \infty$, and also that $f^n(\partial D) \to \infty$. Since $\partial D \subset J(f)$, we have the assertion.

In the case that all the components of $F(f)$ are simply connected, let $\{B_j\}$ be the sequence of domains defined in the proof of Theorem 3.1.15 and we will show that

$$\bigcap_{j=1}^{\infty} B_j \subset J(f).$$

If not, there would be a point $z_0 \in \bigcap_{j=1}^{\infty} B_j$ of which a certain neighborhood V would be contained in $F(f)$. Since the diameters of B_j tend to 0 as $j \to \infty$, we have $B_n \subset V$ for all sufficiently large n. Since $f^{n+1}(B_n)$ contains the annulus

$$\{z \mid e^{-4}M(r_{n+1}) < |z| < e^4 M(r_{n+1})\},$$

$F(f)$ should contain $\{z \mid |z| < e^4 M(r_{n+1})\}$ for all n. But this is a contradiction. □

From Theorems 3.1.8 and 3.1.16, we have the following.

Theorem 3.1.18 *For $f \in \mathcal{B}$,*

$$J(f) = \overline{I_f}.$$

The first fundamental theorem and Theorem 3.1.18 imply that each set of the repelling cyclic points and the escaping points, which have quite different orbits, is dense in $J(f)$. Thus the action of $f \in \mathcal{B}$ on $J(f)$ is really a wild chaos.

Further, we define

$$I'_f = \left\{ z \,\middle|\, \lim_{n\to\infty} \frac{\log|f^{n+1}(z)|}{\log|f^n(z)|} = \infty \right\}.$$

It is clear that $I'_f \subset I_f$. More careful considerations than those in the proofs of Theorems 3.1.15 and 3.1.16 show the following.

Theorem 3.1.19 (Bergweiler 1995a) *For every f in* Ent,

$$I'_f \neq \varnothing \quad \text{and} \quad J(f) = \partial I'_f.$$

3.2 The exponential family

Dynamical properties of exponential maps $E_\lambda(z) = \lambda e^z$ are rather well known and fundamental among those of all transcendental entire functions. The Julia sets of certain exponential maps are the Cantor bouquets. Though the Hausdorff dimension of the Julia set of any exponential function is always

2, the Lebesgue measure of it with an attracting cycle is always 0. On the contrary, we will also show that the Julia set of $\sin \lambda z$ with an attracting cycle has positive Lebesgue measure. Finally, we study the representation space of the exponential family. Each hyperbolic component is unbounded except one, in contrast with the case of the Mandelbrot set.

3.2.1 The Cantor bouquet

In this subsection, we consider the case that λ is real. Since $E_\lambda^n(0) \to \infty$ as $n \to \infty$ for $\lambda > 1/e$, we have $J(E_\lambda) = \mathbb{C}$ by virtue of Theorem 3.1.11. Hence, we fix λ with $0 < \lambda < 1/e$ and abbreviate E_λ as E in this subsection. In this case, E has two fixed points p and q $(0 < p < 1 < q)$ on the real line. As $|E'(p)| = p$, p is an attracting fixed point. Hence we see $F(E) \neq \varnothing$.

Proposition 3.2.1 *Set $E(z) = \lambda e^z$ with $0 < \lambda < 1/e$. Then the immediate basin D of the attracting fixed point p of E coincides with $F(E)$.*

Proof We use the notation above. Since E has the single singular value 0 and has no wandering domains, we have $E^n(z) \to p$ as $n \to \infty$ for all $z \in F(E)$. Set $\nu = \log(1/\lambda) > 1$. Since the half plane $H = \{z \in \mathbb{C} \mid \operatorname{Re} z < \nu\}$ is mapped onto the punctured unit disk $\Delta - \{0\}$ by E and $\Delta \subset H$, we have $H \subset F(E)$. Because $p \in H$, we have $H \subset D$. As the inverse image of Δ by E is H, D is completely invariant. Hence D is the only component of $F(E)$. □

Now we discuss the shape of the Julia set of E, having regard to two kinds of points, those whose orbits are bounded and escaping points. Fix a natural number N and choose ν_1 satisfying $\lambda e^{\nu_1} - \nu_1 > (2N+1)\pi$. We denote the rectangle whose edges are on $x = \nu$, $x = \nu_1$ and $y = (2k \pm 1)\pi$ by $B(k)$ for each integer k with $|k| \leq N$ and also denote $\bigcup_{k=-N}^{N} B(k)$ by B. Each $B(k)$ is mapped onto the annulus with inner radius 1 and outer radius λe^{ν_1} by E. Hence the condition on ν_1 implies that

$$E(B(i)) \supset B(j)$$

for all i and j with $|i|, |j| \leq N$. Now, we set

$$\Lambda_N = \{z \in \mathbb{C} \mid E^n(z) \in B \text{ for all } n = 0, 1, 2, \ldots\}.$$

Then $\Lambda_N \subset J(E)$, because if $z \in \Lambda_N$ then any $E^n(z)$ $(n = 1, 2, 3, \ldots)$ does not belong to H. Further, we set

$$\widetilde{\Sigma_N} = \{\underline{s} = (s_0, s_1, s_2, \ldots) \mid s_j \in \mathbb{Z}, |s_j| \leq N\}$$

and define the shift map $\sigma : \widetilde{\Sigma_N} \to \widetilde{\Sigma_N}$ by

$$\sigma((s_0, s_1, s_2, \ldots)) = (s_1, s_2, \ldots).$$

Proposition 3.2.2 (i) Λ_N *is homeomorphic to* $\widetilde{\Sigma_N}$.

(ii) $E|_{\Lambda_N}$ *is topologically conjugate to* σ *on* $\widetilde{\Sigma_N}$.

Proof Set

$$S(k) = \{z \in \mathbb{C} \mid (2k-1)\pi \le \operatorname{Im} z \le (2k+1)\pi\}$$

for each integer k with $|k| \le N$ and denote by L_k the branch of E^{-1} which maps $\mathbb{C} - \{z \mid \operatorname{Re} z < 0, \operatorname{Im} z = 0\}$ to $S(k)$. Since the upper and the lower edges of each $B(k)$ are mapped by E into the negative real axis, $L_k(B(j))$ is completely contained in the interior of $B(k)$. Because $|E'(z)| > 1$ for $z \in B$,

$$\bigcap_{n \ge 0} L_{s_0} \circ \cdots \circ L_{s_{n-1}}(B)$$

consists of a single point $z_{\underline{s}} \in \Lambda_N$ for $\underline{s} = (s_0, s_1, s_2, \ldots) \in \widetilde{\Sigma_N}$. This is the map we required.

Claim (ii) is clear from the construction above. □

Corollary *If* $\underline{s} \in \widetilde{\Sigma_N}$ *corresponding to* $z \in \Lambda_N$ *is periodic, then* z *is a repelling periodic point of* E.

Definition We say an invariant set C_N of $J(f)$ is an *N-Cantor bouquet* of E, if there is a homeomorphism $h : \widetilde{\Sigma_N} \times [1, \infty) \to C_N$ that satisfies the following four conditions.

(1) $h(\underline{s}, 1) = z_{\underline{s}}$.

(2) For the natural projection $\Pi : \widetilde{\Sigma_N} \times [1, \infty) \to \widetilde{\Sigma_N}$,

$$\Pi \circ h^{-1} \circ E \circ h(\underline{s}, t) = \sigma(\underline{s}).$$

(3) For each $\underline{s} \in \widetilde{\Sigma_N}$, $\lim_{t \to \infty} h(\underline{s}, t) = \infty$.

(4) If $t \ne 1$ then $\lim_{n \to \infty} E^n(h(\underline{s}, t)) = \infty$.

Theorem 3.2.3 *For* $E(z) = \lambda e^z$ $(0 < \lambda < 1/e)$, $J(E)$ *contains an N-Cantor bouquet.*

Proof To define h, first, fix a c so that $\nu < 1 + c < \nu_1$. We consider a sequence of functions

$$h_k(\underline{s}, t) = L_{k, \underline{s}}(E_{1/e}^{k+1}(t) + c) = L_{s_0} \circ L_{s_1} \circ \cdots \circ L_{s_k}(E_{1/e}^{k+1}(t) + c).$$

Since 1 is the fixed point of $E_{1/e}$, $E^n_{1/e}(1) + c = 1 + c \in B$. Hence we have

$$\lim_{k\to\infty} h_k(\underline{s}, 1) = z_{\underline{s}}.$$

Next, since there is a constant $A > 1$ such that

$$|E'_\lambda(z)| \geq A$$

for all z with $\operatorname{Re} z \geq 1 + c$, we can find positive constants M such that

$$\begin{aligned} |h_{n+1}(\underline{s}, t) - h_n(\underline{s}, t)| &= |L_{n,\underline{s}}(E^n_{1/e}(t) - 1 - \log\lambda + i2\pi s_{n+1} + c) \\ &\quad - L_{n,\underline{s}}(E^n_{1/e}(t) + c)| \\ &< M(1/A)^n \end{aligned}$$

for every $t \in [1, \infty)$ and every n. Hence the h_k converge uniformly and the limit function h is the one we required. □

The construction above implies that

$$C_N \subset C_{N+1}$$

and we call

$$\overline{\bigcup_{N\geq 1} C_N}$$

the *Cantor bouquet* of E.

Theorem 3.2.4 *The Cantor bouquet of E coincides with the Julia set of E.*

Proof It is clear that $\overline{\bigcup_{N\geq 1} C_N} \subset J(E)$.

Take $z \in J(E)$ and any neighborhood U of z. Using Proposition 2.4.5, we have, for any Λ_N, $w \in U$ and $k \in \mathbb{N} \cup \{0\}$ such that $E^k(w) \in \Lambda_N$. Hence we obtain the claim, for $w \in \bigcup_{N\geq 1} C_N$. □

Here, each component of $J(E)$ is homeomorphic to $[1, \infty)$.

Definition We call the union A of finite end points of all components of $J(E)$ the *crown* of this Cantor bouquet $J(E)$. We also call each component of $J(E) - A$ a *stem* of $J(E)$.

Actually, the crown A is totally disconnected and contains all repelling periodic points. Hence, the first fundamental theorem implies that A is dense in $J(E)$. Further, we know the following.

Proposition 3.2.5 (Mayer) *The union $A \cup \{\infty\}$ of the crown and infinity is connected.*

Remark Take an $f \in \mathcal{S}$ and denote by D an open disk containing sing(f^{-1}) and the complement by D^c. Any component T of $f^{-1}(D^c)$ is simply connected and its closure contains ∞. The map $f : T \to D^c$ is a universal covering. We call T an exponential tract. If f satisfies some conditions, it is shown that the set of points whose orbits remain in T is a Cantor bouquet (Devaney and Tangerman 1986). For example, $S(z) = \lambda \sin z$ $(0 < \lambda < 1)$ has an attracting fixed point at 0. There are exponential tracts in an upper and a lower half plane respectively, each of which contains a Cantor bouquet.

3.2.2 The size of the Julia set

First, we prepare a useful lemma for estimating Hausdorff dimensions, which we will also use in §4.4.

Lemma 3.2.6 *Let X be a set in $\mathbb{C}$ and μ a probability measure on X. If there are positive constants c, t, r_0 satisfying*

$$\mu(D(z, r)) \leq cr^t$$

for all $z \in X$ and for all positive r with $r < r_0$, then $\dim(X) \geq t$, *where* $D(z, r) = \{\zeta \mid |\zeta - z| < r\}$ *as before.*

Proof Fix a positive number δ with $\delta < r_0/2$ and let $\{D(w_n, r_n)\}_{n=1}^{\infty}$ be any covering of X satisfying $r_n < \delta$ for each n. Choose z_n from each $D(w_n, r_n) \cap X$, then $\{D(z_n, 2r_n)\}$ is again a covering of X, and

$$1 = \mu(X) \leq \sum_{n=1}^{\infty} \mu(D(z_n, 2r_n)) \leq c \sum_{n=1}^{\infty} (2r_n)^t.$$

Hence the t-dimensional Hausdorff measure $H_t(X)$ of X is not less than $1/c$. Letting $\delta \to 0$, we obtain the claim. □

Further, we prove another lemma on a general estimate of Hausdorff dimension. Let Λ_k $(k = 1, 2, 3, \ldots)$ be a finite collection of disjoint compact sets F of $\mathbb{C}$. We denote the set union of the elements of Λ_k by $|\Lambda_k|$. We assume

(1) every $F \in \Lambda_{k+1}$ is contained in an $F' \in \Lambda_k$ and
(2) every $F \in \Lambda_k$ contains at least one element of Λ_{k+1}.

We define

$$\Delta(A, B) = \frac{m(A \cap B)}{m(B)}$$

for two sets A and B, where m means the two-dimensional Lebesgue measure. We also assume

$$\Delta(|\Lambda_{k+1}|, F) \geq \delta_k, \quad \operatorname{diam}(F) \leq d_k$$

for all k and all F in Λ_k. We set $X = \bigcap_{k=1}^{\infty} |\Lambda_k|$.

Lemma 3.2.7 *Let Λ_k, X, δ_k and d_k be as above. Then*

$$\Delta(X, |\Lambda_1|) \geq \prod_{k=1}^{\infty} \delta_k,$$

$$\limsup_{k\to\infty} \frac{\sum_{i=1}^{k} |\log \delta_i|}{|\log d_k|} \geq 2 - \dim(X).$$

Proof The first inequality is clear from the definition.

We will give a proof only in the case that d_k monotonically converges to 0 for the second inequality. We may assume $m(|\Lambda_1|) = 1$, for we can rescale it if necessary. Let μ_1 be the restriction of m to $|\Lambda_1|$. We construct μ_{k+1} on $|\Lambda_{k+1}|$ inductively as follows:

$$\mu_{k+1}(|\Lambda_{k+1}| \cap F) = \mu_k(F), \qquad F \in \Lambda_k.$$

Then $\{\mu_k\}$ has the unique weak limit μ whose support is $X = \bigcap_{k=1}^{\infty} |\Lambda_k|$ and which satisfies

$$\mu(F) \leq \frac{m(F)}{\delta_1 \delta_2 \cdots \delta_{k-1}}$$

for every $F \in \Lambda_k$.

Take $r > 0$ and choose k so that $d_k > r \geq d_{k+1}$. Let D be a disk with radius r and set

$$|\Lambda_{k+1}(D)| = \bigcup_{F \in \Lambda_{k+1} F \cap D \neq \varnothing} F.$$

Then $\operatorname{diam} |\Lambda_{k+1}(D)| < 4r$. From the inequality above, we have for any positive constant δ

$$\mu(D) \leq \mu(|\Lambda_{k+1}(D)|) \leq \frac{m(|\Lambda_{k+1}(D)|)}{\delta_1 \delta_2 \cdots \delta_k} \leq C r^{\delta} \frac{d_k^{2-\delta}}{\delta_1 \delta_2 \cdots \delta_k},$$

where C is a constant depending on δ. If $2-\delta$ is greater than the left side of the second inequality in the proposition, then the right side of the inequality above tends to 0 as $k \to \infty$. We conclude from Lemma 3.2.6 that the Hausdorff dimension of X is at least δ. □

We now consider the size of Julia sets of entire functions following McMullen (1987a). In general, we have $1 \le J(f) \le 2$ for any transcendental entire function f by the corollary to Theorem 3.1.5. Some other estimates are also given by Stallard.

Theorem 3.2.8 *For every $E(z) = \lambda e^z$, the Hausdorff dimension of $J(E)$ is 2.*

Proof To prove this, we show that the Hausdorff dimension of the set of escaping points is 2, because $E(z)$ is an entire function of finite type.

Let $\arg \lambda = \varphi$ and set

$$S = \{z \mid \operatorname{Re} z > 0 \text{ and } \operatorname{Im} z \in I\},$$

where $I = \bigcup_{n\in\mathbb{Z}}[-\pi/4+2n\pi-\varphi, \pi/4+2n\pi-\varphi]$. A point z with $E(z) \in S$ satisfies $\operatorname{Re} E(z) > a \exp(\operatorname{Re} z)$ with some positive constant a. Choose x_0 so that $|E'(z)| \gg 1$ for $\operatorname{Re} z > x_0$ and set

$$T = \{z \in S \mid \operatorname{Re} z > x_0\}.$$

Then we may assume that all z satisfying $E^n(z) \in T$ for every n are escaping points. Let B be a square with side length r. We can choose r so small that $E(B)$ is nearly square-shaped, which follows from boundedness of non-linearity of E, which comes from $E''/E' = 1$. Hence we can pack $E(B) \cap T$ with disjoint squares, the set of which we denote by $\operatorname{pack}(E(B))$, such that $\Delta(|\operatorname{pack}(E(B))|, E(B)) > 1/5$.

Now we define $\{\Lambda_k\}$ inductively. Set $\Lambda_0 = \{B\}$ and define

$$\Lambda_k = \{G \mid G \subset F \in \Lambda_{k-1} \text{ and } E^k(G) \in \operatorname{pack}(E^k(F))\}.$$

Then $A = \cap|\Lambda_k|$ consists of escaping points.

Since

$$\Delta(|\operatorname{pack}(E^k(F))|, E^k(F)) > \frac{1}{5}$$

for each $F \in \Lambda_{k-1}$, and since bounded non-linearity further implies that the distortions of E^{-k} on F are uniformly bounded (cf. the proof of Theorem 1.4.6), we obtain

$$\Delta(|\Lambda_{k+1}|, F) \ge \delta$$

for some $\delta > 0$ independent of k. Further, since $\operatorname{Re} E^k(z) > b$, we see that $\operatorname{diam} F < c/\exp^k(1)$ for every positive k, where b and c are independent of k. And since $1/\exp^k(1) = o(\delta^k)$ as $k \to \infty$, we obtain the claim from Lemma 3.2.7. □

Nevertheless, we still have the following.

Theorem 3.2.9 *Assume that $E(z) = \lambda e^z$ has an attracting cycle. Then the Lebesgue measure of $J(E)$ is* 0.

Proof Since E has only one singular value, the assumption implies that the post singular set $\overline{S^+(E)}$ is compact in $\mathbb{C}$, and $d(S^+(E), J(E)) > 0$.

First, we show $|(E^n)'(z)| \to \infty$ for all $z \in J(E)$ (cf. Lemma 1.4.5). The complement U of $\overline{S^+(E)}$ consists of only one component, which we endow with the hyperbolic metric $\lambda_U(z)|dz|$. Since $E^{-1}(U)$ is a proper subset of U, the Schwarz lemma gives that

$$\|E'(z)\| > 1,$$

for every $z \in J(E)$, where $\|\cdot\|$ denotes the norm of the derivative of E at z in the hyperbolic metric. Hence we have

$$|(E^n)'(z)| = \frac{\|(E^n)'(z)\|\lambda_U(z)}{\lambda_U(E^n(z))}.$$

Since $\lambda_U(z) \to 0$ as $z \to \infty$, we obtain the claim regardless of whether $E^n(z)$ tends to ∞ or $E^n(z)$ return infinitely often to some fixed compact set.

Further, since $\lambda_U(z) = O(1/|z|\log|z|)$ as z tends to infinity, we have, for $z \in J(E)$ with sufficiently large $|z|$,

$$|(E^n)'(z)| \geq C|E^n(z)|\log|E^n(z)|, \tag{3.1}$$

for some constant C.

Let $\{B_\alpha\}$ be a locally finite covering of $J(E)$, each element of which is a disk with center z in $J(E)$ and radius $d(z, \overline{S^+(E)})/2$. Then the sequence of the radii of the B_α tends to infinity.

Since the asymptotic value 0 is contained in $F(E)$, some small disk centered 0 is contained in $F(E)$ and thus so is the inverse image H of it. The set H is of the form $\{z \mid \operatorname{Re} z < x_0\}$ for some x_0. Hence the inverse image of H contains

$$\bigcup_{n\in\mathbb{Z}}\{\operatorname{Re} z > x_1, (2n+1)\pi - \arg\lambda - a < \operatorname{Im} z < (2n+1)\pi - \arg\lambda + a\},$$

for some positive x_1 and a. Thus there is ϵ such that

$$\Delta(J(E), B_\alpha) < 1 - \epsilon,$$

for all B_α.

Fix $z \in J(E)$. Let B_n be an element of $\{B_\alpha\}$ containing $E^n(z)$. Since $S^+(E)$ is disjoint from each B_α, inverses of E^n on B_n are univalent. We denote the inverse which maps $E^n(z)$ to z by E^{-n} and set $U_n = E^{-n}(B_n)$. Using the Koebe distortion theorem, we see that these maps have distortion bounded by a universal constant. It follows that

$$\operatorname{diam}(U_n) = O\left(\frac{\operatorname{diam}(B_n)}{|(E^n)'(z)|}\right).$$

If the $\operatorname{diam}(B_n)$ are bounded, then $\operatorname{diam}(U_n) \to 0$ because of $|(E^n)'(z)| \to \infty$. Unless the $\operatorname{diam}(B_n)$ are bounded, $\operatorname{diam}(B_n) = O(|E^n(z)|)$. By using the inequality (3.1), we have

$$\operatorname{diam}(U_n) = O\left(\frac{1}{\log|E^n(z)|}\right).$$

Hence also in this case, we have $\operatorname{diam}(U_n) \to 0$.

Finally, the Koebe distortion theorem implies that there is a positive ϵ' such that

$$\Delta(J(E), U_n) < 1 - \epsilon',$$

for all n. Hence z is not the Lebesgue density point. □

McMullen (1987a) actually showed that if an entire function is expanding and its Julia set is thin at infinity then the area of its Julia set is zero. Recall that the same (but simpler) argument shows that for every hyperbolic rational function (see §4.4), the area of the Julia set is zero.

Also he gave the following example of a Julia set which does not coincide with the Riemann sphere but whose area is positive.

Theorem 3.2.10 *The Julia set of any element of the sine family has positive Lebesgue measure.*

Remark It is clear that $\sin az$ with $|a| < 1$ has an attracting fixed point at the origin. Hence, its Fatou set is not empty.

Proof of theorem Each element of the sine family is conformally conjugate to a certain entire function of the form

$$f(z) = ae^z + be^{-z}, \qquad a \neq 0 \text{ and } b \neq 0.$$

Hence we consider this function f.

Consider the region

$$R(h) = \{z \mid |\operatorname{Re} z| > h\}.$$

Choose an h large enough so that $|f'(z)|/e^h > c$ for some positive constant c and $|f''(z)/f'(z)|$ is near 1 for $z \in R(h)$. Let B be a square with side length r in $R(h)$. We choose an r so small that $f(B)$ is nearly square-shaped.

Define a sequence $\{h_k\}$ by setting $h_k = 2\exp^k(x)$ for some large x so that $\sum_{k=0}^{\infty}\left(1/\exp^{k+1}(x)\right)$ is sufficiently small. Take a square B in $R(h_k)$. Then we can pack $f(B) \cap R(h_{k+1})$ by squares with side length r, the set of which we denote by pack$(f(B))$, so that

$$\Delta(|\operatorname{pack}(f(B))|, f(B)) > 1 - O\left(\frac{h_{k+1}}{e^{h_k}}\right).$$

Similarly as in the proof of Theorem 3.2.8, define $\{\Lambda_k\}$ inductively by

$$\Lambda_k = \{G \mid G \subset F \in \Lambda_{k-1} \text{ and } f^k(G) \in \operatorname{pack}(f^k(F))\}.$$

Then $A = \cap|\Lambda_k|$ consists of escaping points. Since

$$\Delta(|\operatorname{pack}(f^k(F))|, f^k(F)) \geq 1 - O\left(\frac{h_{k+1}}{e^{h_k}}\right),$$

we complete the proof. □

3.2.3 The representation space

In the rest of this section, we study the representation space $\mathbb{C}$ of the exponential family. In the case that $\lambda = 0$, we consider that E_0 has an attracting fixed point 0 for convenience. The set of parameters which have attracting cycles is an open set. We call each connected component of it a *hyperbolic component*. It is easy to see the following.

Proposition 3.2.11 *Let H be a hyperbolic component. Then there is a unique k such that, for any $\lambda \in H$, E_λ has an attracting cycle with period k.*

Theorem 3.2.12 *There is a unique hyperbolic component H_1 corresponding to an attracting fixed point. Moreover, H_1 is of the form*

$$H_1 = \left\{\lambda = te^{-t} \;\middle|\; |t| < 1\right\}.$$

Proof Let $E(z) = \lambda e^z$. The point t is a fixed point of E if and only if $\lambda = te^{-t}$. Since $E'(z) = \lambda e^z$, t is an attracting fixed point if and only if $|t| < 1$. □

Theorem 3.2.13 *Every hyperbolic component is simply connected.*

Proof Let H be a component corresponding to an attracting cycle with period $p \geq 2$. For every $\lambda \in H$, the immediate basin of the attracting cycle contains the transcendental singularity 0. We denote by $\zeta(\lambda)$ the periodic point contained in the Fatou component containing 0. We will show that $\{E_\lambda^{np}(0)\}$ converges to $\zeta(\lambda)$ uniformly in compact sets of H. Then, $\{E_\lambda^{np+k}(0)\}$ converges to $E_\lambda^k(\zeta(\lambda))$ uniformly in compact sets of H for $1 \leq k \leq p-1$. Hence $\{E_\lambda^n(0)\}$ has precisely p limit functions, which are cyclic. Since the $E_\lambda^n(0)$ are entire functions of λ, the maximal principle shows the claim.

Fix $\lambda_0 \in H$. Set $\mu = |(E_{\lambda_0}^p)'(\zeta(\lambda_0))| < 1$ and take a ν with $\mu < \nu < 1$. Choosing a disk $D = \{z \mid |z - \zeta(\lambda_0)| < r\}$ so that $|(E_{\lambda_0}^p)'(z)| < \nu$ for $z \in D$, we have $E_{\lambda_0}^p(D) \subset D' = \{z \mid |z - \zeta(\lambda_0)| < \nu r\}$. Since $E_{\lambda_0}^{np}(0) \to \zeta(\lambda_0)$, there is m with $E_{\lambda_0}^{mp}(0) \in D'$. We can choose a neighborhood U of λ_0 in H such that $E_\lambda^{mp}(0) \in D$ and $E_\lambda^p(D) \subset D$ for $\lambda \in U$. This implies $\{E_\lambda^n\}$ is a normal family in U. □

Theorem 3.2.14 *Every hyperbolic component except H_1 is unbounded.*

Proof Define a function $M(\lambda)$ on each hyperbolic component corresponding to the period p by

$$M(\lambda) = (E_\lambda^p)'(\zeta(\lambda)).$$

It is analytic in each component. The boundaries of the hyperbolic components are given by the level curve $|M(\lambda)| = 1$. Since $M(\lambda)$ never attains 0 on any component except H_1, each component except for H_1 is unbounded. □

Remark Baker and Rippon (1984) showed that parabolic bifurcations occur at the points te^{-t} on the boundary of H_1 for each primitive root t of unity.

We can formulate the hyperbolic–dense conjecture.

Conjecture The hyperbolic components are dense in the exponential family.

3.3 Topological completeness for decorated exponential families

The exponential family and the sine family are topologically complete in the following sense.

Definition We say that a family of entire functions is *topologically complete* if any entire function topologically conjugate to an element of it is actually conformally conjugate to an element of it.

In this section, we explain two proofs of the following.

Theorem 3.3.1 *For any positive integers k and n, the doubly decorated exponential family*

$$\mathcal{F}_{k,n} = \{P(z) \exp Q(z)\},$$

where $P(z)$ and $Q(z)$ move over all polynomials of degree k and n, respectively, is topologically complete.

Remark The first proof of Theorem 3.3.1 is essentially due to Keen and relies on quasiconformal maps. The second one is more elementary, and gives concrete information on the covering structure induced by an element of the family.

First recall the following fact (cf. the proof of Lemma 2.4.10).

Lemma 3.3.2 *Let g be topologically conjugate to $f(z) = P(z) \exp Q(z) \in \mathcal{F}_{k,n}$. Then g is quasiconformally equivalent to f.*

Proof Let $\varphi(z)$ be a homeomorphism of $\mathbb{C}$ onto itself such that $\varphi \circ f = g \circ \varphi$. Since $\mathrm{sing}(f^{-1})$ is a finite set, there is an isotopy Φ relative to $\mathrm{sing}(f^{-1})$ which connects φ to a quasiconformal map ψ_2.

By the assumption, we can lift Φ to an isotopy $\tilde{\Phi}$ such that $g \circ \tilde{\Phi} = \Phi \circ f$. And $\tilde{\Phi}$ connects φ to a quasiconformal map ψ_1 with $\psi_2 \circ f = g \circ \psi_1$. □

Proof of Theorem 3.3.1 (the first proof) We use the same notation as in the proof of Lemma 3.3.2. We may assume that $\varphi(0) = 0$. Then since $f^{-1}(0)$ and $g^{-1}(0)$ have the same number of points counting multiplicity, we find a polynomial $R(z)$ of degree k such that $g(z)/R(z)$ has no zeros. Hence we can write $g(z)$ as $R(z) \exp h(z)$ with an entire function $h(z)$.

Since quasiconformal maps are Hölder continuous, there are some $K > 1$ and $A > 1$ such that

$$A^{-1}|z|^{1/K} \le |\psi_j(z)| \le A|z|^K$$

for each j. Hence on $\{|z| = r\}$, we have

$$|g(z)| = |\psi_2 \circ f \circ \psi_1^{-1}(z)| \le A|M(f, A^K r^K)|^K,$$

where $M(f,r) = \max_{|z|=r} |f(z)|$ as before. Here since $f \in \mathcal{F}_{k,n}$, the function $\log |M(f, A^K r^K)|^K$ has a polynomial growth order with respect to r. Hence there are some C and N such that

$$|\mathrm{Re}\, h(z)| \leq Cr^N$$

for every z with sufficiently large $|z|$, which implies that $h(z)$ is a polynomial.

Finally, let n' be the degree of $h(z)$. Then $g(z)$ has exactly $k+n'-1$ critical points including multiplicity, which should equal $k+n-1$. Thus we have $n' = n$. □

Next, we give a more elementary proof of Theorem 3.3.1. For this purpose, fix an entire function g which is topologically conjugate to an element f in $\mathcal{F}_{k,n}$, i.e., $\varphi \circ f = g \circ \varphi$ with a homeomorphism φ of $\mathbb{C}$ onto itself. Here, we may assume that $\varphi(0) = 0$.

For the sake of simplicity, we explain here only the case that $Q(z) = z$ and $f(z) = P(z)e^z$ has exactly k non-zero critical values, and let $\mathcal{F}_k$ be the set of all such $P(z)e^z$. Other cases can be treated similarly.

First, we apply a cut–paste surgery to f as follows. Let $\{a_j\}_{j=1}^k$ be critical values of f, and take pairwise disjoint smooth arcs $\{L_j\}_{j=1}^k$ on $\mathbb{C} - \{0\}$ starting at a_j and tending to ∞. Consider f as a branched and incomplete covering of the target, say $\mathbb{C}_w$, of f by the domain, say $\mathbb{C}_z$, of f. Cut $\mathbb{C}_z$ along all lifts of L_j with respect to this covering. Then $\mathbb{C}_z$ is decomposed into $k+1$ components, say $\{R_1, \ldots, R_k, S\}$. Each component can be re-pasted over L_j, and we have $k+1$ Riemann surfaces spreading holomorphically over $\mathbb{C}_w$. Here, k of them corresponding to $\{R_1, \ldots, R_k\}$ are holomorphic universal covering surfaces of $\mathbb{C}_w$, denoted by $\{\widehat{R}_1, \ldots, \widehat{R}_k\}$, while the remaining one, say $\widehat{S}$, is a holomorphic universal covering surface of $\mathbb{C}_w - \{0\}$. Hence, f restricted to a subset S of $\widehat{S}$, which in turn we denote by $\mathbb{C}_\zeta$, can be expressed as

$$f|_S(\zeta) = e^{a\zeta+b}$$

with suitable a and b.

We remark that similar arguments show topological completeness of the exponential family and the sine family.

Definition We call S the *exponential part* of the covering surface $\mathbb{C}_z$, and each R_k a $\mathbb{C}$-*decoration*.

Conversely, such decoration structure characterizes the family $\mathcal{F}_k$. Let $\widehat{S}^0$ be a universal covering surface of $\mathbb{C}_w - \{0\}$. We decorate k copies $\widehat{R}_j^0$ of the whole plane inductively as follows. Take k pairwise disjoint arcs $\{L_j^0\}$ starting from a point in $\mathbb{C}_w - \{0\}$ and tending to ∞. Fix a lift of L_1^0 on $\widehat{S}^0$ and decorate $\widehat{R}_1^0$ by connecting crosswise along this lift. Then, we have a branched incomplete covering of $\mathbb{C}_w$. Next we fix a lift of L_2^0 to this new covering surface, and decorate $\widehat{R}_2^0$ along the lift. Repeating this process, we have a branched and incomplete covering surface D of the whole plane.

Definition We call D a *kth decorated exponential covering surface*, and the projection π to $\mathbb{C}_w$ a *kth decorated exponential function.*

The domain D of π is clearly simply connected. Moreover, the following fact is essentially well known, and hence we omit the proof.

Lemma 3.3.3 *Every kth decorated exponential function π can be regarded as an entire function.*

Roughly speaking, in a neighborhood of ∞, the covering structure of π is the logarithmic one, and also so it is for the non-compact component over a small neighborhood of 0. Thus we can construct a ring domain with arbitrarily large modulus separating a compact part from the boundary.

The covering structure of π depends on the choice of the isotopy class of the arcs $\{L_j^0\}$ relative to end points, and the configuration tree induced from the order of the $\mathbb{C}$-decorations.

Definition The *configuration tree* for a kth decorated exponential function consists of vertices $\{r_1, \ldots, r_k\} \cup \mathbb{Z}$ where each r_j represents $\widehat{R}_j^0$ and each n represents the nth sheet of $\widehat{S}^0$ and they are connected by sides which are given by the following injunctions.

(i) Add one side connecting n with $n+1$ for every n.
(ii) Add one side connecting r_1 with 0 (which determines the 0th sheet of $\widehat{S}^0$).
(iii) Add one side connecting r_2 with r_1 if $\widehat{R}_2^0$ is connected to $\widehat{R}_1^0$. If not, some sheet, say the n_2th sheet, of $\widehat{S}^0$ is connected with $\widehat{R}_2^0$. Then add one side connecting r_2 with n_2.
(iv) Add one side connecting r_j with one of $\{r_1, \ldots, r_{j-1}\} \cup \mathbb{Z}$ by similar injunctions as in (iii), for every $j > 2$.

The configuration tree determines a kth decorated exponential function in the following sense.

Lemma 3.3.4 *Suppose that two kth decorated exponential functions h_1 and h_2 have the same singular values and the same configuration tree with respect to the same cut arcs; then there is a conformal automorphism A of the whole plane such that*

$$h_1 = h_2 \circ A.$$

Proof Let S^m be the exponential part and $\{R_j^m\}$ the $\mathbb{C}$-decorations for each $m = 1, 2$. Then there is a conformal map $A_0 : \widehat{S}^1 \to \widehat{S}^2$ such that the branch point for $\widehat{R}_1^1$ is sent to that for $\widehat{R}_1^2$ and $\widehat{\pi}_2 \circ A_0 \circ \widehat{\pi}_1^{-1}$ is the identity on $\mathbb{C}^*$, where $\widehat{\pi}_m$ is the natural projection of $\widehat{S}^m$. Then A_0 can be extended to a biholomorphic map from $\widehat{S}^1$ decorated with $\widehat{R}_1^1$ to $\widehat{S}^2$ decorated with $\widehat{R}_1^2$, which should send the branch point for $\widehat{R}_2^1$ to that for $\widehat{R}_2^2$. Repeating this, we have a map A as desired. □

On the other hand, every kth decorated exponential covering surface can be realized by an element of $\mathcal{F}_k$. To show this, first we recall that critical values depend biholomorphically on coefficients of monic polynomials.

Lemma 3.3.5 *Let $P_0(z)e^z$ be an element of $\mathcal{F}_k$ with a monic polynomial*

$$P_0(z) = z^k + c_k^0 z^{k-1} + \cdots + c_1^0$$

and pairwise different non-zero critical values $\mathbf{a}_0 = (a_1^0, \ldots, a_k^0)$. Then the map from coefficients $\mathbf{c} = (c_1, \ldots, c_k)$ to critical values $\mathbf{a} = (a_1, \ldots, a_k)$ is biholomorphic in a neighborhood of $\mathbf{c}_0 = (c_1^0, \ldots, c_k^0)$.

Proof First, since critical points $\mathbf{p} = (p_1, \ldots, p_k)$ of $P(z)e^z$ are zeros of $P(z) + P'(z)$, $\mathbf{p}$ is clearly a holomorphic function of $\mathbf{c}$ near $\mathbf{c}_0$, and

$$(P + P')'(p_j)\frac{\partial p_j}{\partial c_\ell} = -p_j^{\ell-1} - (\ell - 1)p_j^{\ell-2}.$$

Here, by the assumption, $(P + P')'(p_j)$ is non-zero for every j and the Jacobian is non-degenerate.

Next, since $\mathbf{a} = (P(p_1)e^{p_1}, \ldots, P(p_k)e^{p_k})$, we have

$$\frac{\partial a_j}{\partial c_\ell} = \left(P'(p_j)\frac{\partial p_j}{\partial c_\ell} + p_j^{\ell-1}\right)e^{p_j} + P(p_j)e^{p_j}\frac{\partial p_j}{\partial c_\ell} = p_j^{\ell-1}e^{p_j}.$$

Hence, we conclude with the assertion. □

Actually, for every point $\mathbf{p} = (p_1, \ldots, p_k)$ in the region

$$\Omega = \left\{\mathbf{p} = (p_1, \ldots, p_k) \in \mathbb{C}^k \;\middle|\; \prod_{i \neq j}(p_i - p_j) \neq 0\right\},$$

it is easy to give an element of $\mathcal{F}_k$ with a monic polynomial which has the critical points $\mathbf{p}$. And we have shown that the map Φ from Ω into $\mathbb{C}^k$, which sends $\mathbf{p}$ to critical values $\mathbf{a}$ of the corresponding element, is locally biholomorphic.

So set

$$\Omega' = \left\{ \mathbf{a} = (a_1, \ldots, a_k) \in \mathbb{C}^k \;\middle|\; a_1 \cdots a_k \prod_{i \neq j} (a_i - a_j) \neq 0 \right\},$$

and consider taking lifts of arcs in Ω' starting from $\mathbf{a}_0$ with respect to Φ, where $\mathbf{a}_0$ is as in Lemma 3.3.5. And we have the following.

Lemma 3.3.6 *The branch of Φ^{-1} which sends $\mathbf{a}_0$ to $\mathbf{p}_0$ can be continued analytically along any arc in Ω' starting from $\mathbf{a}_0$, where $\mathbf{p}_0$ corresponds to $\mathbf{c}_0$ in Lemma 3.3.5.*

Proof Suppose that there were an arc $\gamma : [0, 1] \to \Omega'$ starting from $\mathbf{a}_0$ which could not be lifted from $\mathbf{p}_0$ into Ω with respect to Φ. Further, we may assume that $\gamma : [0, r] \to \Omega'$ has such a lift for every $r < 1$.

If the end of the lift of $\gamma : [0, 1) \to \Omega'$ from $\mathbf{p}_0$ accumulate to a point of Ω, then by the fact noted before Lemma 3.3.6, we would have a lift of the whole given arc γ. So the end should tend to the boundary. Here since γ is in Ω', the end cannot tend to the relative boundary of Ω in $\mathbb{C}^k$. Hence it diverges to ∞. But then, γ should end at $\{a_1 \cdots a_k = 0\}$ or diverge to ∞, which is also impossible. □

Thus perturbation of critical values can be realized by elements of $\mathcal{F}_k$ with monic polynomials, and the following proposition gives the second proof of Theorem 3.3.1.

Proposition 3.3.7 *The family $\mathcal{F}_k$ contains all kth decorated exponential functions modulo conformal conjugation.*

Proof Fix an element h_0 of $\mathcal{F}_k$ with a monic polynomial and a kth decorated exponential function π_0 arbitrarily. First, by starting from h_0 and moving critical values of h_0 so that they are always pairwise distinct, we obtain by Lemma 3.3.6 an element in $\mathcal{F}_k$ whose critical values are just the same as those of π_0 including the order.

When the configuration trees are not the same, we deform that of the element of $\mathcal{F}_k$ by moving each critical value around another critical value or 0. Then again by Lemma 3.3.6, applying a finite number of such deformations along suitable loops, we can get an element h of $\mathcal{F}_k$ which has the same configuration tree as that of π_0.

Then by Lemma 3.3.4, we can find a conformal self-map A of the domain of π_0 such that

$$\pi_0 \circ A = h.$$

This implies that π_0 is conformally conjugate to an element of $\mathcal{F}_k$. □

3.4 Examples

We construct transcendental entire functions having wandering domains or Baker domains.

3.4.1 Wandering domains

The following example due to Baker is the first one which gives a wandering domain.

Theorem 3.4.1 *Fix $r_1 > 1$ and $c > 0$ which satisfy*

$$c \exp\left(\frac{2}{r_1}\right) < \frac{1}{4}, \qquad cr_1 > 1. \tag{3.2}$$

The sequence of positive numbers $\{r_n\}_{n=1}^{\infty}$ is given inductively by the equation

$$r_{n+1} = cr_n^2 \left(1 + \frac{r_n}{r_1}\right)\left(1 + \frac{r_n}{r_2}\right) \cdots \left(1 + \frac{r_n}{r_n}\right). \tag{3.3}$$

Set

$$f(z) = cz^2 \prod_{n=1}^{\infty} \left(1 + \frac{z}{r_n}\right)$$

then $F(f)$ contains a multiply connected component.

Proof From the definition, we see

$$2r_n < r_{n+1}. \tag{3.4}$$

Hence, the infinite product $f(z)$ converges uniformly on compact sets of $\mathbb{C}$, which means that f is a transcendental entire function. The origin is a super-attracting fixed point of f. From (3.2) and (3.4), we have

$$|f(z)| < c \prod_{n=1}^{\infty} \left(1 + \frac{1}{r_n}\right) < c \prod_{n=1}^{\infty} \left(1 + \frac{1}{2^{n-1} r_1}\right) < c \exp\left(\frac{2}{r_1}\right) < \frac{1}{4} \tag{3.5}$$

for $|z| = 1$. This shows that the unit disk is contained in an attracting component.

To verify that $F(f)$ has a multiply connected component, set

$$A_n = \{z \mid r_n^2 < |z| < r_{n+1}^{1/2}\},$$

and by (3.3) and (3.4), we have that

$$4r_n < r_n^2 < r_{n+1}^{1/2} < \frac{1}{4} r_{n+1},$$

and in particular $A_n \neq \varnothing$, for all sufficiently large n.

Now, fix an r with $4r_n < r < r_{n+1}/4$ and set

$$\log \left| \frac{f(r)}{f(-r)} \right| = \sum_{k=1}^{n-1} I_k + I_n + I_{n+1} + \sum_{k=n+2}^{\infty} I_k,$$

where

$$I_k = \log \left| \frac{1 + \frac{r}{r_k}}{1 - \frac{r}{r_k}} \right|.$$

For $k \leq n-1$, we have $I_k < 3r_k/r < 3r_k/(4r_n)$ and hence by (3.4)

$$\begin{aligned} \sum_{k=1}^{n-1} I_k &< \frac{3}{4} \sum_{k=1}^{n-1} \frac{r_k}{r_n} \\ &= \frac{3}{4} \frac{r_{n-1}}{r_n} \left(1 + \frac{r_{n-2}}{r_{n-1}} + \frac{r_{n-3}}{r_{n-1}} + \cdots + \frac{r_1}{r_{n-1}} \right) \\ &< \frac{3}{4} \frac{r_{n-1}}{r_n} \left(1 + \frac{1}{2} + \frac{1}{2^2} + \cdots + \frac{1}{2^{n-1}} \right) < \frac{3r_{n-1}}{2r_n}. \end{aligned}$$

For $k \geq n+2$, we have $I_k < 3r/r_k < 3r_{n+1}/(4r_k)$ and hence by (3.4)

$$\begin{aligned} \sum_{k=n+2}^{\infty} I_k &< \frac{3}{4} \sum_{k=n+2}^{\infty} \frac{r_{n+1}}{r_k} \\ &< \frac{3}{4} \sum_{k=n+2}^{\infty} \frac{r_{n+1}}{r_{n+2}} 2^{n+2-k} < \frac{3r_{n+1}}{2r_{n+2}}. \end{aligned}$$

Further, we have

$$\log \left| \frac{1 + \frac{r}{r_n}}{1 - \frac{r}{r_n}} \right| < \log \frac{5}{3} \quad \text{and} \quad \log \left| \frac{1 + \frac{r}{r_{n+1}}}{1 - \frac{r}{r_{n+1}}} \right| < \log \frac{5}{3}.$$

Since $r_n/r_{n+1} \to 0$ as $n \to \infty$, we get $\log |f(r)/f(-r)| < \log 4$ and hence

$$f(r) < 4|f(-r)| = 4 \min_{0 \leq \theta \leq 2\pi} |f(re^{i\theta})| \tag{3.6}$$

for any r in $(4r_n, r_{n+1}/4)$ with sufficiently large n.

Next, using (3.3), (3.4) and the definition of $f(z)$, we obtain

$$r_{n+1} < f(r_n) = r_{n+1} \prod_{k=n+1}^{\infty} \left(1 + \frac{r_n}{r_k}\right) < r_{n+1} \prod_{k=1}^{\infty} \left(1 + \frac{1}{2^k}\right) < e r_{n+1}. \tag{3.7}$$

Since $\max_{|z|=r} |f(z)| = f(r)$, Hadamard's three-circle theorem implies that

$$\log f(e^{2s}) > 2 \log f(e^s) - \log f(e^0)$$

for $s > 0$. Set $e^s = r_n^{1/2}$, and by adding (3.5) and (3.7) we have

$$f(r_n^{1/2}) < \left\{ \frac{f(r_n)}{4} \right\}^{1/2} < \left(\frac{e}{4}\right)^{1/2} r_{n+1}^{1/2} < r_{n+1}^{1/2}. \tag{3.8}$$

Hence

$$|f(z)| \le f(|z|) < f(r_{n+1}^{1/2}) < r_{n+2}^{1/2}$$

for $z \in A_n$. Set $e^s = r_n$, and similarly we have

$$\frac{1}{4} f(r_n^2) > f(r_n)^2 > r_{n+1}^2. \tag{3.9}$$

From (3.6) and (3.9), we have

$$|f(z)| \ge f(-|z|) > \frac{1}{4} f(|z|) > \frac{1}{4} f(r_n^2) > r_{n+1}^2,$$

for every $z \in A_n$. Thus $f(A_n) \subset A_{n+1}$.

We have shown that $\{f^m\}_{m=1}^{\infty}$ converges uniformly to ∞ on A_n. Let D_n be the Fatou component containing A_n. Then every D_n is multiply connected because the origin is a superattracting fixed point of f. From Theorem 3.1.5, we see that all D_n are pairwise disjoint. Thus D_n is the component we required. □

Remark For the transcendental entire function given in Theorem 3.4.1, the connectivity of the multiply connected component constructed is unknown. Baker (1985) also constructed a transcendental entire function whose Fatou set contains an infinitely connected component.

Next, we give some transcendental entire functions whose Fatou sets contain simply connected wandering domains. As a preparation, we give a theorem on coincidence of the Julia sets.

Theorem 3.4.2 *Assume that a transcendental entire function g satisfies $g(z + c) = g(z) + c$ for some constant c. Then*

(i) *if $h(z) = z + c$, then $h(F(g)) = h^{-1}(F(g)) = F(g)$, and*

(ii) *if* $f(z) = g(z) + c$, *then* $F(f) = F(g)$.

Proof (i) Assume that $z \in F(g)$ and choose a neighborhood $U \subset F(g)$ of z. By induction, we have $g^n(z + c) = g^n(z) + c$. Hence $\{g^n\}_{n=0}^{\infty}$ is a normal family in $h(U)$ and thus $h(z) \in h(U) \subset F(g)$.

Also, we have $g(z - c) = g(z) - c$. By the same argument as above, we have $h^{-1}(z) \in h^{-1}(U) \subset F(g)$.

(ii) By (i), we have $f(F(g)) = h(g(F(g))) \subset h(F(g)) = F(g)$. Since $f(z+c) = f(z)+c$ by the definition, applying (i) for f, we have $g(F(f)) = h^{-1}(f(F(f))) \subset h^{-1}(F(f)) = F(f)$. The inclusion relation $f(F(g)) \subset F(g)$ implies that $F(f) \supset F(g)$. Similarly, we have $F(g) \supset F(f)$. Therefore we have $F(f) = F(g)$. □

Remark In the case that f and g are rational functions, from the uniform continuity of rational functions with respect to the spherical metric, $f \circ g = g \circ f$ implies that $g(F(f)) \subset F(f)$ and $f(F(g)) \subset F(g)$ and hence $F(f) = F(g)$.

Remark Let $g(z)$ be a non-constant entire function and set $f(z) = g(z) + c$. Two functions f and g commute if and only if g satisfies the functional equation $g(z + c) = g(z) + c$.

Example 3.4.1 (Baker) *Let* $f(z) = z - e^z + 1 + 2\pi i$. *Then* $F(f)$ *contains a simply connected wandering domain (see Figure 3.2).*

Verification In the rest of this section, we denote by F_z the Fatou component which contains $z \in F(f)$.

Set $g(z) = z - e^z + 1$. Since all $z = 2n\pi i$ $(n \in \mathbb{Z})$ are superattracting fixed points of g, by Theorem 3.1.7 each $F_{2n\pi i}$ is simply connected. Because $g(z + 2\pi i) = g(z) + 2\pi i$ and $f(z) = g(z) + 2\pi i$, we have $F(f) = F(g)$ by Theorem 3.4.2. Moreover since $f(2n\pi i) = 2(n + 1)\pi i$, $F_{2n\pi i}$ is a wandering domain of f.

Example 3.4.2 *Let* λ *satisfy the conditions that* $|\lambda| = 1$ *and that every* $\phi(z) = \lambda z + c_2 z^2 + \cdots$ *is linearizable at the origin. Set* $f(z) = z + (\lambda - 1)(e^z - 1) + 2\pi i$. *Then* $F(f)$ *contains a simply connected wandering domain* D *such that* f *is univalent on* D.

Verification Set $g(z) = z + (\lambda - 1)(e^z - 1)$; then the assumption on λ implies that each $2n\pi i$ $(n \in \mathbb{Z})$ is the fixed point in the Siegel disk $F_{2n\pi i}$. Hence g is

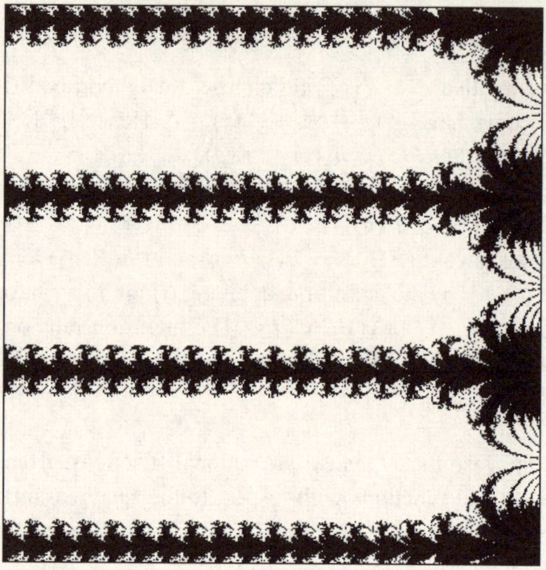

Fig. 3.2. The Julia set of $z - e^z + 1 + 2\pi i$

univalent on $F_{2n\pi i}$. Since $g(z + 2\pi i) = g(z) + 2\pi i$ and $f(z) = g(z) + 2\pi i$, we have $F(f) = F(g)$ by Theorem 3.4.2. Our claim follows from $f(2n\pi i) = 2(n+1)\pi i$.

Finally, we construct a transcendental entire function whose Fatou set contains many grand orbits of wandering domains. As in §2.4.2, we denote, for any $z \in \mathbb{C}$ and any subset $U \subset \mathbb{C}$, by $[z]$ and $[U]$ the grand orbits of z and U, respectively.

Example 3.4.3 (Baker) *Assume that* $\{a_j\}$ *satisfies* $0 < 2a_j < a_{j+1}$ *and set*

$$\varphi(z) = \prod_{j=1}^{\infty} \left(1 - \frac{z}{a_j}\right).$$

Then the Fatou set of

$$f(z) = z + \{\varphi(e^z)\}^2 + 2\pi i$$

contains countably many wandering domains whose grand orbits are pairwise disjoint.

Verification The assumption on $\{a_j\}_{j=1}^{\infty}$ implies that $\varphi(z)$ is a transcendental entire function. Set $g(z) = z + \{\varphi(e^z)\}^2$ and $\alpha_j = \log a_j \in \mathbb{R}$. Then every $\beta_{(j,k)} = \alpha_j + 2\pi ik$ is a parabolic fixed point of g. We denote by $F_{(j,k)}$ the immediate basin of $\beta_{(j,k)}$. Since $g(z + 2\pi i) = g(z) + 2\pi i$ and $f(z) = g(z) + 2\pi i$, we have $F(f) = F(g)$ by Theorem 3.4.2. Because $g^n(f(z)) = g^{n+1}(z) + 2\pi i$, the immediate basin of $\beta_{(j,k)}$ is mapped to that of $\beta_{(j,k+1)}$. Hence each $F_{(j,0)}$ is a wandering domain of $F(f)$. It is easy to see that the $[F_{(j,0)}]$ do not intersect each other.

Though all the limit functions on the wandering domains given in the above are the constant ∞, Eremenko and Lyubich (1987) showed the following.

Theorem 3.4.3 *There is an entire function f with wandering domains on which infinitely many limit functions of $\{f^n\}$ exist.*

Proof Take ϵ and q satisfying $0 < \epsilon < 1/2$ and $1 < q < 2^{1/3}$ and set

$$s = \min\left\{\frac{1}{qr}\log q, \frac{1}{2q}\right\},$$

where r (> 1) satisfies $1 - x \geq \exp(-rx)$ for every $0 < x < 1/2$. Consider a sequence $\{R_m\}_{m=0}^{\infty}$ such that

$$0 < R_m < \frac{1}{2}R_{m-1}, \quad \epsilon\sum_{k=0}^{m-1}\frac{R_m}{R_k} < \min\left\{s, \frac{1}{8}\pi\right\}$$

for all $m = 1, 2, \ldots$ and choose $\{a_m\}_{m=0}^{\infty}$ with $a_0 = 0$ and $a_{m+1} - a_m > 2R_0$. Set

$$\begin{aligned} B_m &= \left\{z \,\middle|\, |z - a_m| \leq \frac{1}{2}R_m\right\}, \\ D_m &= \left\{z \,\middle|\, q^{-2}R_m < |z| < q^{-1}R_m, |\arg z| < \frac{1}{4}\pi\right\}, \\ Q_m &= \left\{z \,\middle|\, q^{-3}R_m \leq |z - a_m| \leq R_m, |\arg z| \leq \frac{1}{2}\pi\right\}. \end{aligned}$$

Then $G = \bigcup_{m=0}^{\infty}(B_m \cup Q_m)$ is the union of the disjoint compact sets. We define an analytic function ϕ as follows.

$$\phi(z) = \begin{cases} z + a_{m+1} - a_m, & z \in B_m, \\ q^{-3/2}R_{m+1}, & z \in Q_m. \end{cases}$$

By using Runge's theorem, we can find an entire function f satisfying

$$f(a_m) = \phi(a_m) = a_{m+1}, \quad f'(a_m) = 1,$$

$$|f(z) - \phi(z)| < \frac{1}{2}\epsilon R_m \quad (z \in B_m),$$

and $f(Q_m) \subset D_{m+1}$. Then, we have

$$f^m(D_m) \subset Q_m.$$

Indeed, clearly we have $D_0 \subset Q_0$. To see this for $m \geq 1$, for every z satisfying $|z| < q^{-1}R_m$ with $m > 0$, we set $r_k = |f^k(z) - a_k|$ for every k and show

$$q^{-1}r_0 \leq r_k \leq qr_0 \quad (k = 0, \ldots, m).$$

Assume that the inequalities above hold for i, $0 \leq i \leq k-1 < m$. Then we have $f^i(z) \in B_i$ for $0 \leq i \leq k-1$. We can put

$$f(z + a_i) = z + g_i(z) + a_{i+1}$$

with $g_i(0) = g_i'(0) = 0$ for $z \in B_i' = \{z \mid |z| < R_i/2\}$. The definition of ϕ implies $|g_i(z)| < \epsilon R_i/2$ for $z \in B_i'$. Applying the maximum modulus theorem for $g(z)/z^2$, we have

$$r_i\left(1 - \frac{\epsilon}{R_i}r_i\right) \leq r_{i+1} \leq r_i\left(1 + \frac{\epsilon}{R_i}r_i\right).$$

Hence, owing to the choice of s, we have the inequalities. Further, we have $f^k(z) \in B_k$ for $0 \leq k \leq m-1$. Moreover, considering $g(z)$ again, we have

$$|\arg(f^{k+1}(z) - a_{k+1}) - \arg(f^k(z) - a_k)| \leq 2\epsilon R_m/R_k.$$

Thus, we obtain

$$|\arg(f^m(z) - a_m) - \arg z| \leq 2\epsilon \sum_{k=0}^{m-1} \frac{R_m}{R_k} \leq \pi/4.$$

These imply that $f^m(D_m) \subset Q_m$.

Thus we have

$$f^{m+1}(D_m) \subset f(Q_m) \subset D_{m+1}$$

and have

$$f^{m(m+1)/2}(D_0) \subset D_m.$$

This means that $D_0 \subset F(f)$ and $a_0 = 0$ is a limit function. Since $f^n(a_0) = a_n$, a_n is also a limit function. □

3.4.2 Logarithmic lifts and Baker domains

We now turn to examples of transcendental entire functions whose Fatou sets contain Baker domains. Fatou has already given an example $f(z) = z+e^{-z}+1$ whose Fatou set contains a Baker domain. Indeed, take $\alpha \geq 1$, then we have $\operatorname{Re} f(z) > \operatorname{Re} z + (1 - \frac{1}{\alpha})$ for $\operatorname{Re} z > \log \alpha$. It follows that there is a Baker domain containing the right half plane.

A Baker domain of an entire function f resembles a parabolic basin in the sense that f is eventually univalent and semiconjugate to a translation of either $\mathbb{C}$, or the upper half plane, or a strip, which come from results due to Cowen (1981). Nevertheless, we construct examples, considering a relation between Baker domains and singular values.

Now to obtain example, we consider lifting an element h of $\mathrm{Ent} \cap \mathrm{End}(\mathbb{C}^*)$ to an entire function f which satisfies

$$h \circ \exp = \exp \circ f$$

and we call such an entire function f a *logarithmic lift* of h.

Remark Here, h can be expressed as

$$h(z) = z^m \exp(\varphi(z)) \quad (m \in \mathbb{N})$$

by some entire function φ.

Also note that such a function in $\mathrm{End}(\mathbb{C}^*)$ as

$$z^m \exp(az + b/z + c)$$

has a logarithmic lift. When $h = \exp(az+b/z)$, the logarithmic lifts are in the sine family. And if $h = z\exp(a + bz - b/z)$, then the logarithmic lifts are in the standard family of circle maps

$$z - ia + 2b \sin z.$$

First, we prove the following proposition on singular values.

Proposition 3.4.4 *Let f be a logarithmic lift of $h \in \mathrm{Ent} \cap \mathrm{End}(\mathbb{C}^*)$. Then e^z is in $\mathrm{sing}(h^{-1})$ if and only if z is in $\mathrm{sing}(f^{-1})$.*

Proof Since

$$\exp(f(z)) = h(e^z),$$

it is clear that e^z is a singular value of h if and only if z is one of f. □

The following fact is also fundamental.

Theorem 3.4.5 *Let f be a logarithmic lift of $h \in \mathrm{Ent} \cap \mathrm{End}(\mathbb{C}^*)$. Then e^z is in $F(h)$ if and only if z is in $F(f)$.*

Proof By induction, we have $h^n(e^z) = \exp(f^n(z))$. Hence if z is a periodic point of f, then e^z is one of h. First, we assume that $z \in J(f)$. The first fundamental theorem shows that the set of repelling periodic points is dense in the Julia set. Thus we have $e^z \in J(h)$.

Conversely, we assume that $\zeta = e^z \in J(h)$. For any neighborhood U of z, there are points z_1 and z_2 such that $\zeta_1 = e^{z_1}$ is a repelling periodic point with period p of h and that $\zeta_2 = e^{z_2}$ is in I_h', which is defined at the end of §3.1. Now suppose that there is a neighborhood U of z such that $U \cap J(f) = \varnothing$. Then z_1 and z_2 are contained in the same Fatou component of f. Hence both of them are in I_f.

First, since $f(ix) = O(x)$ with $x \in \mathbb{R}$, a generalization of Proposition 3.1.4 gives that every Fatou component is simply connected (cf. Bergweiler 1995a). Hence by Theorem 2.1.2 and the Schwarz lemma, we can find a constant C such that

$$|f^n(z_2)| \le C|f^n(z_1)|,$$

for all sufficiently large n.

Next, since ζ_1 is a periodic point of period p, there is a constant $m \in \mathbb{Z}$ such that

$$f^{np}(z_1) = z_1 + m\left(\sum_{j=0}^{n-1} l^{jp}\right) 2\pi i.$$

Take $M > \max\{1, |l|\}$; then we have $|f^{np}(z_1)| = o(M^{np})$ and hence

$$|f^{np}(z_2)| = o(M^{np})$$

as $n \to \infty$.

On the other hand, since ζ_2 is in I_h', we have $\mathrm{Re}\, f^n(z_2) \to \infty$ and $(\mathrm{Re}\, f^{n+1}(z_2))/(\mathrm{Re}\, f^n(z_2)) \to \infty$ as $n \to \infty$ and thus

$$|f^n(z_2)| \ge |\mathrm{Re}\, f^n(z_2)| \ge M^n.$$

This is a contradiction. Hence we conclude that $z \in J(f)$. □

Example 3.4.4 *Let $f(z) = z + e^z + \log\lambda$ $(0 < |\lambda| < -\log|\lambda|)$. Then $F(f)$ contains a Baker domain D such that f is not univalent on D (see Figure 3.3).*

Verification Set $h(z) = \lambda z e^z$. Then the origin is an attracting fixed point of h and -1 is a critical point of h. Since $0 < |\lambda| < -\log|\lambda|$, the open disk D_0

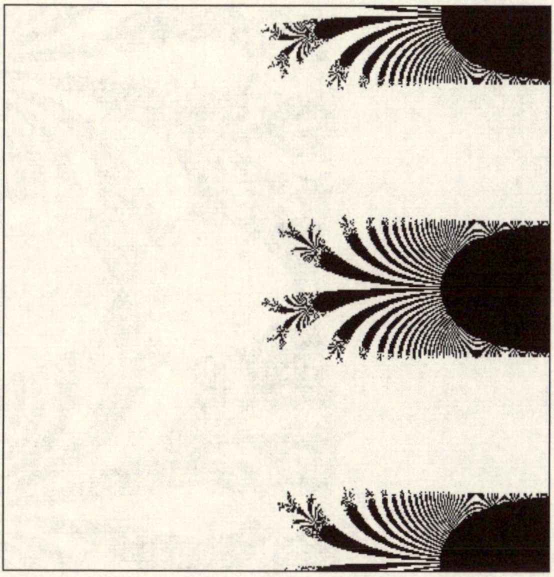

Fig. 3.3. The Julia set of $z + e^z + \log \lambda$

centered at the origin of radius λ is contained in the attracting component F_0. Denoting the segment from the origin to -1 by l, we have $h(l) \subset D_0$. Hence $-1 \in F_0$. The entire function f is a logarithmic lift of h and the lift of F_0 is a Baker domain as required.

Example 3.4.5 *Let λ satisfy the conditions that $|\lambda| = 1$ and that every $\phi(z) = \lambda z + c_2 z^2 + \cdots$ is linearizable at the origin. Set $f(z) = z + e^z + \log \lambda$. Then $F(f)$ contains a Baker domain D such that f is univalent on D and whose boundary is contained in $\overline{S^+(f)}$.*

Verification Set $h(z) = \lambda z e^z$. Then the origin is the fixed point of the Siegel disk of h. The entire function f is a logarithmic lift of h. Let D be a lift of the Siegel disk. Then f is univalent on D. From Proposition 3.4.4 and Theorem 2.4.6, we have $\partial D \subset \overline{S^+(f)}$.

Example 3.4.6 (Bergweiler) *Let $f(z) = 2 - \log 2 + 2z - e^z$. Then $F(f)$ contains a Baker domain D such that f is univalent on D and the Euclidean distance between D and $\overline{S^+(f)}$ is positive (see Figure 3.4).*

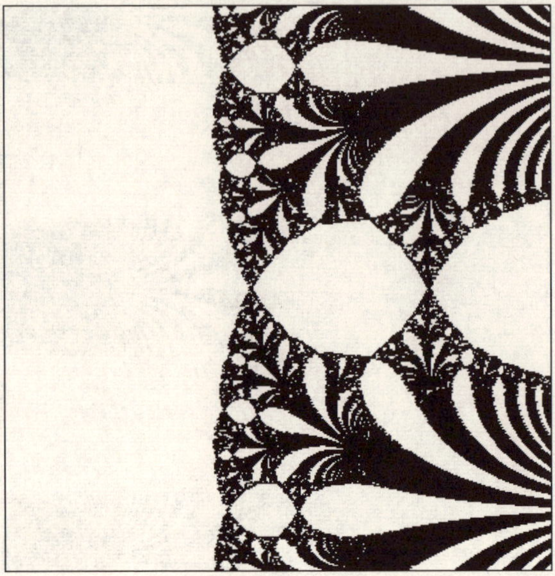

Fig. 3.4. The Julia set of $2 - \log 2 + 2z - e^z$

Verification Set $h(z) = \frac{1}{2}z^2 e^{2-z}$. Then the critical points of h are 0 and 2, both of which are superattracting fixed points of h. The entire function f is a logarithmic lift of h. Note that $\mathrm{sing}(f^{-1}) = \{\log 2 + 2n\pi i \mid n \in \mathbb{Z}\}$, each of whose points is a superattracting fixed point of f. We can choose the disks each of which is centered at $\log 2 + 2n\pi i$, of the same radius, and contained in $F(f)$. A component D of $F(f)$ which is a lift of the component F_0 of $F(h)$ is a Baker domain. Since $\mathrm{sing}(h^{-1}) - \{0\} = \exp\{\mathrm{sing}(f^{-1})\}$, the Euclidean distance between D and $\overline{S^+(f)} = \mathrm{sing}(f^{-1})$ is positive. Because D is simply connected and does not contain $\mathrm{sing}(f^{-1})$, f is univalent on D.

Finally, we give an example of periodic Baker domains.

Example 3.4.7 (Morosawa) *Let $f(z) = -z(2 + \exp(-z^2))$; then $F(f)$ has Baker domains with period* 2.

Verification First, we show that $g(z) = z(2+\exp(-z^2))$ has at least two Baker domains. Take θ with $0 < \theta < \pi/4$ and sufficiently large $R_1 > 3/(2\cos 2\theta)$ and set

$$A = \left\{ z \,\middle|\, |\arg z| < \theta \text{ and } |z| > R_1 \right\}.$$

For $z \in A$, we have

$$|\arg(2+\exp(-z^2))| < c|\exp(-z^2)|$$

for some c. We take $R_2 > R_1$ such that

$$c\sum_{n=0}^{\infty} \exp\left(-\left(\frac{3}{2}\right)^n R_2\right) < \theta.$$

Choose φ satisfying

$$0 < \varphi < \theta - c\sum_{n=0}^{\infty} \exp\left(-\left(\frac{3}{2}\right)^n R_2\right)$$

and set

$$A' = \{z \mid |\arg z| < \varphi \text{ and } |z| > R_2\}.$$

Elementary calculation shows that $\{g^n(z)\}_{n=0}^{\infty} \subset A$ for every $z \in A'$, which implies that $\{g^n\}$ converges to ∞ uniformly on compact sets in A'. Similarly, it converges to ∞ uniformly on compact sets in

$$B' = \{z \mid |\arg z - \pi| < \varphi \text{ and } |z| > R\}.$$

The Fatou components containing A' and B' are simply connected for they are unbounded. Since $J(g)$ is symmetric with respect to the real axis and the origin belongs to the Julia set, A' and B' are contained in different Fatou components.

Since $f^2(z) = g^2(z)$, we have $J(f) = J(g)$. Hence the components containing A' and B' are periodic Baker domains.

4
Dynamics of rational functions

The Newton method, which is the method to find zeros of meromorphic functions, is an origin of complex dynamics. In §4.1, we summarize the Newton method and explain some computational troubles of it. These facts imply, in a sense, the complexity of Julia sets and the Mandelbrot set.

The study of the Newton method was succeeded by the work of Fatou and Julia on the iteration of rational and entire functions. In §4.2, we explain almost independently of Chapters 1 and 2 some fundamental results on the iteration of rational functions. Sullivan's no wandering domain theorem is the answer to one of the most important problems posed by Fatou and its method indicates the next step in complex dynamics. In §4.3, we explain the answers of other important problems proved by Shishikura, which show some estimates of the number of non-repelling cycles.

In §4.4, we show some fundamental properties of hyperbolic rational functions.

4.1 Topics on the Newton method

4.1.1 The Newton method

In general, let a meromorphic function $f(z)$ be given. Then, one way to find a zero of $f(z)$ is the so-called *Newton method*. That is, consider the *Newton map*

$$N(z) = N_f(z) = z - \frac{f(z)}{f'(z)}$$

for f, and we conclude that the orbit $N^n(z)$ of a suitable seed z converges to a zero of $f(z)$.

Proposition 4.1.1 *Let $f(z)$ be a meromorphic function on $\mathbb{C}$. Then, a point z_0 is a zero of $f(z)$ if and only if it is a finite attracting fixed point of $N_f(z)$. Here a simple zero of f corresponds to a finite superattracting fixed point.*

Proof In the domain where $f(z)$ is holomorphic, it is easy to see that $f(z_0) = 0$ if and only if $N(z_0) = z_0$. Moreover, if z_0 is a zero of order ν, then $N'(z_0) = 1 - 1/\nu$, for

$$N'(z) = \frac{f(z)f''(z)}{f'(z)^2}. \tag{4.1}$$

In particular, if z_0 is a simple zero, then $N'(z_0) = 0$.

Next, if z_0 is a pole of order ν, then similarly we have that z_0 is a fixed point of N, and that $N'(z_0) = 1 + 1/\nu$, hence z_0 is a repelling fixed point of $N(z)$, which shows the assertion. □

Now we consider the case that f is a polynomial P of degree k. Then simple computations show that

$$N_P(\infty) = \infty, \quad N'_P(\infty) = \frac{k}{k-1}.$$

That is, ∞ is a repelling fixed point of N_P, and hence the orbit $(N_P)^n(z)$ never converges to ∞ for any given $z \in \mathbb{C} - (\bigcup_{n \geq 1} N_P^{-n}(\infty))$.

Further, if z_0 is a simple zero of P, and if z belongs to the superattracting basin for z_0, then the orbit $(N_P)^n(z)$ converges very rapidly to z_0. For instance, compare the speeds of convergence to 0 between the cases that $P(z) = z^3$ and $P(z) = z^3 + z$.

Thus, the Newton method is an effective way to know approximate values of zeros. An obstacle is that we cannot use every value as the starting point. The set of prohibited values as above is always simple in the case of quadratic polynomials.

Theorem 4.1.2 (Schröder–Cayley) *Suppose that a quadratic polynomial P has two distinct zeros. Then the Newton map $N_P(z)$ is conjugate to z^2 by a Möbius transformation.*

Remark When a quadratic polynomial P has a double zero, the Newton map reduces to a linear function.

Proof of theorem Letting α and β be zeros of P, we set $T(z) = \frac{z-\alpha}{z-\beta}$. Then $\widehat{N} = T \circ N_P \circ T^{-1}$ is a rational function of degree 2 and has superattracting fixed points at 0 and ∞. Since N_P fixes ∞, we also see that $\widehat{N}(1) = 1$, which implies that $\widehat{N}(z) = z^2$. $\square$

The proposition above shows that, in the case of a quadratic polynomial P, for every value z not on a line which passes through $(\alpha+\beta)/2$ and is orthogonal to the segment between α and β, the orbit $(N_P)^n(z)$ always converges to a zero (α or β) of P. Moreover, each critical point of N_P is a zero of P.

But, when we turn to the case of polynomials of higher degrees, serious difficulties occur. First, the set of prohibited values, by which we mean the Julia set of the Newton map, becomes very complicated and in general a fractal. Next, even if the starting point lies in the Fatou set, there may be other components not in the attractive basin of any attracting fixed point.

Such difficulties were already pointed out by Julia. We explain these by some examples.

Example 4.1.1 *A critical point of N_P may not be a zero of P. Equation (4.1) shows that inflection points, that is zeros of P'', are also critical points of N_P.*

Let

$$P(z) = \frac{z^3}{2} - z + 1.$$

Then

$$N_P(z) = z - \frac{z^3 - 2z + 2}{3z^2 - 2} = \frac{2z^3 + 2}{3z^2 - 2},$$

and hence 0 *is a superattracting periodic point of order* 2*, but not a fixed point. As a consequence, starting from a point in the immediate basin for* 0*, we cannot reach a zero of P.*

Now it is important to know the character of Newton maps induced from polynomials of degree $k \geq 2$. Clearly, N_P is a rational function of degree k. Hence, as will be shown in §4.2, it has at most $2k - 2$ attracting cycles. This bound is still sharp for Newton maps.

Theorem 4.1.3 (Hurley 1986) *For every $k \geq 2$, there exists a polynomial P of degree k such that the Newton map N_P has $2k - 2$ attracting periodic cycles.*

Remark Attracting fixed points of N_P corresponding to zeros of P are at most k in number. Hence each of at least $k - 2$ cycles of attracting periodic points is never a fixed point.

The Newton method still works well on the Fatou set for some kinds of entire functions. The following result is such a one.

Theorem 4.1.4 (Bergweiler) *Let $f(z)$ be an entire function represented as*

$$f(z) = \int_0^z P(t)e^{Q(t)}dt + c$$

with polynomials P and Q, but $P(z)z^{Q(z)}$ is not coincident with e^{az+b} ($a, b \in \mathbb{C}$). Further, suppose that $N_f^n(\alpha)$ converges for every zero α of f'' that is not a zero of f'. Then for every z in the Fatou set $F(N_f)$, the orbit $N_f^n(z)$ converges to one of the zeros of f.

On the other hand, Newton maps are special in the following sense.

Theorem 4.1.5 (Shishikura preprint) *For every Newton map induced from a polynomial, the Julia set is connected.*

Actually, Shishikura showed that the Julia set is connected for every rational function having only one weakly repelling fixed point. Here we say that a fixed point z_0 of a meromorphic function f is *weakly repelling* if the multiplier λ at z_0 satisfies $|\lambda| > 1$ or $\lambda = 1$.

Corollary *Every Newton map induced from a polynomial has no Herman rings.*

Remark Bergweiler and Terglane (1996) showed that every transcendental entire function with a multiply connected Fatou component has infinitely many weakly repelling fixed points.

Contrary to Example 4.1.1, there is a universal set for finding roots. Recently, the following theorem has been announced by Schleicher.

Theorem 4.1.6 (Hubbard, Schleicher and Sutherland preprint) *For every $d \geq 2$, there is a finite set S_d of points in $\mathbb{C}$ such that for every root of every polynomial P with degree d, all of whose zeros are in the unit disk, there is a point in S_d converging to this root under the Newton method.*

4.1.2 Iterative algorithms and computability over the reals

A basic problem is whether there exists any purely iterative algorithm to find roots of polynomials.

Definition We call a rational map $\tilde{N}$ of the space Poly_k of all polynomials of degree k to the space Rat_ℓ of all rational function of degree ℓ a *purely iterative algorithm*.

We say that a rational function $R(z) : \widehat{\mathbb{C}} \to \widehat{\mathbb{C}}$ is *convergent* for a polynomial P if for every z in a dense open subset of $\mathbb{C}$, $R^n(z)$ converges to a zero of P.

And we say that a purely iterative algorithm $\tilde{N} : \mathrm{Poly}_k \to \mathrm{Rat}_\ell$ is *generally convergent* if the rational function $\tilde{N}(P)$ is convergent for every polynomial P in a dense open subset of Poly_k.

Example 4.1.2 *Theorem 4.1.2 implies that the Newton method $N : \mathrm{Poly}_2 \to \mathrm{Rat}_2$ is generally convergent.*

But, in the case of higher degree, we cannot find any such algorithm.

Theorem 4.1.7 (McMullen 1987b) *For $k \geq 4$, there exists no generally convergent, purely iterative algorithm of Poly_k to any Rat_ℓ.*

Example 4.1.3 *When $k = 3$, the purely iterative algorithm $\tilde{N}$ which maps*

$$P(z) = z^3 + az + b$$

to

$$\tilde{N}(P)(z) = z - \frac{(z^3 + az + b)(3az^2 + 9bz - a^2)}{3az^4 + 18bz^3 - 6a^2z^2 - 6abz - 9b^2 - a^3}$$

is generally convergent.

Actually, $\tilde{N}(P)$ is the Newton function N_Q with

$$Q(z) = \frac{P(z)}{3az^2 + 9bz - a^2}.$$

Now we have restricted ourselves to $\tilde{N} : \mathrm{Poly}_k \to \mathrm{Rat}_\ell$ such that $\tilde{N}(P)$ is determined from the coefficients of $P \in \mathrm{Poly}_k$ by using only addition, multiplication and their inverse operations.

If we allow taking the square root and the complex conjugate, we can always construct a generally convergent algorithm.

Theorem 4.1.8 (Shub–Smale 1986) *For every $k \geq 2$, let G_k be the open dense set of Poly_k consisting of all polynomials whose zeros are all simple.*

Next, for every

$$P(z) = a_k z^k + \cdots + a_1 z + a_0,$$

set $M_P = \max_j |a_j|$ and

$$\rho(z) = \min\left\{1, \frac{\phi(|z|)|P'(z)|^2}{2\phi'(|z|)^2|P(z)|M_P}\right\}$$

with

$$\phi(r) = r^k + \cdots + r + 1.$$

Finally set

$$\widehat{N}(P)(z) = z - \rho(z)\frac{P(z)}{P'(z)}.$$

Then $\widehat{N}(P)^n(z)$ *converges to one of the zeros of* P *for almost all* $z \in \widehat{\mathbb{C}}$.

Now, in the above definition of a generally convergent algorithm, we did not care about the size of the excluded set. Even if the given algorithm works well not only on a dense open subset but also almost everywhere, there may still occur serious trouble in the computation level. Here we explain such computational troubles in the case of the classical Newton method. First we introduce the concept of computability over the real, following Blum, Shub and Smale (1989).

Definition A *machine over* $\mathbb{R}$, by definition, consists of one input space $\bar{I}$, several output spaces $\bar{O}_j$, several state spaces $\bar{S}_k$, and directed connectors between these spaces, each of which represents a process of computation. Each space is a finite-dimensional real number space $\mathbb{R}^m$. From $\bar{I}$, there exists only one directed connector, which represents a linear map for input to a state space. To each $\bar{O}$, there exists only one directed connector from a state space, which represents a linear map for output. A connector from a state space is either to another state space or to two state spaces. The former represents a rational map for calculation and the latter an indication map for branching.

Remark In general, rational maps for calculation need not be defined on the whole state space. But by suitable change of definition, we can assume that they are always defined on the whole space.

When we construct the universal machine, we should consider the real number space $\mathbb{R}^\infty$ of infinite dimension. But we can still use the same definitions as below.

Definition For a machine M, we denote by Ω_M the set of all points in the initial space of M, for which the output is determined in a finite number of processes. We call Ω_M the *halting set* of M.

We say that a map ϕ defined on a subset Ω_ϕ of $\mathbb{R}^j$ to another $\mathbb{R}^k$ is *computable over* $\mathbb{R}$ if there exists a machine M over $\mathbb{R}$ such that the halting set Ω_M coincides with Ω_ϕ and that the output by M for any $x \in \Omega_\phi$ equals $\phi(x)$. That is, ϕ is computable if and only if there is a machine over $\mathbb{R}$ which realizes ϕ.

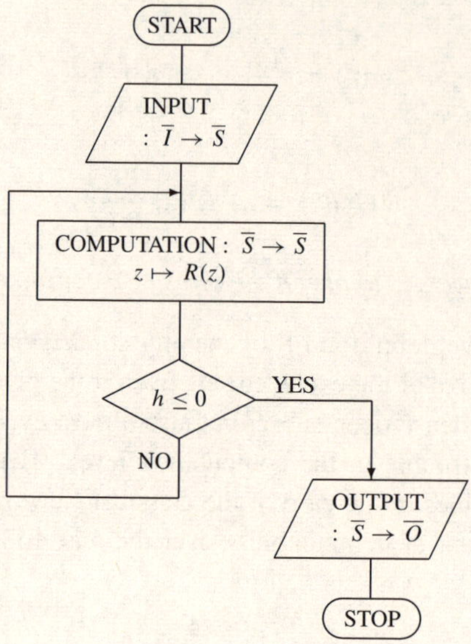

Fig. 4.1. The flow chart

Here, computability of a map can be relaxed to decidability of the domain of it.

Definition We say that a subset E of $\mathbb{R}^k$ is *recursively enumerable over* $\mathbb{R}$ if there exists a machine M over $\mathbb{R}$ such that $\Omega_M = E$.

When both E and the complement $\mathbb{R}^k - E$ are recursively enumerable, we say that E is *decidable over* $\mathbb{R}$.

Remark Computable functions and decidable sets over $\mathbb{R}$ are coincident with recursive partial functions and recursive sets over $\mathbb{R}$, respectively.

Example 4.1.4 *The union E of all attractive basins of a rational function is recursively enumerable over* $\mathbb{R}$.

Indeed, considering $\mathbb{C} = \mathbb{R}^2$, we can find a suitable polynomial $h(x, y)$ of the real coordinates x, y such that the set $\{h(x, y) \leq 0\}$ is contained in the Fatou set of the given rational function R and is a closed neighborhood of all attracting periodic points of R. Then the halting set of the machine defined by the flow chart in Figure 4.1 is clearly the union of all attractive basins for R.

The argument in the above example also indicates the following fact.

Theorem 4.1.9 *Every recursively enumerable set over $\mathbb{R}$ is a union of countably many fundamental semi-algebraic sets.*

Here, a *fundamental semi-algebraic set* is one represented as

$$\{h_j < 0 \ (\text{or } h_j \leq 0) \mid j = 1, \ldots, m\}$$

with suitable m and polynomials $\{h_j\}_{j=1}^{m}$.

Corollary *If the Julia set $J(P)$ of a polynomial P is totally disconnected, then $J(P)$ is not recursively enumerable over $\mathbb{R}$.*

In this case, the Fatou set $F(P)$ is recursively enumerable but undecidable over $\mathbb{R}$.

Proof The first assertion is clear since $J(P)$ has uncountably many connected components. The second one follows by the fact that $F(P)$ is the attractive basin of ∞. □

Remark From the viewpoint of computability, we may call a set E a fractal set if E cannot be represented as a union of countably many fundamental semi-algebraic sets.

Actually, the Julia set of a fairly general rational function is undecidable over $\mathbb{R}$. Compare with Theorem 1.4.7.

Theorem 4.1.10 (Blum–Shub–Smale 1989) *Let $R(z) : \widehat{\mathbb{C}} \to \widehat{\mathbb{C}}$ be a rational function. Then the Julia set $J(R)$ is recursively enumerable over $\mathbb{R}$ if and only if one of the following conditions holds.*

(i) *$J(R)$ consists of at most one point.*
(ii) *$J(R)$ is either a real analytic arc or a real analytic simple closed curve.*
(iii) *$J(R) = \widehat{\mathbb{C}}$.*

Proof First, note that, if $J(R)$ is recursively enumerable over $\mathbb{R}$, $J(R)$ should be connected, for the number of connected components is countable.

If $J(R) = \widehat{\mathbb{C}}$, then $J(R)$ is clearly recursively enumerable over $\mathbb{R}$. Further, if $J(R)$ consists of at most one point, then the degree of R is at most 1, and $J(R)$ is recursively enumerable over $\mathbb{R}$.

Otherwise, R is of degree not less than 2, and $J(R)$ has no interior. Suppose that $J(R)$ is recursively enumerable over $\mathbb{R}$, or equivalently is a union of countably many semi-algebraic sets. Then Baire's category theorem implies that one of these sets has an interior in $J(R)$. Hence we can find an analytic arc I in $J(R)$ such that $R^n(I) = J(R)$ for some n. Moreover, if $R^n(I)$ has either a branching point or a self-crossing point, then such a point should be dense in $R^n(I)$, which contradicts the choice of I. Thus condition (ii) holds. □

Corollary *If a rational function R has at least three attracting fixed points, then $J(R)$ is not recursively enumerable over $\mathbb{R}$.*

In particular, for every polynomial P with at least three simple zeros, the Newton map N_P has Julia set not recursively enumerable over $\mathbb{R}$.

Thus even if the Fatou set $F(P)$ of such a polynomial as in the above corollary is recursively enumerable over $\mathbb{R}$, $F(P)$ is never decidable over $\mathbb{R}$.

Proof of corollary The assertion follows by noting that, in all cases stated in Theorem 4.1.10, the Fatou set has at most two connected components. □

Finally, recall that, since the boundary of the Mandelbrot set is a fractal set by Shishikura's theorem, we also reach the following conclusion.

Theorem 4.1.11 (Blum–Smale 1993) *The Mandelbrot $\mathcal{M}$ set is undecidable over $\mathbb{R}$.*

Proof If not, $\mathcal{M}$ is a union of countably many semi-algebraic sets S_j. Since $\mathcal{M}$ is closed, we may assume that each S_j is closed. Here $\partial\mathcal{M}$ should not intersect with the interior of any S_j. Hence, $\partial\mathcal{M} \cap S_j = \partial\mathcal{M} \cap \partial S_j$ has dimension at most 1, which implies that $\dim \partial\mathcal{M} \leq 1$. This contradicts Shishikura's Theorem 1.4.4. □

4.2 Fundamental theorems of Fatou, Julia and Sullivan

Let $f(z)$ be a holomorphic function on a neighborhood of ζ, which is a fixed point of f. Then the Taylor expansion of f at ζ is

$$f(z) = \zeta + \lambda(z - \zeta) + a_2(z - \zeta)^2 + \cdots .$$

Here, λ is a multiplier of f at ζ. We say that f has a fixed point with order k at ζ, or ζ is a fixed point with *multiplicity* k, if $f(z) - z$ has a zero with order k at ζ. Hence multiplicity of a fixed point is more than 2 if and only if its multiplier is 1.

Hereafter we will reconstruct some of the fundamental results in §2.3 for the case of rational functions, in a simpler manner.

Proposition 4.2.1 *A rational function with degree k except for the identity has $k+1$ fixed points with multiplicity.*

Proof Assume that ∞ is not a fixed point of a rational function f and set $f(z) = P(z)/Q(z)$, where P and Q are mutually prime polynomials. Then $\deg Q = k$. From the equation $P(z)/Q(z) = z$, we have $zQ(z) - P(z) = 0$, which has $k+1$ roots in $\mathbb{C}$ with multiplicity. □

For a fixed point $\zeta \in \mathbb{C}$ of f, take a sufficiently small circle C with center ζ and set

$$I(f, \zeta) = \frac{1}{2\pi i} \int_C \frac{dz}{z - f(z)},$$

where the direction of C is positive with respect to ζ.

Theorem 4.2.2 *Any rational function $f(z)$ except for the identity satisfies*

$$\sum_f I(f, z) = 1,$$

where $\sum_f$ means the sum over all fixed points of f.

Proof We may assume that ∞ is not a fixed point. Take a sufficiently large r so that all the fixed points are in $|z| < r$. From the residue theorem, we have

$$\sum_f I(f, z) = \frac{1}{2\pi i} \int_{|z|=r} \frac{dz}{z - f(z)} = \lim_{r\to\infty} \frac{1}{2\pi i} \int_{|z|=r} \frac{dz}{z - f(z)} = 1.$$

□

Proposition 4.2.3 *A fixed point ζ of f whose multiplier λ is not equal to 1 is attracting if and only if* $\mathrm{Re}(I(f, \zeta)) > 1/2$.

Proof The inequality

$$\mathrm{Re}(I(f, \zeta)) = \frac{1}{2}\left(\frac{1}{1-\lambda} + \frac{1}{1-\bar{\lambda}}\right) > \frac{1}{2}$$

holds if and only if $\lambda\bar{\lambda} < 1$. □

Corollary 1 *A rational function with degree at least 2 has either a repelling fixed point or an indifferent fixed point whose multiplier is* 1.

Proof For an indifferent fixed point whose multiplier is not 1, $\mathrm{Re}(I(f, \zeta)) = 1/2$. Hence if there are $k+1$ attracting fixed points or indifferent fixed points whose multipliers are not 1, then we have $\sum_f \mathrm{Re}(I(f,\zeta)) \geq (k+1)/2 > 1$ and this contradicts Theorem 4.2.2. □

Corollary 2 *The Julia set of a rational function with degree at least two is not empty.*

Proof Since repelling fixed points and indifferent fixed points with multiplier 1 are contained in Julia sets, we obtain the claim from Corollary 1. □

In what follows, we consider rational functions with degree at least 2, that is, elements of Rat. First, we consider periodic points of elements of Rat.

Proposition 4.2.4 (Beardon 1991) *A rational function f with degree k (≥ 2) has cycles of periodic points with every prime period except for at most $k+1$ prime.*

Proof Assume that f does not have periodic points with some prime period, say p. Hence if $f^p(\zeta) = \zeta$, then there is a divisor q ($< p$) of p such that $f^q(\zeta) = \zeta$. Since p is a prime number, we have $q = 1$ and hence ζ is a fixed point of f. Let

$$f(z) = \zeta + a(z-\zeta) + b(z-\zeta)^{r+1} + \cdots \qquad (b \neq 0)$$

be the Taylor expansion of f at ζ. By calculation, we have

$$f^p(z) = \zeta + a^p(z-\zeta) + b'(z-\zeta)^{r+1} + \cdots,$$

where

$$b' = a^{p-1} b \sum_{k=0}^{p-1} a^{rk}.$$

Since f and f^p have $k+1$ and k^p+1 fixed points, respectively, there is a fixed point ζ of f whose multiplicity of f^p is greater than that of f. For such ζ, we see that $a \neq 1$ and $a^p = 1$ from the two Taylor expansions above. Hence, we obtain the claim by this fact and Proposition 4.2.1. □

A much stronger result on the existence of periodic points is known. Compare the following theorem with Proposition 4.2.4.

Theorem 4.2.5 (Baker) *Assume that a rational function with degree k has no periodic points with period n. Then the pair (k, n) is one of the following.*

$$(2,2), \quad (2,3), \quad (3,2), \quad (4,2).$$

Remark In fact, rational functions corresponding to each pair above exist. The normal forms for such rational functions are determined by Kisaka (1995). For example $f(z) = z^2 - z$ represents (2, 2).

As was seen in §2.4.1, there are relations between singular values, and non-repelling periodic points and cycles of Herman rings. Hence, from the observation of singular orbits, we have obtained information on Julia sets.

If we consider the case of rational functions, singular values are just critical values. Moreover, if $\zeta \in \mathbb{C}$ is a critical point of f with $f(\zeta) \in \mathbb{C}$, then $f'(\zeta) = 0$. Further, if $\zeta = \infty$, then, by using $\phi(z) = 1/z$, $(f \circ \phi^{-1})'(0) = 0$ when $f(\infty) \neq \infty$ and $(\phi \circ f \circ \phi^{-1})'(0) = 0$ when $f(\infty) = \infty$. We say that the *multiplicity* of a critical point ζ is the multiplicity of the zero of these derivatives at ζ. Then we have the following theorem from the Riemann–Hurwitz relation.

Theorem 4.2.6 *A rational function with degree k has* $2k - 2$ *critical points with multiplicity.*

The following theorem follows from Theorems 2.4.1, 2.4.2 and 4.2.6.

Theorem 4.2.7 *Denote the number of attracting cycles and that of parabolic cycles of a rational function with degree k by* $N(A)$ *and* $N(P)$*, respectively. Then the following inequality holds.*

$$N(A) + N(P) \leq 2k - 2.$$

Remark We do not use the first fundamental theorem to prove Theorems 2.4.1 and 2.4.2. Hence we can use Theorem 4.2.7 to give another proof of the first fundamental theorem for Rat (see Theorem 4.2.10).

Now we estimate the number of remaining non-repelling cycles.

Theorem 4.2.8 *The number of indifferent cycles whose multipliers are not* 1 *of a rational function* f *with degree* k *is at most* $4k - 4$.

Proof Setting $f(z) = P(z)/Q(z)$, we define the perturbation

$$f(z, t) = \frac{(1 - t)P(z) + tz^k}{(1 - t)Q(z) + t},$$

for $t \in \mathbb{C}$. We see that $f(z, 0) = f(z)$ and $f(z, 1) = z^k$ and that $f(z, t)$ are rational functions with degree k except for a finite number of t. Assume that f has m indifferent cycles whose multipliers are not 1. We have the cycles

$\{z_i(t), \ldots, f^{p(i)-1}(z_i(t), t)\}$ $(i = 1, 2, \ldots, m)$ with multiplier $\lambda_i(t)$ from the cycles $\{z_{i1}, \ldots, z_{ip(i)}\}$ of f with multiplicity $\lambda_i = \lambda_i(0)$ by the perturbation above. Then the $\lambda_i(t)$ are holomorphic with respect to t and not constant. Since $\lambda_i(0) \neq 1$, we have

$$\lambda_i(t) = \lambda_i(0) + a_{ir}t^r + \cdots,$$

$a_{ir} \neq 0$ near $t = 0$. Set

$$\epsilon_i(t) = \operatorname{sgn} \log |\lambda_i(t)|;$$

then the limit

$$\epsilon_i^*(\theta) = \lim_{\rho \to 0} \epsilon_i(\rho e^{i\theta})$$

exists and is a sum of step functions with value 1 or -1 of intervals with length π/r except at at most a finite number of points. Moreover, since

$$\int_0^{2\pi} \epsilon_i^*(\theta) d\theta = 0,$$

setting $\eta(\theta) = \sum_{i=1}^m \epsilon_i^*(\theta)$, we have

$$\int_0^{2\pi} \eta(\theta) d\theta = 0,$$

which implies that at least $[(m + 1)/2]$ of the ϵ_i are negative for sufficiently small $|t|$, that is, they correspond to attracting cycles. We obtain the claim by Theorem 4.2.7. □

We have the following theorem from Theorems 4.2.7 and 4.2.8.

Theorem 4.2.9 *The number of non-repelling cycles of a rational function with degree k is at most* $6k - 6$.

The best possible estimate of the number of non-repelling cycles was given by Shishikura, which is proved in the next section. However, if we know only that it is finite, we give another proof for the first fundamental theorem for Rat.

Theorem 4.2.10 (First fundamental theorem for Rat) *The Julia set of a rational function is the closure of the set of the repelling periodic points.*

Proof The same argument as in the proof of Theorem 2.3.7 shows that periodic points are dense in the Julia set. Because the number of non-repelling periodic points is finite, we have the assertion. □

After proving the first fundamental theorem, we can show Proposition 2.4.5. Since $J(f)$ is compact in $\widehat{\mathbb{C}}$ and any exceptional point of f is not contained in $J(f)$ for $f \in \mathrm{Rat}$, we have the following variant of Proposition 2.4.5.

Proposition 4.2.11 *Let f be in* Rat *and take any open set U intersecting with $J(f)$. Then $f^n(U) \supset J(f)$ holds for all sufficiently large n.*

Thus we conclude that the action of f on $J(f)$ is chaotic. Here we recall the definition of chaos given by Devaney.

Definition Let (X, ρ) be a metric space and f be a map from X onto itself. The dynamical system (f, X) is said to be *chaotic* if it satisfies the following three conditions.

(i) The set of periodic points of f is dense in X.
(ii) The function f has a transitivity property, i.e., for any two open sets U and V of X, there is a k such that $f^k(U) \cap V \neq \emptyset$.
(iii) The function f has sensitive dependence on initial conditions, i.e., there is a constant δ such that, for any $x \in X$ and any neighborhood N, there are $y \in N$ and n satisfying

$$\rho(f^n(x), f^n(y)) > \delta.$$

Now we consider the dynamical system $(f, J(f))$, where $J(f)$ is a metric space endowed with the restricted spherical metric ρ.

Proposition 4.2.12 *For every $f \in$* Rat, *the dynamical system $(f, J(f))$ is chaotic.*

Proof Condition (i) is the first fundamental theorem and condition (ii) follows from Proposition 4.2.11.

Choose periodic points p_1 and p_2 of f satisfying

$$\delta_0 = \rho(O(p_1), O(p_2)) > 0,$$

where $O(p_i)$ is a cycle containing p_i. Hence, for any $z \in J(f)$, the distance between z and one of $O(p_1)$ and $O(p_2)$ is at least $\delta_0/2$. We show that $\delta = \delta_0/8$ satisfies condition (iii).

Choose any point $z \in J(f)$ and any neighborhood N of it. From condition (i), we can take a periodic point p in $U = N \cap D(z, \delta)$ and we assume its period is n. From the definition of δ_0, we can also take a periodic point q such that the distance between $O(q)$ and z is at least 4δ. Set

$$V = \bigcap_{i=0}^{n} f^{-i}(D(f^i(q), \delta)).$$

Then V is open and further is not empty for $q \in V$. From condition (ii), there are $y \in D(z, \delta)$ and k satisfying $f^k(y) \in V$. Set $j = [(k/n) + 1]$; then since $1 \leq nj - k \leq n$ we have

$$f^{nj}(y) \in D(f^{nj-k}(q), \delta).$$

Since $f^{nj}(p) = p$, we obtain

$$\rho(f^{nj}(p), f^{nj}(y)) \geq \rho(z, f^{nj-k}(q)) - \rho(f^{nj-k}(q), f^{nj}(y)) - \rho(p, z) > 2\delta.$$

By using the triangle inequality, at least one of

$$\rho(f^{nj}(z), f^{nj}(y)) \quad \text{and} \quad \rho(f^{nj}(z), f^{nj}(p))$$

is greater than δ. □

Remarks (1) Similarly, the dynamical system $(f, J(f))$ is chaotic for every $f \in$ Ent.

(2) After Devaney gave the definition, it was proved as above that conditions (i) and (ii) imply (iii) for a general dynamical system. Nevertheless, condition (iii) is the essence of chaos. For example, by condition (iii), it is difficult to predict the orbit of a point in the Julia set even by computer.

Next Sullivan answered the problem given by Fatou.

Theorem 4.2.13 (Second fundamental theorem for Rat**)** *There are no wandering domains in the Fatou set of a rational function.*

Recalling the weak 'no wandering domains theorem' (Theorem 2.4.11), we need only show the following lemma.

Lemma 4.2.14 *Assume that* $f \in$ Rat *has wandering domains. Then there is a wandering domain* D_0 *such that each* $D_n = f^n(D_0)$ $(n \geq 0)$ *is simply connected and contains no critical points.*

Proof As the number of the critical points of f is finite, there is a wandering domain D_0 such that each $D_n = f^n(D_0)$ contains no critical points. Hence if D_0 is simply connected, so is D_n because $f^n : D_0 \to D_n$ is a smooth covering.

Let γ be a closed curve in $\widehat{\mathbb{C}}$ contained in a hemisphere of $\widehat{\mathbb{C}}$. In this proof, the complementary component of γ containing a hemisphere is called the exterior of γ and is denoted by $\mathrm{Ext}(\gamma)$, and the union of the other complementary components is called the interior of γ and is denoted by $\mathrm{Int}(\gamma)$. We denote the diameter of a set W with respect to the spherical metric σ_0 by $\mathrm{diam}_0(W)$. Since a rational function is Lipschitz continuous with respect to the spherical metric, we have

$$\mathrm{diam}_0(f(W)) \leq M\,\mathrm{diam}_0(W),$$

where M is a Lipschitz constant of f.

First, we prove that $\lim_{n\to\infty} \mathrm{diam}_0(f^n(K)) = 0$ for any compact set K in D_0. Assume the contrary:

$$\mathrm{diam}_0(f^{n(j)}(K)) \geq \epsilon$$

for some compact set K, a positive number ϵ and a subsequence $\{f^{n(j)}\}$. Since $\{f^n\}$ is a normal family in D_0, a certain subsequence of $\{f^{n(j)}\}$ converges uniformly in K to a limit function g, which is non-constant by the assumption. We denote the subsequence again by $\{f^{n(j)}\}$. Choose a $\zeta \in D_0$ and take a sufficiently small circle C contained in D_0 with center ζ, where g is injective. Then we have

$$|f^{n(j)}(z) - g(z)| < \inf_{w\in C} |g(w) - g(\zeta)| \leq |g(z) - g(\zeta)| \quad (z \in C)$$

for all sufficiently large j. Denoting the interior of C by D, we have $f^{n(j)}(D) \ni g(\zeta)$ from Rouché's theorem. This contradicts that D_0 is a wandering domain. Hence we have the assertion.

Now suppose that D_0 is not simply connected; then there is a closed curve γ_0 which is not contractible in D_0. According to the monodromy theorem, $\gamma_n = f^n(\gamma_0)$ is also not contractible in D_n. On the other hand,

$$\mathrm{diam}_0(\gamma_n) \to 0.$$

If necessary changing D_0, we may assume

$$\mathrm{diam}_0(\gamma_n) \leq \pi/(M+1)$$

for all $n \geq 0$. Hence each γ_n is contained in a hemisphere and satisfies

$$\mathrm{diam}_0(f(\mathrm{Int}(\gamma_n))) \leq M\,\mathrm{diam}_0(\gamma_n) < \pi.$$

This implies that $f(\mathrm{Int}(\gamma_n))$ does not contain $\mathrm{Ext}(\gamma_{n+1})$. Moreover, since

$$\partial(f(\mathrm{Int}(\gamma_n))) \subset f(\partial\,\mathrm{Int}(\gamma_n)) \subset f(\gamma_n),$$

we have $\mathrm{Ext}(f(\gamma_n)) \cap f(\mathrm{Int}(\gamma_n)) = \varnothing$, which implies

$$f(\mathrm{Int}(\gamma_n)) \subset \mathrm{Int}(\gamma_{n+1}).$$

Hence we have

$$\mathrm{diam}_0\left(\bigcup_{n\geq 0} f^n(\mathrm{Int}(\gamma_0))\right) \leq \pi,$$

which contradicts the three-point condition because $\mathrm{Int}(\gamma_0)$ intersects $J(f)$. □

The second fundamental theorem can be paraphrased as follows.

Theorem 4.2.15 *Let f be a rational function whose Fatou set is not empty. Then the orbit of every $z \in F(f)$ is eventually contained in either an immediate attracting or parabolic basin, or a cycle of Siegel disks or of Herman rings.*

Here we recall two well-known results as direct applications of the second fundamental theorem.

Theorem 4.2.16 *For $f \in$ Rat, $F(f)$ has at most two completely invariant components. If there are two, then each of them is simply connected and there are no other Fatou components.*

Proof If there are two completely invariant Fatou components, then by the second corollary of Theorem 2.3.5, every Fatou component is simply connected. Hence, the Riemann–Hurwitz formula (Lemma 1.1.5) implies that every completely invariant Fatou components contains $k - 1$ critical points including multiplicity. It follows that there can be at most two completely invariant components. If there are two, all critical points are contained in them. Hence there are no other Fatou components by the theorems in §2.4.1 and the second fundamental theorem. □

Corollary *For $f \in$ Rat, the number of the components of $F(f)$ is* 0, 1, 2 *or* ∞.

Proof If the number of the components of $F(f)$ is finite then there is a natural number n such that all the components of $F(f)$ are completely invariant with respect to f^n. Since $F(f) = F(f^n)$, we obtain the claim. □

In the case of Ent, the first half of Theorem 4.2.16 is as follows.

Theorem 4.2.17 (Baker 1970) *For $f \in \mathrm{Ent}$, $F(f)$ has at most one completely invariant component.*

Theorem 4.2.18 *If all the critical points of a rational function are preperiodic, then its Julia set coincides with $\widehat{\mathbb{C}}$.*

Proof There are no wandering domains by Theorem 4.2.13. A cycle of a superattracting periodic point contains critical points. If a Fatou component of another kind exists, then there is a critical point whose orbit consists of infinitely many points. □

Remark Theorem 4.2.18 can be applied to Example 2.3.2. The Julia set of the rational function

$$R(z) = -\frac{i}{2}\left(z - \frac{1}{z}\right)$$

coincides with $\widehat{\mathbb{C}}$.

Indeed, the critical points are i and $-i$ and ∞ is the fixed point. And we have

$$R^3(i) = R^3(-i) = \infty,$$

which implies $J(R) = \widehat{\mathbb{C}}$ by Theorem 4.2.18.

4.3 Theorems of Shishikura

In this section we give proofs of the sharp estimates due to Shishikura, following his paper. For the more precise details and related results, see that paper (Shishikura 1987). In what follows, we suppose that the degree k of a rational function f considered is $k \geq 2$.

Definition We denote the number of critical points in a set A by $N_c(A)$, and if A is the Fatou set, we denote it simply by N_c. We also denote the number of cycles of irrationally indifferent periodic points in a set A by $N_{ir}(A)$, and in the case $A = \widehat{\mathbb{C}}$, we denote it simply by N_{ir}.

The numbers of cycles of non-repelling periodic points and of Herman rings are denoted by $N(NR)$ and $N(HR)$, respectively.

Theorem 4.3.1 *For any rational function f,*

$$N_c + N_{ir} + 2N(HR) \leq 2k - 2$$

holds. Moreover, this estimate is the best possible.

First we state some direct corollaries. We denote the numbers of immediate attracting and parabolic basins by $N(AB)$ and $N(PB)$. Also we denote the numbers of cycles of Siegel disks and of Cremer points by $N(SD)$ and $N(Cr)$, respectively.

Corollary 1

$$N(NR) + 2N(HR) \leq 2k - 2.$$

Proof Since the number N_c is not less than $N(AB) + N(PB)$, the assertion follows. □

Corollary 2

$$N(AB) + N(PB) + N(SD) + 2N(HR) + N(Cr) \leq 2k - 2.$$

Proof Since $N_{ir} = N(SD) + N(Cr)$, the assertion is clear. □

Now the proof of Theorem 4.2.8 shows that, even in the worst case, at least half the number of non-repelling cycles can be deformed to attracting cycles by general holomorphic perturbation. Hence Shishikura used quasiconformal mappings to perturb them, using the idea originated by Douady and Hubbard and made sophisticated by him. It is called *quasiconformal surgery*.

To show Theorem 4.3.1, first we suppose that a rational function f has one or more Herman rings, all of which we include in a set $\mathcal{A}_{HR}$. We consider how to eliminate all Herman rings of f. Recall that $\mathcal{A}_{HR}$ is finite in number. For each $A \in \mathcal{A}_{HR}$, we associate an oriented analytic simple closed dividing curve γ_A passing through no orbits of critical points, so that

$$f(\gamma_A) = \gamma_{f(A)}.$$

Let $\mathcal{A}$ be the set of connected components of $A - \gamma_A$ with $A \in \mathcal{A}_{HR}$ and $\mathcal{X}$ the union of $\mathcal{A}$ and all non-repelling periodic points. Then every element $X \in \mathcal{X}$ is contained in a connected component of $\widehat{\mathbb{C}} - \Gamma_n$, where we denote by Γ_n the set of all connected components of $f^{-n}(\gamma_A)$ with $A \in \mathcal{A}_{HR}$. We assign orientations to all elements of Γ_n so that f respects them.

Definition We call a component of $\widehat{\mathbb{C}} - \Gamma_n$ an *n-piece*.

Since $\mathcal{X}$ is a finite set, we can see that there is a non-negative integer N such that, if $X_1, X_2 \in \mathcal{X}$ are in the same N-piece, then so are $f(X_1), f(X_2)$. Fix such an N, and divide the set $\mathcal{D}$ of all N-pieces into the two classes $\mathcal{D}_I$ and $\mathcal{D}_{II}$ according as $D \in \mathcal{D}$ contains some element of $\mathcal{X}$ or not. Since every $X \in \mathcal{X}$ is periodic as a set, the set $\mathcal{D}_I$ can be decomposed into cycles $\{D_{i,j}\}_{j=0}^{m_i-1}$ $(i = 1, \ldots, L)$ so that, if $X \in \mathcal{X}$ is contained in $D_{i,j}$, then $f(X)$ is contained in $D_{i,j+1}$, where we set $D_{i,m_i} = D_{i,0}$.

For the sake of simplicity, fix i and we abbreviate $D_{i,j}$ and m_i as D_j and m.

Lemma 4.3.2 *Let $\gamma \in \Gamma_N$ be a boundary component of D_j. Suppose that D_j is contained in* $\operatorname{Int}\gamma$ *or* $\operatorname{Ext}\gamma$, *then D_{j+1} is contained in* $\operatorname{Int} f(\gamma)$ *or* $\operatorname{Ext} f(\gamma)$, *respectively.*

Proof Assume that $D_j \subset \operatorname{Int}\gamma$. If $N \geq 1$, then we see that $f(D_j)$ is disjoint from $f(\gamma)$ but not from $\operatorname{Int} f(\gamma)$. Hence

$$\operatorname{Int} f(\gamma) \supset f(D_j) \supset D_{j+1}.$$

If $N = 0$, some $A \in \mathcal{A}$ having γ as a boundary component is contained in D_j. Since $f(A) \subset D_{j+1}$ and D_{j+1} is disjoint from $f(\gamma)$, we have the assertion. □

Let $m\widehat{\mathbb{C}}$ be the disjoint union of m copies of $\widehat{\mathbb{C}}$, or more precisely, $(\mathbb{Z}/m\mathbb{Z}) \times \widehat{\mathbb{C}}$. We write $\{j\} \times \widehat{\mathbb{C}}$ as $\widehat{\mathbb{C}}_j$. We say that a map $g : \widehat{D}\ (\subset m\widehat{\mathbb{C}}) \to m\widehat{\mathbb{C}}$ is a *cyclic map* if $g(\widehat{D} \cap \widehat{\mathbb{C}}_j) \subset \widehat{\mathbb{C}}_{j+1}$ for all j, and that a cyclic map g is rational if so are all $g|_{\widehat{\mathbb{C}}_j}$. Basic facts still hold even in the case of rational cyclic maps, and we will use the same terminologies as in the case of rational functions.

Let $\widehat{D}_j$ be D_j considered as a subset of $\widehat{\mathbb{C}}_j$, and $\widehat{D} = \bigcup \widehat{D}_j$. Define a cyclic map $\hat{f} : \widehat{D} \to m\widehat{\mathbb{C}}$ by setting $\hat{f}(j, z) = (j+1, f(z))$.

Proposition 4.3.3 (Reduction to the no HR case) *There exist a rational cyclic map $F : m\widehat{\mathbb{C}} \to m\widehat{\mathbb{C}}$ and a quasiconformal map $\varphi : m\widehat{\mathbb{C}} \to m\widehat{\mathbb{C}}$ such that*

(i) *$\varphi \circ \hat{f} = F \circ \varphi$ on $\widehat{D}$,*
(ii) *$F(m\widehat{\mathbb{C}} - \varphi(\widehat{D})) \subset m\widehat{\mathbb{C}} - \varphi(\widehat{D})$,*
(iii) *for every $A \in \widehat{\mathcal{A}}$, $\varphi(A)$ is contained in a Siegel disk of F,*
(iv) *there is $M > 0$ such that $F^M(m\widehat{\mathbb{C}} - \varphi(\widehat{D}))$ is contained in the union of attractive basins and Siegel disks of F.*

Here $\widehat{\mathcal{A}}$ is the set of $A \in \mathcal{A}$ considered as a subset in $\widehat{D}_j$ when $A \subset D_j$.

We call a continuous map *quasiregular* if it is a post-composition of a holomorphic map to a quasiconformal map. The crucial fact for the proof is the following generalization of the straightening theorem (Theorem 1.5.1).

Lemma 4.3.4 (Fundamental lemma for qc-surgery) *Let g be a quasiregular cyclic map from $m\widehat{\mathbb{C}}$ onto itself. Assume that there are an open set E of $m\widehat{\mathbb{C}}$,* *a quasiconformal map Φ of E and a non-negative integer M satisfying the following conditions:*

(i) $g(E) \subset E$,

(ii) $\Phi \circ g \circ \Phi^{-1}$ *is analytic in* $E' = \Phi(E)$ *and*

(iii) $g_{\bar{z}} = 0$ *a.e. on* $m\widehat{\mathbb{C}} - g^{-M}(E)$.

Then there is a quasiconformal map φ such that $\varphi \circ g \circ \varphi^{-1}$ is a rational cyclic map.

Proof Let μ be the Beltrami coefficient of Φ on E; then μ is invariant on E under the action of g. Pulling back the complex structure induced by μ by g and noting condition (iii), we have an invariant Beltrami coefficient $\tilde{\mu}$ on the whole of $\mathbb{C}$ by setting it to be 0 outside $\bigcup_{n\geq 0} g^{-n}(E)$. Then the assertion follows by the measurable Riemann mapping theorem. □

Proof of Proposition 4.3.3 We will extend $\hat{f}$ to a cyclic map g of the whole of $m\widehat{\mathbb{C}}$. First we can extend $\hat{f}$ continuously to all boundary components of $\widehat{D}$. These belong to the set $\widehat{\Gamma}_N$ of all $\gamma \in \Gamma_N$ with $\gamma \subset \partial D$ considered as a subset of $\widehat{\overline{D}}_j$ when $\gamma \subset \overline{D}_j$. Set

$$\widehat{\Gamma}'_N = \widehat{\Gamma}_N \cup \{\hat{f}(\gamma) \mid \gamma \in \widehat{\Gamma}_N\}.$$

Here the orientations are taken so that $\hat{f} : \widehat{\Gamma}_N \to \widehat{\Gamma}'_N$ respects them. And by Lemma 4.3.2, we may also assume, changing some of the orientations if necessary, that

$$\widehat{D}_j \cap \operatorname{Ext}\gamma = \emptyset \quad \text{for every } \gamma \in \widehat{\Gamma}'_N,$$

where $\operatorname{Ext}\gamma$ means the exterior of γ in $\widehat{\mathbb{C}}_j$ if $\gamma \subset \widehat{\mathbb{C}}_j$. Note that, setting $\mathcal{E} = \{E_\gamma = \overline{\operatorname{Ext}\gamma} \mid \gamma \in \widehat{\Gamma}'_N\}$, we see that $m\widehat{\mathbb{C}}$ is the disjoint union of $\widehat{D}$ and $\bigcup_{\gamma\in\mathcal{E}} E_\gamma$. Also, for every $E \in \mathcal{E}$, there is a unique $E' \in \mathcal{E}$ such that $\hat{f}(\partial E) \subset E'$. Then we write $E \to E'$, and we define g on E so that $g(E) \subset E'$ and $g = \hat{f}$ on ∂E as follows.

When $E = E_\gamma$ belongs to $\mathcal{E}_1 = \{E_\gamma \mid \gamma \in \widehat{\Gamma}_0\}$, and if $E \to E'$, then $E' \in \mathcal{E}_1$ and $\hat{f}(\partial E) = \partial E'$. Hence we have a cycle

$$E = E_0 \to E_1 \to \cdots \to E_q = E$$

with some q, where we may assume that $\{E_0, \ldots, E_{q-1}\}$ are distinct. Then we can find an orientation-preserving diffeomorphism $\psi_0 : \gamma_0 = \partial E_0 \to \partial\Delta$ such that $\psi_0 \circ \hat{f}^q \circ \psi_0^{-1}(z) = \lambda z$, where $\Delta = \{|z| < 1\}$. For other j, set $\psi_j = \psi_0 \circ \hat{f}^{q-j}$. Then we can extend these ψ_j to quasiconformal maps of E_j

onto $\overline{\Delta}$. Hence we set

$$g|_{E_0} = \psi_1^{-1} \circ \lambda \circ \psi_0$$

and

$$g|_{E_j} = \psi_{j+1}^{-1} \circ \psi_j$$

for positive j. We set $\Phi|_{E_j} = \psi_j$.

Next assume that $E = E_\gamma$ does not belong to $\mathcal{E}_1$. If E is not periodic, then it is preperiodic, for $\mathcal{E}$ is a finite set. And if $E \to E'$, define g so that $g = \hat{f}$ on ∂E and that $g(E) \subset E'$. Hence it suffices to consider only the case that E is periodic. Then we have the cycle $\{E_j\}$ as above. We denote the set of all such periodic E by $\mathcal{E}_2$.

Here we can see that $\hat{f}(\gamma_{j_0}) \neq \gamma_{j_0+1}$ for some j_0 $(< q)$, since $E \notin \mathcal{E}_1$. Take smaller open disks E'_j such that $\overline{E'_j} \subset \operatorname{Int} E_j$ for every j, $\hat{f}(\gamma_{j0}) \subset E'_{j_0+1}$ and $E'_q = E'_0$. Then it is easy to construct quasiregular maps g on E_j such that $g = \hat{f}$ on γ_j and is analytic in E'_j,

$$g(E_j) \subset E_{\hat{f}(\gamma_j)} \subset E_{j+1}, \quad \text{and} \quad g(E'_j) \subset E'_{j+1}.$$

We set $\Phi|_{E'_j}$ as the identity map.

Thus by setting $g = \hat{f}$ on $\widehat{D}$, we have a quasiregular map g on the whole of $m\widehat{\mathbb{C}}$. Set

$$E^1 = \bigcup_{E \in \mathcal{E}_1} E, \quad E^2 = \bigcup_{E \in \mathcal{E}_2} E',$$

where the E' are as above for $E \in \mathcal{E}_2$, and set $E^* = E^1 \cup E^2$. Then g, E^*, Φ and a suitable M satisfy the conditions in the fundamental lemma (Lemma 4.3.4) for the cyclic maps, and we have a quasiconformal map φ such that $F = \varphi \circ g \circ \varphi^{-1}$ is analytic. □

By using Proposition 4.3.3, we reduce the proof of Theorem 4.3.1 to the case of cyclic rational maps without Herman rings. In this case, we have the following.

Proposition 4.3.5 (Perturbation lemma) *Let F be a rational cyclic map without Herman rings. There is a family of quasiconformal maps $\{\phi_\epsilon\}_{0 \le \epsilon < \epsilon_0}$ converging to the identity as $\epsilon \to 0$ such that $F_\epsilon = \phi_\epsilon \circ F \circ \phi_\epsilon^{-1}$ are rational cyclic maps and $z_\epsilon = \phi_\epsilon(z)$ is an attractive periodic point of F_ϵ for every non-repelling periodic point z of F.*

Proof For the sake of simplicity, we describe a construction of f_ϵ for a rational function f, and concentrate on a single cycle $\{z_0, \ldots, z_{p-1}\}$ of non-repelling periodic points of f. So fix such a cycle $\{z_0, \ldots, z_{p-1}\}$. First suppose that the z_j are parabolic. Then by the petal theorem (Theorem 1.2.6) there is a union W of attracting petals of f at z_j such that

$$f(W) \subset W, \qquad f^p(\overline{W}) \subset W \cup \{z_0, \ldots, z_{p-1}\}.$$

We may assume that $f(\infty) \in W - f^p(\overline{W})$.

Take a polynomial $h(z)$ such that

$$h(z_j) = 0, \quad h'(z_j) = -1.$$

Then, we can take a neighborhood U_ϵ of $\{z_0, \ldots, z_{p-1}\}$ depending on ϵ such that

$$\tilde{f}_\epsilon(W \cup U_\epsilon) \subset W \cup U_\epsilon,$$

where we set $\tilde{f}_\epsilon(z) = f(z + \epsilon h(z))$ with small ϵ.

Now, we can patch $z + \epsilon h(z)$ with small $\epsilon > 0$ and the identity smoothly and in an invariant manner in a small neighborhood of z_j (cf. the proof of Theorem 1.5.1) so that the resulting maps Φ_ϵ are quasiconformal maps with maximal dilatation tending to 0 which are conformal outside a small neighborhood of ∞ in $g_\epsilon^{-1}(W)$ and tend to the identity uniformly, where $g_\epsilon = f \circ \Phi_\epsilon$. Then the fundamental lemma gives quasiconformal maps φ_ϵ of $\widehat{\mathbb{C}}$ such that

$$f_\epsilon = \varphi_\epsilon \circ g_\epsilon \circ \varphi_\epsilon^{-1}$$

is rational and converges to f uniformly. Also, since $g_\epsilon^p(\overline{W \cup U_\epsilon}) \subset W \cup U_\epsilon$, the $(z_j)_\epsilon$ are attractive periodic points of f_ϵ. Thus we have finished in this case.

Next suppose that z_j are the centers of Siegel disks S_j. Letting w_j be a conformal map of S_j onto the unit disk Δ such that $w(z_j) = 0$, we set $B_j = w_j^{-1}(\{|z| \leq r\})$. Here we take $r < 1$ so that $S_j - \operatorname{Int} B_j$ are disjoint from the forward orbits of critical points. Again we may assume that $\infty \in f^{-1}(\operatorname{Int} B_0) - \operatorname{Int} B_{p-1}$. Set

$$h_j = -\frac{w_j(z)}{w_j'(z)}$$

on S_j. Then by using Runge's theorem, we can find a polynomial h such that

$$h(z_j) = 0, \quad h'(z_j) = -1, \quad |h - h_j| < \delta \text{ on } B_j$$

for any $\delta > 0$. Using this h, we can construct Φ_ϵ and g_ϵ as above. And, if δ is sufficiently small, we can see that

$$\Phi_\epsilon(B_j) \subset B_j, \quad g_\epsilon(B_j) \subset B_{j+1}.$$

Hence, as above, we have the desired φ_ϵ and f_ϵ.

Finally, suppose that the z_j are Cremer points. We may assume that $f(\infty) = z_0$ and $\infty \neq z_{p-1}$. Again let h be a polynomial such that $h(z_j) = 0$ and $h'(z_j) = -1$. Then we can take a suitable neighborhood E_ϵ of $\{z_0, \ldots, z_{p-1}\}$ so small that

$$\tilde{f}_\epsilon(E_\epsilon) \subset E_\epsilon$$

with $\tilde{f}_\epsilon(z) = f(z + \epsilon h(z))$. Though the E_ϵ shrink to points in this case, we can construct g_ϵ as before so that, for a sufficiently small neighborhood V_ϵ of ∞ with $V_\epsilon \cap E_\epsilon = \emptyset$,

$$g_\epsilon(\overline{V_\epsilon}) \subset E_\epsilon, \quad g_\epsilon^p(\overline{E_\epsilon}) \subset E_\epsilon$$

and g_ϵ is analytic outside V_ϵ. Then we can apply again the fundamental lemma and we have the desired function f_ϵ.

Since the case that $\{z_j\}$ are (super)attracting is easy, we have proved the assertion for a single cycle of non-repelling periodic points. In the general case, we can take a polynomial h so that it satisfies the conditions stated above for each cycle of non-repelling periodic points of f, and we obtain the perturbation which makes all non-repelling periodic points attracting ones simultaneously. □

Proof of Theorem 4.3.1 We can see that for f_ϵ constructed above, the number of critical points of these f_ϵ on $\tilde{F}(f_\epsilon)$ is not less than

$$N_c(\tilde{F}(f)) + N_{ir},$$

where $\tilde{F}(f_\epsilon)$ and $\tilde{F}(f)$ are the unions of Fatou components of f_ϵ and f, respectively, never mapped to Herman rings. Since the number of critical points is not greater than $2k - 2$ in general, we have the desired estimate by Proposition 4.3.5 if f has no Herman rings.

So assume that f has a Herman ring, and we use the same notation as before. By construction, the rational cyclic map F as in Proposition 4.3.3 has no Herman rings, and by (iv) of Proposition 4.3.3, all critical points of F in $m\widehat{\mathbb{C}} - \varphi(\widehat{D})$ are contained in the Fatou set, which is denoted here by $\widehat{F}(F)$, of F. Hence each irrationally indifferent cycle of F is either entirely contained in $\varphi(\widehat{D})$ or the cycle of centers of Siegel disks containing some $\varphi(A)$ with $A \in \widehat{\mathcal{A}}$.

Let $N_{\mathcal{A}}(D^{(i)})$ be the number of cycles of $\mathcal{A}$ in $D^{(i)} = \bigcup_{j=1}^{m_i} D_{i,j}$. Then by the construction we have

$$N_c(\varphi(\widehat{D}) \cap \widehat{F}(F)) + N_{\mathcal{A}}(D^{(i)}) + N_{ir}(\varphi(\widehat{D})) \leq N_c(\varphi(\widehat{D})),$$

for $N_{\mathcal{A}}(D^{(i)})$ equals the number $N_{\widehat{\mathcal{A}}}$ of cycles of $\widehat{\mathcal{A}}$ in our setting.

Next, let $z \in D^{(i)} \cap F(f)$, and $\hat{z} \in m\widehat{\mathbb{C}}$ be the point corresponding to z. Set $\tilde{z} = \varphi(\hat{z})$ and let d_0 be the distance between $J(F)$ and $m\widehat{\mathbb{C}} - \varphi(\widehat{D}^{(i)})$. If some forward orbit of $\tilde{z}$ by F has a distance less than d_0 from $m\widehat{\mathbb{C}} - \varphi(\widehat{D}^{(i)})$, then $\tilde{z} \in \widehat{F}(F)$. And if not, condition (iv) of Proposition 4.3.3 implies again $\tilde{z} \in \widehat{F}(F)$. Hence we conclude that

$$N_c(D^{(i)} \cap F(f)) \leq N_c(\varphi(\widehat{D}) \cap \widehat{F}(F)).$$

Also since it is easy to see that z is irrationally indifferent if and only if so is $\tilde{z}$, we have

$$N_{ir}(D^{(i)}) \leq N_{ir}(\varphi(\widehat{D})).$$

Thus we have

$$N_c(D^{(i)} \cap F(f)) + N_{\mathcal{A}}(D^{(i)}) + N_{ir}(D^{(i)}) \leq N_c(D^{(i)}). \tag{4.2}$$

Here, by the definition of $\mathcal{D}_I$, we have

$$N_{ir} = \sum_i N_{ir}(D^{(i)}).$$

And since every Herman ring is divided into two elements of $\mathcal{A}$, we have

$$\sum_i N_{\mathcal{A}}(D^{(i)}) = 2N(HR).$$

Hence summing up the inequality (4.2) gives that

$$N_c(D_I \cap F(f)) + 2N(HR) + N_{ir} \leq N_c(D_I).$$

And it is obvious that

$$N_c(D_{II} \cap F(f)) \leq N_c(D_{II}), \tag{4.3}$$

Here D_I and D_{II} are unions of all D belonging to $\mathcal{D}_I$ and $\mathcal{D}_{II}$, respectively.

Thus we conclude with the desired inequality. The sharpness follows from Theorem 4.3.7 at the end of this section. □

As to the number $N(HR)$, we further have the following.

Theorem 4.3.6

$$N(HR) \leq k - 2.$$

In particular, rational functions of degree 2 *have no Herman rings. Moreover, this estimate is the best possible.*

Proof If $N_c \neq 0$ or $N_{ir} \neq 0$, the assertion follows by Theorem 4.3.1. So assume that $N_c = N_{ir} = 0$.

We say that $A \in \mathcal{A}_{HR}$ is *innermost* if $\operatorname{Int}\gamma_A \cap A' = \emptyset$ for any other $A' \in \mathcal{A}_{HR}$. Reversing the orientation if necessary, we can find an innermost ring, say A_0. Set $A_j = f^j(A_0)$, and let p be the order of A_0. Also set $A'_j = \operatorname{Int}\gamma_{A_j} \cap A_j$.

Then there exists some k such that the component C_k of $\widehat{\mathbb{C}} - f^{-1}(\gamma_{A_{k+1}})$ containing A'_k is not simply connected, and that A_{k+1} is innermost. Indeed, if all A_j are innermost, then some $f(\operatorname{Int}\gamma_{A_k})$ cannot be contained in $\operatorname{Int}\gamma_{A_{k+1}}$, and hence C_k is not simply connected. And if one of the A_j is not innermost, then for some k, A_{k+1} is innermost but A_k is not. Then C_k should not be simply connected, for $\operatorname{Int}\gamma_{A_k}$ contains an element of $\mathcal{A}_{HR}$, which is mapped to $\operatorname{Ext}\gamma_{A_k}$.

Now since $f : C_k \to \operatorname{Int}\gamma_{A_{k+1}}$ is proper and C_k is not simply connected, C_k should contain at least two critical points. Let $D_{i,j}$ be the element of $\mathcal{D}_I$ containing A'_k. Since A_{k+1} is innermost, C_k contains no element of $\mathcal{A}$ but A'_k and no element of $\mathcal{D}_I$ but $D_{i,j}$. Hence $D_{i,j} \subset C_k \subset \overline{D_{i,j} \cup D_{II}}$, which implies that

$$N_c(D_{i,j}) + N_c(D_{II}) \geq 2.$$

If $N_c(D^{(i)}) \geq N_c(D_{i,j}) \geq 2$ then the equality in (4.2) cannot hold, for $N_{\mathcal{A}}(D^{(i)}) = 1$. On the other hand, if $N_c(D_{II}) \geq 1$, then formula (4.3) cannot hold by the assumption. Thus we have the assertion. □

Finally, we state the best possible result without proof.

Theorem 4.3.7 *Let* $m_{AB}, m_{PB}, m_{SD}, m_{HR}, m_{Cr}$ *be non-negative integers satisfying*

$$m_{AB} + m_{PB} + m_{SD} + 2m_{HR} + m_{Cr} \leq 2k - 2, \quad m_{HR} \leq k - 2.$$

Then there exists a rational function of degree k such that

$$N(AB) = m_{AB}, N(PB) = m_{PB}, N(SD) = m_{SD},$$

$$N(HR) = m_{HR}, N(Cr) = m_{Cr}.$$

4.4 The Julia sets of hyperbolic rational functions

In this section, we consider a subfamily of Rat whose Julia sets have a relatively tame structure. To do this, the concept of hyperbolicity, which is defined for polynomials in Chapter 1, is extended to the case of rational functions.

Definition A rational function f is called *hyperbolic* if it satisfies $J(f) \cap \overline{C^+(f)} = \emptyset$.

In the following, we see nice properties which hyperbolic rational functions have. Note also the following conjecture.

Density conjecture on rational functions (HD conjecture) The set of all hyperbolic rational functions of degree k is open and dense in rational functions of degree k.

One of the nice properties of hyperbolic rational functions on the Julia sets is expandingness.

Definition A rational function f is *expanding* on $J(f)$ if there are a Riemannian metric $\sigma(z)^2|dz|^2$ on some neighborhood of $J(f)$ and positive constants c and $\lambda > 1$ such that

$$\sigma(f^n(z))|(f^n)'(z)| \geq c\lambda^n \sigma(z)$$

for all n and $z \in J(f)$. This λ is called an *expanding constant* of f on $J(f)$.

Theorem 4.4.1 *If $f \in$ Rat is expanding with respect to a Riemannian metric as above, then so is it with respect to all such Riemannian metrics. An expanding constant is independent of choice of metrics.*

Proof Let $\omega(z)^2|dz|^2$ and $\mu(z)^2|dz|^2$ be metrics on a neighborhood of $J(f)$. Since $\omega(z)/\mu(z)$ is continuous on $J(f)$ and $J(f)$ is compact, there are positive constants m and M such that

$$m \leq \frac{\omega(z)}{\mu(z)} \leq M.$$

Hence, we have

$$\frac{m}{M} \frac{\mu(f^n(z))|(f^n)'(z)|}{\mu(z)} \leq \frac{\omega(f^n(z))|(f^n)'(z)|}{\omega(z)} \leq \frac{M}{m} \frac{\mu(f^n(z))|(f^n)'(z)|}{\mu(z)},$$

which implies the claim. □

We show that, in the case of rational functions, hyperbolicity coincides with expandingness.

Theorem 4.4.2 *For $f \in$ Rat, the following three conditions are equivalent to each other.*

(i) $J(f) \cap \overline{C^+(f)} = \varnothing$.
(ii) *All the critical points are in $F(f)$, each of whose orbits converges towards an attracting cycle.*

(iii) *The function f is expanding on $J(f)$.*

Proof (i)⇒(iii) With an assumption $\infty \in F(f)$, the proof is the same as that of Lemma 1.4.5.

(iii)⇒(ii) Assume that $J(f) \not\ni \infty$ and take $\sigma(z) \equiv 1$. The critical point ζ is not contained in $J(f)$ because $f'(\zeta) = 0$. Hence, from Theorems 2.4.6 and 2.4.7, there are neither Siegel disks, Herman rings nor Cremer points. If a parabolic periodic point ζ with period p, which is in $J(f)$, exists, then we have $|(f^{np})'(\zeta)| = 1$. This contradicts that f is expanding on $J(f)$. Thus all non-repelling cycles are attracting, and the orbits of all critical points converge toward one of these cycles.

(ii)⇒(i) This is clear. □

In particular, we have the following.

Theorem 4.4.3 *The two-dimensional Lebesgue measure of the Julia set of a hyperbolic rational function is* 0.

The proof follows by the same argument as that for Theorem 1.4.6 with the assumption that ∞ is contained in the Fatou set. See also §5.4.

Next we show a topological property of the Julia sets as boundaries of Fatou components of hyperbolic rational functions.

Theorem 4.4.4 *Assume that the Fatou set of a hyperbolic rational function f contains a simply connected component D. Then ∂D is a closed curve. In particular, ∂D is a locally connected.*

Corollary *For a hyperbolic polynomial P, if $J(P)$ is connected, then $J(P)$ is locally connected.*

Proof If $J(P)$ is connected, then the completely invariant component $I_P = F_\infty$ is simply connected and $\partial F_\infty = J(P)$. Hence, we obtain the claim. □

Proof of Theorem 4.4.4 We may assume that the period of each attracting cycle is 1 by replacing f by some f^n. From Theorems 4.2.15 and 4.4.2, for a component D of $F(f)$, there are an attracting fixed point ζ and a natural number N such that $f^N(D) = D_\zeta$ and $f^n(D) \neq D_\zeta$ for all $n < N$, where D_ζ is an immediate basin of ζ. If D is simply connected, then so is D_ζ by the Riemann–Hurwitz relation.

If we take ϵ sufficiently small, then $U_0 = \left\{z \,\middle|\, |z-\zeta| < \epsilon\right\}$ satisfies $f(\overline{U_0}) \subset U_0$. The component of $f^{-n}(U_0)$ containing ζ is denoted by U_n. Then each U_n

is simply connected, $\overline{U_n} \subset U_{n+1}$, and $\bigcup_{n\geq 0} U_n = D_\zeta$. Fix a point η contained in $f^{-N}(\zeta) \cap D$. The component of $f^{-N}(U_n)$ containing η is denoted by V_n. Then each V_n is simply connected and $\bigcup_{n\geq 0} V_n = D$. If necessary by replacing U_0 and V_0, we may assume V_0 contains all the critical points in D. Let $A = \{z \mid 1 < |z| < r\}$ be conformally equivalent to a doubly connected domain B bounded by ∂U_0 and ∂U_1, and ∂U_0 corresponds to $\{z \mid |z| = 1\}$. The curve in B corresponding to an external segment

$$t \mapsto (1 + t(r-1))e^{i\theta} \qquad (0 \leq t \leq 1)$$

is called an external segment of B and is denoted by $l_\theta(t)$. We denote points on ∂U_0 and ∂U_1 by $z_0(\theta) = l_\theta(0)$ and $z_1(\theta) = l_\theta(1)$, respectively. Since $f(z_1(\theta))$ is in ∂U_0, there is the external segment $l_{\theta'}(t)$ of B starting at $z_0(\theta') = f(z_1(\theta))$. Using the branch φ of f^{-1} mapping $z_0(\theta')$ to $z_1(\theta)$, we define the point $z_2(\theta) = \varphi(z_1(\theta'))$ on ∂U_2. Repeating this procedure, we define the point $z_n(\theta)$ on ∂U_n. Finally, suppose the local degree of f^N on D is k. Since each ∂V_n covers ∂U_n k-fold, we also define the point $\gamma_n(\tilde{\theta})$ on ∂V_n from $z_n(\theta)$ by f^{-N}, where $\tilde{\theta}$ is one of $(\theta + 2\pi j i)/k$ $(0 \leq j \leq k-1)$.

Since f is hyperbolic, there are a Riemannian metric $\sigma(z)^2|dz|^2$ on some neighborhood K of $J(f)$ and a constant $\lambda > 1$ satisfying the condition in the above definition. Assume that each $\gamma_n(\tilde{\theta})$ is in K, and we have

$$d_\sigma(\gamma_{n+1}(\tilde{\theta}), \gamma_n(\tilde{\theta})) \leq \frac{1}{c\lambda^n} d_\sigma(\gamma_1(\tilde{\theta}), \gamma_0(\tilde{\theta})),$$

where d_σ is the distance with respect to $\sigma(z)^2|dz|^2$. This inequality shows that $\{\gamma_n(\tilde{\theta})\}$ converges uniformly with respect to $\tilde{\theta}$. We denote the limit function by $\Gamma(\tilde{\theta})$. Then $\Gamma([0, 2\pi))$ is a closed curve and $\Gamma([0, 2\pi)) = \partial D$. Hence ∂D is locally connected. □

Further, we have the following theorem on local connectedness of the whole Julia set.

Theorem 4.4.5 *Let f be hyperbolic and $J(f)$ connected. Then $J(f)$ is locally connected.*

Before proving this, we include here two useful criteria on local connectedness.

Proposition 4.4.6 (Whyborn 1942) *A compact set X in $\mathbb{C}$ is locally connected if and only if it satisfies the following two conditions.*

(i) *The boundary of each complementary component of X is locally connected.*

(ii) *The number of the complementary components of X whose diameters exceed ϵ is finite for all $\epsilon > 0$.*

Proposition 4.4.7 *A compact set X in $\mathbb{C}$ is locally connected if and only if for all $z_0 \in X$ and all sequences $\{z_n\}_{n=1}^{\infty}$ tending to z_0, there are connected sets L_n containing z_0 and z_n whose diameters tend to 0.*

Proof of Theorem 4.4.5 Assume $\infty \in F(f)$. Condition (i) in Proposition 4.4.6 has already been obtained in Theorem 4.4.4. Hence the proof will be completed after proving the following, Lemma 4.4.8. □

Lemma 4.4.8 *Let f be a hyperbolic rational function. Assume that $J(f)$ is connected and $\infty \in F(f)$. Then $J(f)$ satisfies condition (ii) in Proposition 4.4.6.*

Proof Assume that there are infinitely many components of $F(f)$ whose diameters exceed ϵ for some ϵ. We choose a sequence $\{D_n\}_{n=1}^{\infty}$ of distinct such components and a component D_0 of $F(f)$ such that $f^{j(n)}(D_n) = D_0$ and $(\bigcup_{n\geq 0} D_n) \cap C^+(f) = \varnothing$. Let S_n be the single valued branch of $f^{-j(n)}$ on D_0 with $S_n(D_0) = D_n$. Lemma 2.4.3 shows that $\{S_n\}$ is a normal family. Since the D_n are pairwise disjoint, any limit function of it is a constant. The hyperbolicity of f ensures existence of a simply connected neighborhood U_z of z with $C^+(f) \cap U_z = \varnothing$ for every $z \in \partial D_0$. Let $\widehat{S}_{z,n}$ be the single valued branch of $f^{-j(n)}$ on U_z which coincides with S_n on D_0. Using Lemma 2.4.4, we see that any limit function of the sequence is also a constant.

Take two points ζ_n and η_n in D_n with $|\zeta_n - \eta_n| > \epsilon$. Since $\zeta_n' = f^{j(n)}(\zeta_n)$ and $\eta_n' = f^{j(n)}(\eta_n)$ are in D_0, choosing a subsequence if necessary, we may assume $\zeta_n' \to \zeta'$ and $\eta_n' \to \eta'$. In the case that $\zeta' \in D_0$ or $\eta' \in D_0$, we take as $U_{\zeta'}$ or $U_{\eta'}$ an open set U in D_0 such that $\overline{U} \subset D_0$ and $U \supset \{\zeta', \eta'\} \cap D_0$. In any case, take another open set U_0 in D_0 such that $\overline{U_0} \subset D_0$ and $U_0 \cap U_{\zeta'}$ and $U_0 \cap U_{\eta'}$ are non-empty. Then the limit constants of $\{\widehat{S}_{\zeta',n}\}$ and $\{\widehat{S}_{\eta',n}\}$ should be equal to each other, which is a contradiction. □

Local connectedness is still valid for wider classes of rational functions.

Definition We say that a rational function is *subhyperbolic* if the orbit of each critical point in the Fatou set accumulates at an attracting cycle and each critical point in the Julia set is preperiodic.

Further we say that a rational function is *geometrically finite* simply if each critical point in the Julia set is preperiodic.

Actually, a subhyperbolic rational function f is not expanding on the Julia set with respect to any Riemannian metric. Nevertheless, if we allow a metric on some neighborhood of $J(f)$ to have a finite number of singularities with behavior like $|d\sqrt[n]{z}|^2$ at the origin, then it is expanding on the neighborhood of $J(f)$ with respect to such a metric. Such a metric is called an orbifold metric.

Theorem 4.4.9 *A rational function is subhyperbolic if and only if it is expanding with respect to some orbifold metric on a neighborhood of its Julia set.*

Once we obtain the expandingness of subhyperbolic rational functions, we have the following theorem by the same argument as in the proof of Theorem 4.4.4.

Theorem 4.4.10 *Assume that the Fatou set of a subhyperbolic rational function f contains a simply connected component D. Then ∂D is a closed curve. In particular, ∂D is a locally connected.*

Corollary *For a subhyperbolic polynomial P with connected $J(P)$, $J(P)$ is locally connected.*

For example, quadratic polynomials with a parabolic cycle are geometrically finite. We have already used in Chapter 1 the following theorem. See also §5.4.

Theorem 4.4.11 (Douady–Hubbard) *For a geometrically finite polynomial P with connected $J(P)$, $J(P)$ is locally connected.*

We now show an advantage of local connectedness of boundaries of Fatou components.

There is a conformal map from a simply connected hyperbolic domain onto a unit disk Δ by the Riemann mapping theorem. We call such a map a Riemann map, and the following theorem is fundamental.

Theorem 4.4.12 (Carathéodory) *Let D be a domain conformally equivalent to the unit disk Δ and $\varphi : \Delta \to D$ be an inverse map of a Riemann map. Then φ can be extended continuously to $\overline{\Delta}$ if and only if ∂D is locally connected.*

Proof Assume that φ can be extended continuously to $\overline{\Delta}$. Fix $z \in \partial D$ and a sequence $\{z_n\}$ on ∂D converging to z. There is a sequence consisting of ζ_n in $\varphi^{-1}(z_n)$ which converges to ζ in $\varphi^{-1}(z)$. Let L_n be an arc $\overline{\zeta_n\zeta}$ on $\partial\Delta$ which is not longer than π. Since L_n converges to 0, ∂D is locally connected at z by Proposition 4.4.7.

Conversely, suppose that ∂D is locally connected. Assume $\varphi(0) = \infty$ by using a Möbius transformation. Fix $\zeta_0 \in \partial\Delta$ and set

$$L(\rho) = \int_{\Delta\cap\{|\zeta-\zeta_0|=\rho\}} |\varphi'(\zeta)|\,|d\zeta|$$

for $\rho > 0$. By the Cauchy–Schwarz inequality, we have

$$\int_0^{\rho_0} \frac{L(\rho)^2}{\rho} d\rho < \pi \iint_{\Delta\cap\{|\zeta-\zeta_0|=\rho_0\}} |\varphi'(\zeta)|^2 d\xi d\eta \qquad (\zeta = \xi + i\eta)$$

for a fixed $\rho_0 < 1$. Since the right hand side of the inequality is not greater than the area of D, it is finite. Hence there is a sequence $\{\rho_n\}$ converging to 0 such that $\{L(\rho_n)\}$ also converges to 0.

Since the length $L(\rho_n)$ of the image Γ_n of $\Delta\cap\{|\zeta-\zeta_0| = \rho_n\}$ by φ is finite, its end points α_n and β_n are well defined on ∂D. Further $L(\rho_n) \to 0$ implies $|\alpha_n - \beta_n| \to 0$. Hence we may assume $\{\alpha_n\}$ and $\{\beta_n\}$ converge to the same point $w_0 \in \partial D$.

The assumption and Proposition 4.4.7 show existence of connected sets L_n, containing w_0, α_n and β_n, whose diameters tend to 0. Here, for every large n, the image D_n of each $\Delta \cap \{|\zeta - \zeta_0| < \rho_n\}$ by φ is contained in a bounded component of $\widehat{\mathbb{C}} - (\Gamma_n \cup L_n)$. Indeed, otherwise $z_0 \in D_n$ would be connected to ∞ by a curve in $\widehat{\mathbb{C}}-(\Gamma_n \cup L_n)$. On the other hand, z_0 is connected to ∞ by a curve in D. Thus there is a simple closed curve C in $\widehat{\mathbb{C}} - L_n$ which crosses Γ_n at only one point. This means that C separates α_n and β_n, which contradicts connectedness of L_n.

From the construction above, the diameters of $\Gamma_n \cup L_n$ converge to 0. Hence φ has a limit at ζ_0, which we denote by $\varphi(\zeta_0)$. Then it is easy to see that $\varphi : \overline{\Delta} \to \overline{D}$ is continuous. □

Remark By introducing the concept of prime ends, the inverse of any Riemann mapping is extended to a homeomorphism from Δ onto the union of D and the set of all prime ends with natural topology. This is the original Carathéodory theorem. The cluster set of φ at $\zeta \in \partial\Delta$ is said to be an *impression* of the prime end e_ζ and the non-tangential limit at ζ is said to be the principal point of e_ζ. It is known that the principal point exists for almost all ζ. Theorem 4.4.12 shows ∂D is locally connected if and only if the impression of every prime end consists of a single point.

Using Carathéodory's theorem, we can extend the action of a rational function on a Fatou component to its boundary continuously and study the Julia set as the boundary of an invariant component.

Proposition 4.4.13 *Assume that the Fatou set of a hyperbolic rational function f contains a simply connected invariant component D whose local degree of f is k. Set $S^1 = \{z \mid |z| = 1\}$ and $h(z) = z^k$. Then there is a continuous map φ from S^1 onto ∂D such that*

$$\varphi \circ h(z) = f \circ \varphi(z).$$

Moreover, h is injective on $\varphi^{-1}(\zeta)$ for all $\zeta \in \partial D$.

Proof Let ϕ be a Riemann map from D onto the unit disk. Since D is an attracting invariant component, $\phi \circ f \circ \phi^{-1}$ is a finite Blaschke product of degree k with attracting fixed point. Thus there are a compact set E of D, an annulus $A = \{z \mid r < |z| < 1\}$ and a homeomorphism φ from A onto $D - E$ which satisfies

$$f \circ \varphi(z) = \varphi(z^k)$$

for $z \in \{z \mid r^{1/k} < |z| < 1\}$. Since f is hyperbolic, φ can be extended continuously to a map from $A \cup S^1$ onto $\overline{D} - E$. We again denote this by φ.

For $t \in \mathbb{R}/\mathbb{Z}$, defining $e(t) = \{le^{2\pi it} \mid r^{1/k} < l \leq 1\}$, we have $f \circ \varphi(e(t)) = \varphi(e(kt))$. Suppose that h is not injective on $\varphi^{-1}(\zeta)$ for some $\zeta \in \partial D$, there are $s, t \in \mathbb{R}$ such that

$$e(s) \neq e(t), \quad \varphi(e^{2\pi is}) = \varphi(e^{2\pi it}) \qquad \text{and} \qquad e(ks) = e(kt).$$

Since $\varphi(e(s)) \cap \varphi(e(t)) = \{\zeta\}$, we have

$$f \circ \varphi(e(s)) = \varphi(e(ks)) = \varphi(e(kt)) = f \circ \varphi(e(t)),$$

which shows ζ is a critical point. This is a contradiction. □

Though Theorem 4.4.4 shows that the boundary of a simply connected Fatou component of a hyperbolic rational function is a closed curve, it is not necessarily a Jordan curve.

Proposition 4.4.14 *Let f, D and φ be as in Proposition 4.4.13. Then, for all $\zeta \in \partial D$, the number of $\varphi^{-1}(\zeta)$ is bounded from above by a constant independent of ζ.*

Proof Set $d(e^{it}, e^{is}) = |e^{it} - e^{is}|$ for $e^{it}, e^{is} \in S^1$ and let k and $h(z) = z^k$ be as in Proposition 4.4.13. Take sufficiently small $\epsilon > 0$; then there is $\mu > 1$ such that

$$d(h(z), h(w)) \geq \mu d(z, w)$$

for all $z, w \in S^1$ with $d(z, w) \leq \epsilon$.

Let $\zeta \in \partial D$, and set $X = \varphi^{-1}(\zeta)$ and $\tilde{X} = \varphi^{-1}(f(\zeta)) = h(\varphi^{-1}(\zeta))$. Since $h : X \to \tilde{X}$ is injective by Proposition 4.4.13 for all $\epsilon > 0$, there is $\delta = \delta(\zeta, \epsilon) > 0$ such that $d(z, w) < \epsilon$ for all $z, w \in X$ with $d(h(z), h(w)) < \delta$. Indeed, otherwise there would be sequences $\{z_n\}$ and $\{w_n\}$ and $\epsilon > 0$ satisfying $z_n \to z$, $w_n \to w$, $d(z_n, w_n) \geq \epsilon$ and $d(h(z_n), h(w_n)) \to 0$. Hence we have $d(z, w) \geq \epsilon$ and $h(z) = h(w)$. Compactness of X implies $z, w \in X$, which is a contradiction.

Further, we show $\inf_{\zeta \in S^1} \delta(\zeta, \epsilon) > 0$. Assume that $\inf_{\zeta \in S^1} \delta(\zeta, \epsilon) = 0$. Then there is a sequence $\{\zeta_j\}$ satisfying $\zeta_j \to \zeta$ and $\delta(\zeta_j, \epsilon) \to 0$. Hence, we can take z_j and w_j in $\varphi^{-1}(\zeta_j)$ such that $d(z_j, w_j) \geq \epsilon$ and $d(h(z_j), h(w_j)) \to 0$ as $j \to \infty$. Choosing subsequences if necessary, we may assume that $z_j \to z$ and $w_j \to w$. Hence, we have

$$h(z) = \lim_{j \to \infty} h(z_j) = \lim_{j \to \infty} h(w_j) = h(w).$$

This contradicts Proposition 4.4.13, because $z, w \in \varphi^{-1}(\zeta)$ and $z \neq w$ for $d(z_j, w_j) \geq \epsilon$.

Now we set $\inf_{\zeta \in S^1} \delta(\zeta, \epsilon) = \delta > 0$. Fix a positive integer n and a point $\zeta \in \partial D$, and set $X_n = \varphi^{-1}(f^n(\zeta)) = h^n(\varphi^{-1}(\zeta))$. Since $X_n \subset S^1$, there is a natural number N independent of ζ such that each X_n is covered by N sets each of whose diameters is less than δ, say

$$X_n \subset U_1^n \cup \cdots \cup U_N^n$$

with diam $U_i^n \leq \delta$. Setting $U_i^{n-1} = h^{-1}(U_i^n) \cap X_{n-1}$, we have

$$X_{n-1} \subset U_1^{n-1} \cup \cdots \cup U_N^{n-1}$$

with diam $U_i^{n-1} \leq \delta/\mu < \delta$. By repeating this procedure, we can cover X_0 by N sets each of whose diameter is less than δ/μ^n. Since n is arbitrary, $\varphi^{-1}(\zeta)$ consists of at most N points. □

Especially, we obtain the following result on orbits of points on boundaries.

Theorem 4.4.15 *Let f be a hyperbolic rational function and D be a simply connected invariant component of $F(f)$. Then each one of the following three sets is dense in ∂D.*

$$\begin{aligned} A_1 &= \{\zeta \in \partial D \mid \zeta \text{ is a periodic point of } f\}, \\ A_2 &= \{\zeta \in \partial D \mid \{f^n(\zeta)\}_{n=0}^{\infty} \text{ is dense in } \partial D\}, \\ A_3 &= \left\{\zeta \in \partial D - \bigcup_{n=0}^{\infty} f^{-n}(A_1) \,\middle|\, \{f^n(\zeta)\}_{n=0}^{\infty} \text{ is not dense in } \partial D\right\}. \end{aligned}$$

Proof Consider the action of $h(z) = z^k$ on S^1. For $z = e^{2\pi i t}$ $(0 \le t < 1)$, we write t by the k-adic representation

$$t = \sum_{j=1}^{\infty} a_j k^{-j} \qquad (a_j \in \{0, 1, \ldots, k-1\}),$$

where we assume that infinitely many a_j's are not 0. The map ϕ from S^1 to the sequence space on k symbols is defined by

$$\phi(z) = (a_1, a_2, \ldots).$$

Denoting by σ the shift map on Σ_k, we have

$$\phi \circ h = \sigma \circ \phi.$$

Let A_1', A_2' and A_3' be the subsets of Σ_k as follows.

$$\begin{aligned} A_1' &= \{u \in \Sigma_k \mid u \text{ is cyclic}\}, \\ A_2' &= \{u \in \Sigma_k \mid u \text{ contains all the finite sequences of } k \text{ symbols}\}, \\ A_3' &= \left\{ u \in \Sigma_k - \bigcup_{n=0}^{\infty} \sigma^{-n}(A_1') \,\middle|\, u \text{ does not contain some finite} \right. \\ &\qquad \left. \text{sequence of } k \text{ symbols} \right\}. \end{aligned}$$

It is easy to see that $\phi^{-1}(A_i')$ is dense in S^1 for each i. Using σ and Proposition 4.4.14, we have

$$\varphi \circ \phi^{-1}(A_i') = A_i \qquad (i = 1, 2, 3).$$

Thus, by Proposition 4.4.14, A_i is dense in ∂D for each i. □

Recall that the boundary of a completely invariant component coincides with the Julia set. The converse is also true in the case of hyperbolic rational functions.

Theorem 4.4.16 *Let f be a hyperbolic rational function and assume that $J(f)$ is connected. If there is an invariant component D of $F(f)$ with $\partial D = J(f)$, then D is completely invariant.*

Proof Assume that the degree of f is k, the local degree of f on D is p and φ is the function obtained in Proposition 4.4.13. Fix $\eta_0 \in \partial D$. Since there is no critical point of f on $J(f)$, $f^{-1}(\eta_0)$ consists of k distinct points, say $\{\zeta_0^1, \ldots, \zeta_0^k\}$. Denoting the cardinality of $\varphi^{-1}(\eta_0)$ and that of $\varphi^{-1}(\zeta_0^i)$ by t_0 and t_0^i, respectively, we have $t_0^i \le t_0$ from Propositions 4.4.13 and 4.4.14. If the inequality $t_0^i < t_0$ holds for some i, then we set $\eta_1 = \zeta_0^i$ and $t_1 = t_0^i$. Let $f^{-1}(\eta_1) = \{\zeta_1^1, \zeta_1^2, \ldots, \zeta_1^k\}$ and t_1^i be the cardinality of $\varphi^{-1}(\zeta_1^i)$. Then we have $t_1^i \leq t_1$. Repeating this procedure, we obtain n which satisfies $t_n^i = t_n$ for all i. For this n, we set $\eta = \eta_n$ and $t = t_n$. The cardinality of $(f \circ \varphi)^{-1}(\eta)$ is kt and that of $(\varphi \circ h)^{-1}(\eta)$ is tp. Hence

$$f \circ \varphi(z) = \varphi \circ h(z)$$

implies $p = k$, and D is completely invariant. □

In general, points in Julia sets are not necessarily in the boundaries of Fatou components. We denote the set of all the points of $J(f)$ which are not in the boundaries of any Fatou components by $J_0(f)$ and call it the *residual Julia set* of f (see Morosawa 1997 and cf. Abikoff 1993).

Theorem 4.4.17 *Let f be a rational function. Assume that $J_0(f)$ is not empty. Then $J_0(f)$ is a completely invariant subset of $J(f)$ and dense in $J(f)$. Moreover, it contains uncountably many points.*

Proof We assume $F(f) \neq \varnothing$. First, for a component D of $F(f)$, $f(D)$ and components of $f^{-1}(D)$ are those of $F(f)$. Hence $J_0(f)$ is a completely invariant subset of $J(f)$.

Next, since any $\zeta \in J_0(f)$ is not an exceptional point, $\{f^n(\zeta)\}_{n=-\infty}^{\infty}$ is an infinite set. Thus $\overline{J_0(f)}$ is a closed completely invariant set with at least three points. By minimality and complete invariance of $J(f)$, $J_0(f)$ is dense in $J(f)$.

Finally, we can construct a Cantor set as a subset of $J_0(f)$ because $J(f)$ and $J_0(f)$ are completely invariant and $J_0(f)$ is dense in $J(f)$. Hence J_0 is an uncountable set. □

Remark If $F(f)$ has a completely invariant component D, then $J_0(f) = \varnothing$ because $\partial D = J(f)$. Further, if $F(f)$ consists of precisely two components, each of them is completely invariant under f^2. Since $J(f) = J(f^2)$, we obtain $J_0(f) = \varnothing$.

Proposition 4.4.18 *Let f be a rational function and $\{D_i\}_{i=1}^M$ be the set of all immediate (super)attracting or parabolic basins, all Siegel disks and all Herman rings of f. Then $J_0(f)$ is empty if and only if $\bigcup_{i=1}^M \partial D_i = J(f)$.*

Proof From the definition of $J_0(f)$, it is clear that $\bigcup_{i=1}^M \partial D_i = J(f)$ implies $J_0(f) = \varnothing$.

We assume $\bigcup_{i=1}^{M} \partial D_i \neq J(f)$ and set $D = \bigcup_{i=1}^{M} D_i$. Let ζ_1 be a point in $J(f) - \bigcup_{i=1}^{M} \partial D_i$. We choose an open disk U_1 with center ζ_1 such that $\overline{U}_1 \cap \overline{D} = \varnothing$. Since, for sufficiently large n, $f^n(U_1) \supset J(f)$, there exists N such that $f^n(U_1) \cap \overline{D} = \varnothing$ for all $n < N$ and that $f^N(U_1) \cap \overline{D} \neq \varnothing$. If $f^N(U_1) \cap J(f) \subset \overline{D}$, complete invariance of $J(f)$ and Proposition 4.4.14 imply that $J(f) = f^n(U_1) \cap J(f) \subset \overline{D}$ for all sufficiently large n, which contradicts the assumption. Hence $(f^N(U_1) - \overline{D}) \cap J(f) \neq \varnothing$. Let V_1 be a connected open set satisfying $V_1 \subset f^N(U_1) - \overline{D}$ and $V_1 \cap J(f) \neq \varnothing$. Further we choose a component V_1' of $f^{-N}(V_1)$ with $V_1' \subset U_1$ and take $\zeta_2 \in V_1' \cap J(f)$. Defining ζ_n by induction, we have the limit $\zeta = \lim_{n\to\infty} \zeta_n$ which is in $J_0(f)$ □

In the case of a disconnected Julia set, the converse of the first statement of the remark before Proposition 4.4.18 holds.

Theorem 4.4.19 *Let f be a rational function and assume that $J(f)$ is disconnected. If $J_0(f)$ is the empty set, then $F(f)$ contains a completely invariant component.*

Proof There is a component D of $F(f)$ whose connectivity is greater than 1 by the assumption. Unless $F(f)$ contains a completely invariant component, $\{f^{-n}(D)\}_{n=0}^{\infty}$ is an infinite set each of whose elements has a connectivity which is greater than 1. Hence, for $\{D_i\}_{i=1}^{M}$ defined in Proposition 4.4.18, we have $J(f) \neq \bigcup_{i=1}^{M} \partial D_i$ since M is finite. Thus we obtain $J_0(f) \neq \varnothing$. □

Corollary *If the Fatou set of $f \in$ Rat contains a cycle of Herman rings, then $J_0(f) \neq \varnothing$.*

Proof If there is a completely invariant component D, there are no Herman rings because of $\partial D = J(f)$. □

Remark In the case of a transcendental entire function f, if f has a multiply connected Fatou component, then $J_0(f) \neq \varnothing$. See Dominguez (1997).

Here we give an example of a rational function whose Fatou set is disconnected, and contains neither Herman rings, nor completely invariant components. Hence its residual Julia set is not empty.

Example 4.4.1 (Morosawa 1995) *Let*

$$f(z) = \frac{a(z-2a)}{z^2(z-4a)},$$

where a is a constant with $0 < a < 1/58$. Then f is hyperbolic, $J(f)$ is disconnected and $J_0(f) \neq \varnothing$.

Verification Four critical points of f are 0, ∞, $\alpha = a(5 + i\sqrt{7})/2$ and $\overline{\alpha}$. Moreover, $\{0, \infty\}$ is a superattracting cycle.

We now show that F_0 is infinitely connected. Let $r > 2$ and C_1 be the circle given by $|z| = r$. Then we have

$$|f(z)| \leq \frac{a(r+2a)}{r^2(r-4a)} < a$$

on C_1. Next, let $r < a$ and C_2 be the circle given by $|z| = r$. Then we have

$$|f(z)| \geq \frac{2a-r}{a(r+4a)} > 2$$

on C_2. We set $D_a = \{z \mid |z| < a\}$. The two inequalities above imply $f^2(D_a) \subset D_a$. Further, since $F(f^2) = F(f)$, we have $D_a \subset F_0$. Let C_3 be the circle given by $|z - 2a| = 3a/2$. We have

$$|f(z)| \geq \frac{1}{a}\frac{12}{7^3} > 2$$

on C_3. This shows $f^2(C_3) \subset D_a \subset F_0$. Since $D_a \cap C_3 \neq \varnothing$, F_0 contains C_3. Elementary calculation shows that a fixed point x_0 of f exists in an interval $(a, 2a)$. Because F_0 is the immediate basin of the superattracting fixed point 0 of f^2, F_0 does not contain any fixed point of f^2 except 0. Thus F_0 is not simply connected, and should be infinitely connected. Hence $F(f)$ does not contain any completely invariant component.

Finally, we show that f is hyperbolic. Let C_4 be the circle given by $|z - 2a| = \sqrt{2}a$, on which two critical points α and $\overline{\alpha}$ exist. By an argument similar to the above, we can show that $C_4 \cap D_a \neq \varnothing$ and $f^2(C_4) \subset D_a$. Hence we have $\{\alpha, \overline{\alpha}\} \subset F_0$.

In the case of hyperbolic rational functions, we have the following characterization of those with non-empty residual Julia set.

Theorem 4.4.20 *Let f be a hyperbolic rational function. Then $J_0(f)$ is empty if and only if $F(f)$ has a completely invariant component or consists of only two components.*

Proof Since the if part is shown in the remark before Proposition 4.4.18, we show the only-if part. If $J(f)$ is disconnected, then Theorem 4.4.19 gives the proof. So assume that $J(f)$ is connected and $F(f)$ has infinitely many components but no completely invariant ones, and we will deduce a contradiction.

By $\{D_i\}_{i=1}^M$ we denote the set defined in Proposition 4.4.18. Suppose that $J(f) = \bigcup_{i=1}^M \partial D_i$. Because $J(f)$ is the Baire space, for some component of $\bigcup_{i=1}^M D_i$, say D, ∂D has non-empty interior as a subset of $J(f)$. That is, there exists an open set U of $\widehat{\mathbb{C}}$ such that $\partial D \supset U \cap J(f) \neq \varnothing$. Hence, for a sufficiently large m, $f^m(\partial D) \supset f^m(U \cap J(f)) = J(f)$. In particular, let p be the period of D, then for some n, $f^{np}(\partial D) = J(f)$, and since $f^p(\partial D) = \partial D$, we have $\partial D = J(f)$.

Since D is simply connected, there exists a Riemann mapping from D onto the unit disk Δ. Let φ be the inverse of the Riemann mapping. Since ∂D is locally connected, φ can be extended continuously to the map from $\overline{\Delta}$ onto $\overline{D}$. We again denote it by φ.

If $\partial D = J(f)$ is a simple closed curve, then $F(f)$ should consist of only two components, which contradicts the assumption. If not, there exist two disjoint open interval I_1 and I_2 on $\partial\Delta$ such that each end point of I_i $(i = 1, 2)$ is mapped to the same point ζ in ∂D by φ. From the theorem of F. and M. Riesz, each I_i $(i = 1, 2)$ contains η_i such that $\varphi(\eta_i) \neq \zeta$ and $\varphi(\eta_1) \neq \varphi(\eta_2)$. Let γ be a simple arc in Δ whose end points coincide with the end points of I_1. Then $\gamma' = \varphi(\gamma) \cup \{\zeta\}$ is a Jordan curve with two complementary components V_1 and V_2. Assume that one of them, say V_1, contains both $\varphi(\eta_1)$ and $\varphi(\eta_2)$. Take a simple arc α in Δ whose end points are η_1 and η_2 and which crosses γ exactly once. Then $\varphi(\alpha)$ is a simple arc which crosses γ' exactly once. Since there is a simple arc in V_1 whose end points are $\varphi(\eta_1)$ and $\varphi(\eta_2)$, there exists a Jordan curve which crosses γ' exactly once. This is impossible, and hence one of $\varphi(\eta_1)$ and $\varphi(\eta_2)$ is contained in V_1 and the other in V_2. Each component of $F(f)$ except D is contained in either V_1 or V_2. Thus the boundary of each component of $F(f)$ except D contains at most one of $\varphi(\eta_1)$ and $\varphi(\eta_2)$. This shows that $p = 1$. By means of Theorem 4.4.16, D is a completely invariant component. Again this contradicts our assumption.

Thus we conclude that $J(f) \neq \bigcup_{i=1}^M \partial D_i$. Then Proposition 4.4.18 implies that $J_0(R) \neq \varnothing$. Thus we have completed the proof. □

5

Kleinian groups and Sullivan's dictionary

The landmark of the modern study of complex dynamics is Sullivan's no wandering domain theorem. In the paper where he proved it, he also showed, by the same argument, Ahlfors' finiteness theorem, which is a fundamental theorem in the theory of Kleinian groups. In the same paper, he enumerated analogous properties between iteration of rational functions and action of Kleinian groups. This is the first version of what is called Sullivan's dictionary.

In §5.1, bearing in mind iteration theory, we give a survey of the theory of Kleinian groups.

In §5.2, we construct our version of Sullivan's dictionary based on the contents of the previous chapters.

In §5.3, we discuss the score-cards for stability, which afford the central conjectures. Finally, in §5.4, we explain some topics concerning probability measures on the Julia sets. We construct conformal densities on the Julia sets as an analogy of those on the limit sets of Kleinian groups. Further, using these probability measures, we estimate the Hausdorff dimensions of Julia sets.

5.1 The limit sets and the ordinary sets

Recall that the group consisting of Möbius transformations is denoted by $\mathrm{Möb}(\widehat{\mathbb{C}})$, i.e.

$$\mathrm{Möb}(\widehat{\mathbb{C}}) = \left\{ z \mapsto \frac{az+b}{cz+d} \;\middle|\; a, b, c, d \in \mathbb{C}, ad - bc = 1 \right\}.$$

The group $\mathrm{Möb}(\widehat{\mathbb{C}})$ is identified with $\mathrm{PSL}(2, \mathbb{C}) = \mathrm{SL}(2, \mathbb{C})/\{\pm I\}$. Every Möbius transformation except for the identity conjugates in $\mathrm{Möb}(\widehat{\mathbb{C}})$ to

(i) $z \mapsto z + 1$,
(ii) $z \mapsto \lambda z$ ($|\lambda| = 1, \lambda \neq 1$), or
(iii) $z \mapsto \lambda z$ ($|\lambda| > 1$),

and then is called *parabolic*, *elliptic*, or *loxodromic*, respectively. Further, if the multiplier λ in the loxodromic case is real, then the transformation is called *hyperbolic*.

The points z in $\widehat{\mathbb{C}}$ satisfying $\gamma(z) = z$ for Möbius transformation γ are called fixed points of γ. A parabolic transformation has only one fixed point in $\widehat{\mathbb{C}}$ and an elliptic or loxodromic transformation has two. Representing $\gamma(z)$ as

$$\gamma(z) = \frac{az+b}{cz+d} \qquad (ad-bc=1, \ \ a,b,c,d \in \mathbb{C}),$$

we call $a+d$ the *trace* of γ and denote it by $\mathrm{tr}(\gamma)$. Note that $\mathrm{tr}(\gamma)$ is determined up to sign and is invariant under conjugation by elements of $\mathrm{Möb}(\widehat{\mathbb{C}})$. Easy calculation shows the following proposition.

Proposition 5.1.1 *Let γ be a Möbius transformation not the identity. Then:*

(i) *In the case that* $\mathrm{tr}(\gamma)$ *is real, if* $|\,\mathrm{tr}(\gamma)| < 2$, $|\,\mathrm{tr}(\gamma)| = 2$ *or* $|\,\mathrm{tr}(\gamma)| > 2$ *then γ is elliptic, parabolic or hyperbolic, respectively.*

(ii) *If* $\mathrm{tr}(\gamma)$ *is not real, then γ is non-hyperbolic loxodromic.*

Moreover, the converse is true.

Definition We say that a subgroup of $\mathrm{Möb}(\widehat{\mathbb{C}})$ is a *Kleinian group*, if it is a discrete subgroup of $\mathrm{PSL}(2, \mathbb{C})$. That is, it is a discrete set in $\mathrm{PSL}(2, \mathbb{C})$.

From the definition, a Kleinian group is countable and does not contain any elliptic element with infinite order.

Definition For a Kleinian group Γ, we say that $z \in \widehat{\mathbb{C}}$ is an *ordinary point* of Γ, if Γ is a normal family on some neighborhood of z, otherwise we call z a *limit point* of Γ. The set of ordinary points is called the *ordinary set* of Γ and that of limit points is called the *limit set* of Γ. We denote them by $\Omega(\Gamma)$ and $\Lambda(\Gamma)$, respectively.

Proposition 5.1.2 *Let Γ be a normal family of Möbius transformations on a domain D, and γ_1 and γ_2 any Möbius transformations. Then $\gamma_1 \Gamma \gamma_2^{-1}$ is a normal family on $\gamma_2(D)$.*

Proof A Möbius transformation is Lipschitz continuous with respect to the spherical metric σ. More explicitly, representing a Möbius transformation γ by

$$\gamma(z) = \frac{az+b}{cz+d}, \qquad ad-bc=1,$$

we have

$$\sigma(\gamma(z), \gamma(w)) \leq (|a|^2 + |b|^2 + |c|^2 + |d|^2)\sigma(z, w)$$

for all z and w in $\widehat{\mathbb{C}}$. Hence the assertion is easily obtained. □

Corollary 1 *Let Γ be a Kleinian group. The ordinary set $\Omega(\Gamma)$ is a Γ-invariant open set, i.e., $\gamma(\Omega(\Gamma)) = \Omega(\Gamma)$ for all $\gamma \in \Gamma$. Hence $\Lambda(\Gamma)$ is a Γ-invariant closed set.*

Corollary 2 *Let Γ be a Kleinian group. Set $\tilde{\Gamma} = \gamma\Gamma\gamma^{-1}$ for any $\gamma \in Möb(\widehat{\mathbb{C}})$. Then $\Omega(\tilde{\Gamma}) = \gamma(\Omega(\Gamma))$ and $\Lambda(\tilde{\Gamma}) = \gamma(\Lambda(\Gamma))$.*

Proposition 5.1.3 *The fixed points of a loxodromic element of Γ and that of a parabolic element of Γ are contained in $\Lambda(\Gamma)$.*

Proof It suffices to consider $\gamma(z) = \lambda z$ with $|\lambda| > 1$ as a loxodromic element and 0 as a loxodromic fixed point of it. Since $|\gamma'(0)| = |\lambda|$, we have $|(\gamma^n)'(0)| = |\lambda|^n \to \infty$ as $n \to \infty$. Hence $\{\gamma^n\}_{n=1}^{\infty}$ is not a normal family in a neighborhood of 0.

Next it suffices to consider $\gamma(z) = z + 1$ as a parabolic element. Take a neighborhood $\{z \mid |z| > R\}$ of ∞. For sufficiently large n, $-n$ is contained in this neighborhood. On the other hand, $\gamma^n(-n) = 0$. This shows $\{\gamma^n\}_{n=1}^{\infty}$ is not a normal family in the neighborhood of ∞. □

Now we give a sufficient condition for the normality of a family of Möbius transformations.

Theorem 5.1.4 *Let a and b be distinct points in $\widehat{\mathbb{C}}$ and D a domain. Assume that a family $\mathcal{F}$ consists of Möbius transformations, each of which does not take the value either a or b in D. Then it is normal in D.*

Proof It is enough to prove that for each $z_0 \in D$, $\mathcal{F}$ is a normal family on some disk with center z_0. If necessary, we consider a subdomain D' of D such that D' contains at most one of a and b. We may assume that $a = 0$, $b = \infty$ and $D' \not\ni \infty$. Take z_0 in D' and suppose the open disk $K = D(z_0, r)$ is contained in D'. Considering the set $\delta(\Gamma)$ with $\delta(z) = z - z_0$ we may assume $z_0 = 0$.

We take an arbitrary subsequence

$$\gamma_n(z) = \frac{a_n z + b_n}{c_n z + d_n}, \qquad a_n d_n - b_n c_n = 1,$$

of $\mathcal{F}$. Since neither $\gamma_n^{-1}(0) = -b_n/a_n$ nor $\gamma_n^{-1}(\infty) = -d_n/c_n$ is contained in K, we have $|b_n/a_n| > r$ and $|d_n/c_n| > r$. Further, if $|z| < r' < r$, then we obtain

$$\left(1 - \frac{r'}{r}\right)\left(1 + \frac{r'}{r}\right)^{-1} \le \left|\frac{d_n}{b_n}\gamma_n(z)\right| \le \left(1 + \frac{r'}{r}\right)\left(1 - \frac{r'}{r}\right)^{-1}.$$

If the $|b_n/d_n|$ are bounded, then there is a constant M_1 depending on only r and r' such that $|\gamma_n(z)| < M_1$ for all $|z| < r'$. Thus $\{\gamma_n\}$ is normal on K. If not, there is a subsequence $\{n(i)\}$ such that $|1/\gamma_{n(i)}(z)| \to 0$ uniformly on $\{|z| < r'\}$. Thus we can get a convergent subsequence of $\{\gamma_n\}$. Hence $\mathcal{F}$ is again normal on K. □

Corollary *Let z be a point in $\Lambda(\Gamma)$. For each point w except at most one point in $\widehat{\mathbb{C}}$, there is a sequence $\{\gamma_n\}_{n=1}^{\infty}$ of Γ mutually distinct elements of Γ such that $\gamma_n(w) \to z$ as $n \to \infty$.*

Proof Since Γ is not normal in any neighborhood U of z, there is a γ of Γ with $\gamma(U) \ni w$ for all w but at most one point. We choose γ_1 of Γ such that $\gamma_1^{-1}(D(z, 1)) \ni w$. Since $\Gamma - \{\gamma_1\}$ is not normal on $D(z, 1/2)$, we can choose γ_2 in $\Gamma - \{\gamma_1\}$ such that $\gamma_2^{-1}(D(z, 1/2)) \ni w$. Repeating this procedure, we have the sequence we need. □

Theorem 5.1.5 *For an arbitrary Kleinian group Γ, either $\Lambda(\Gamma) = \widehat{\mathbb{C}}$ or the interior of $\Lambda(\Gamma)$ is empty.*

Proof Suppose $\Lambda(\Gamma) \neq \widehat{\mathbb{C}}$. From the corollary above, for each $z \in \Lambda(\Gamma)$ there are a point w of $\Omega(\Gamma)$ and a sequence $\{\gamma_n\}$ of Γ such that $\gamma_n(w) \to z$ as $n \to \infty$. Since $\Omega(\Gamma)$ is Γ-invariant, $\{\gamma_n(w)\} \subset \Omega(\Gamma)$. Hence z is a boundary point of $\Omega(\Gamma)$. □

A Kleinian group whose limit set consists of at most two points is called *elementary*. Properties of elementary groups are well known.

Example 5.1.1 *Choose $\gamma(z) = \lambda z$ with $|\lambda| \neq 1$. The group $\Gamma = \langle\gamma\rangle$ generated by γ is an elementary Kleinian group and $\Lambda(\Gamma) = \{0, \infty\}$. On the other hand, the Julia set $J(\gamma)$ of γ is $\{0\}$ for $|\lambda| > 1$.*

From the definition, a finite group is elementary. Further we have the following.

Theorem 5.1.6 *A Kleinian group consisting only of elliptic transformations apart from the identity is a finite group.*

Proof If two Möbius transformations γ and η have only one common fixed point, then elementary calculation shows that $\gamma \circ \eta \circ \gamma^{-1} \circ \eta^{-1}$ is parabolic. Hence any two elements of Γ have either two common fixed points or none. If Γ is an infinite group, then there are infinitely many elliptic elements, any two of which have no common fixed points. We assume an elliptic element γ of Γ fixes 0 and ∞. The representation of γ by a matrix in $\mathrm{SL}(2, \mathbb{C})$ is

$$\begin{pmatrix} e^{i\theta} & 0 \\ 0 & e^{-i\theta} \end{pmatrix},$$

where $\theta \neq n\pi$. We show that all other elements of Γ are represented by

$$\begin{pmatrix} a & b \\ -\overline{b} & \overline{a} \end{pmatrix}, \qquad |a|^2 + |b|^2 = 1,$$

under a suitable conjugacy, which contradicts the discreteness of Γ.

Choose any η in Γ which does not fix 0 or ∞, and represent it by

$$\begin{pmatrix} a & b \\ c & d \end{pmatrix},$$

where $ad - bc = 1$ and $bc \neq 0$. Since $\mathrm{tr}(\gamma) = a + d$ and $\mathrm{tr}(\gamma\eta) = ae^{i\theta} + de^{-i\theta}$ are real, we have $d = \overline{a}$. Further, we have $bc < 0$, because $\gamma\eta\gamma^{-1}\eta^{-1}$ is the identity or elliptic and its trace is $2(1 + 2bc\sin^2\theta)$. Hence we have

$$-\frac{c}{\overline{b}} = -\frac{|c|^2}{\overline{bc}} > 0.$$

Set $\mu^4 = -c/\overline{b}$ with $\mu > 0$ and consider $\Gamma' = \rho\Gamma\rho^{-1}$ with ρ represented by

$$\begin{pmatrix} \mu & 0 \\ 0 & \mu^{-1} \end{pmatrix}.$$

Then $\eta' = \rho \circ \eta \circ \rho^{-1}$ is represented by

$$\begin{pmatrix} a & e \\ -\overline{e} & \overline{a} \end{pmatrix}$$

with $e = b\mu^2$, for $c\mu^{-2} = -\overline{b\mu^2}$.

Take any element σ of Γ' and represent it by

$$\begin{pmatrix} x & u \\ v & \overline{x} \end{pmatrix}.$$

Then $\eta' \circ \sigma$ is represented by

$$\begin{pmatrix} ax + eu & * \\ * & -\bar{e}u + \overline{ax} \end{pmatrix}.$$

Since $\overline{ax + ev} = -\bar{e}u + \overline{ax}$, we have $v = -\bar{u}$. Thus we obtain the claim. □

Theorem 5.1.7 *Let Γ be a non-elementary Kleinian group. Then $\Lambda(\Gamma)$ is a perfect set. In particular, $\Lambda(\Gamma)$ is an uncountable set.*

Proof Since $\Lambda(\Gamma)$ is a closed set, it is sufficient to show that $\Lambda(\Gamma)$ has no isolated point. Fix $z \in \Lambda(\Gamma)$. First, assume that z is a fixed point of infinitely many distinct elements of Γ. Since Γ is a discrete group, z is a fixed point of a parabolic or loxodromic element γ. There is a point $w \in \Lambda(\Gamma)$ which is not a fixed point of γ, because $\Lambda(\Gamma)$ contains at least two points other than z. Then $\{\gamma^n(w)\}_{n=-\infty}^{\infty} \subset \Lambda(\Gamma)$, the $\gamma^n(w)$ are all distinct, and $\gamma^n(w) \to z$ as $n \to +\infty$ or $-\infty$.

Next assume that z is a fixed point of finitely many distinct elements of Γ. There are a point w of $\Lambda(\Gamma)$ and a sequence $\{\gamma_n\}$ of distinct elements of Γ such that $\gamma_n(w) \to z$ by the corollary to Theorem 5.1.4. Unless z is an accumulation point of $\{\gamma_n(w)\}$, there is N such that $\gamma_n(w) = z$ for $n \geq N$. This implies that $\gamma_n \circ \gamma_N^{-1}(z) = z$ and $\gamma_n \circ \gamma_N^{-1} \neq \mathrm{id}$ for all $n \geq N$, which contradicts the assumption. □

In the case of non-elementary groups, the following theorem also holds.

Theorem 5.1.8 *A non-elementary group Γ contains infinitely many loxodromic elements, any two of which have no common fixed points.*

We need the following lemma to prove this theorem.

Lemma 5.1.9 *Let γ be a loxodromic transformation and assume that a Möbius transformation η has only one common fixed point with γ. Then the group $\langle \gamma, \eta \rangle$ generated by γ and η is not Kleinian.*

Proof Since conjugation preserves discreteness, we may assume that the common fixed point is ∞. We represent γ and η by

$$\begin{pmatrix} \alpha & 0 \\ 0 & \alpha^{-1} \end{pmatrix}, \quad |\alpha| > 1, \qquad \text{and} \qquad \begin{pmatrix} a & b \\ 0 & a^{-1} \end{pmatrix}, \quad a \neq 0,$$

respectively. Here, the assumption implies that $b \neq 0$. Hence the $\gamma^{-n} \circ \eta \circ \gamma^n$ converge:

$$\begin{pmatrix} a & b\alpha^{-2n} \\ 0 & a^{-1} \end{pmatrix} \to \begin{pmatrix} a & 0 \\ 0 & a^{-1} \end{pmatrix},$$

as $n \to \infty$. Hence the group $\langle \gamma, \eta \rangle$ is not Kleinian. □

Proof of Theorem 5.1.8 If Γ contains no loxodromic elements, it contains a parabolic element γ by Theorem 5.1.6. We may represent γ by

$$\begin{pmatrix} 1 & 1 \\ 0 & 1 \end{pmatrix}.$$

If there is a parabolic element η whose fixed point is not ∞, then a calculation shows that $\eta \circ \gamma$ or $\eta \circ \gamma \circ \eta^{-1} \circ \gamma^{-1}$ is a loxodromic element. Hence the fixed point of every parabolic element of Γ must be ∞. Further, unless ∞ is a fixed point of some elliptic element η of Γ, $\eta(\infty) \neq \infty$ is a fixed point of the parabolic element $\eta \circ \gamma \circ \eta^{-1}$. Thus one of the fixed points of every elliptic element of Γ is ∞, and we have $\Lambda(\Gamma) = \{\infty\}$, which contradicts the fact that Γ is non-elementary. Thus Γ contains a loxodromic element.

Next take a loxodromic element γ, whose fixed points are z_1 and z_2. Suppose that $\{\eta(z_1), \eta(z_2) \mid \eta \in \Gamma\} \subset \{z_1, z_2\}$ for all $\eta \in \Gamma$. Then for any point w distinct from z_1 and z_2, the corollary to Theorem 5.1.4 gives a sequence $\{\gamma_n\}$ of Γ such that $\gamma_n(z_1)$ or $\gamma_n(z_2)$ tends to w as $n \to \infty$, which is a contradiction. If $\{z_1, z_2\}$ and $\{\eta(z_1), \eta(z_2)\}$ have a single common point for some $\eta \in \Gamma$, it should be a fixed point of $\eta \circ \gamma \circ \eta^{-1}$. This contradicts the discreteness of Γ by Lemma 5.1.9.

Thus there is an element η such that $\{\eta(z_1), \eta(z_2)\}$ does not intersect $\{z_1, z_2\}$. Moreover, the $\{\gamma^n \circ \eta(z_1), \gamma^n \circ \eta(z_2)\}$ are pairwise disjoint and each of them is the pair of fixed points of $\gamma^n \circ \eta \circ \gamma \circ \eta^{-1} \circ \gamma^{-n}$. □

Actually we have a more precise statement than that of the corollary to Theorem 5.1.4.

Theorem 5.1.10 *Let Γ be a non-elementary Kleinian group, $z \in \Lambda(\Gamma)$ and $w \in \widehat{\mathbb{C}}$. Then there is a sequence $\{\gamma_n\}$ of Γ such that $\gamma_n(w) \to z$ as $n \to \infty$.*

Proof Assume that there were $z \in \Lambda(\Gamma)$ and $w \in \widehat{\mathbb{C}}$ such that the $\gamma_n(w)$ do not converge to z for any $\{\gamma_n\}$. There is a loxodromic element η of Γ such that $\eta(w) \neq w$ by Theorem 5.1.8. Hence from the corollary to Theorem 5.1.4, there is a sequence $\{\omega_n\}$ of Γ such that $\omega_n \eta(w) \to z$ as $n \to \infty$. This is a contradiction. □

Corollary 1 *Let Γ be a non-elementary Kleinian group. For $z \in \Lambda(\Gamma)$, $\Gamma(z) = \{\gamma(z) \mid \gamma \in \Gamma\}$ is dense in $\Lambda(\Gamma)$.*

Proof From Theorem 5.1.10, $\Lambda(\Gamma)$ is contained in the derived set of $\Gamma(z)$. We obtain the claim because of Γ-invariance of $\Lambda(\Gamma)$. □

Corollary 2 *Let Γ be a non-elementary Kleinian group. Then the set of fixed points of all loxodromic elements is dense in $\Lambda(\Gamma)$.*

Proof From Theorem 5.1.8, Γ contains a loxodromic element γ and its fixed point z is in $\Lambda(\Gamma)$. The point $\eta(z)$ is a fixed point of a loxodromic transformation of $\eta \circ \gamma \circ \eta^{-1}$ for every η of Γ. Thus we obtain the claim by the corollary above. □

Corollary 3 *Let Γ be a non-elementary Kleinian group containing parabolic transformations. Then the set of fixed points of all parabolic elements is dense in $\Lambda(\Gamma)$.*

Proof We can prove this by an argument similar to that in the proof above. □

Corollary 4 *Let Γ be a non-elementary Kleinian group and U an open set intersecting with $\Lambda(\Gamma)$. Then there are finitely many elements $\gamma_1, \gamma_2, \ldots, \gamma_N$ of Γ such that*

$$\bigcup_{n=1}^{N} \gamma_n(U) \supset \Lambda(\Gamma).$$

Proof Let γ be a loxodromic element with the fixed points z_1 and z_2. If V is an open set containing z_1, then we have

$$\bigcup_{n=-\infty}^{\infty} \gamma^n(V) \supset \widehat{\mathbb{C}} - \{z_2\}.$$

Hence $\bigcup_{\gamma\in\Gamma} \gamma(U) \supset \Lambda(\Gamma)$ by Corollary 2 above. Since $\Lambda(\Gamma)$ is compact, we obtain the claim. □

Now we study limit functions of a uniformly convergent sequence of Möbius transformations.

Proposition 5.1.11 *Suppose that a sequence $\{\gamma_n\}$ of Γ converges uniformly on a domain D. Then the limit function is constant.*

Proof Since each γ_n is univalent on D, the limit function f is constant or univalent. Suppose it is univalent. Fix three distinct points z_1, z_2 and z_3 in D and set $f(z_i) = w_i$ $(i = 1, 2, 3)$. Take Möbius transformations ω_1 and ω_2 so that $\omega_2 \circ f \circ \omega_1$ fixes 0, 1 and ∞, and set $\eta_n = \omega_2 \circ \gamma_n \circ \omega_1$. Then we have $\eta_n(0) \to 0$, $\eta_n(1) \to 1$ and $\eta_n(\infty) \to \infty$. We represent η_n as

$$\begin{pmatrix} a_n & b_n \\ c_n & d_n \end{pmatrix}, \qquad a_n d_n - b_n c_n = 1.$$

Then we obtain $a_n/c_n \to \infty$, $b_n/d_n \to 0$, and $(a_n + b_n)/(c_n + d_n) \to 1$. Moreover, we have $c_n d_n \to 0$ by $a_n d_n - b_n c_n = 1$. Since

$$\frac{d_n(a_n + b_n)}{d_n(c_n + d_n)} = \frac{1}{d_n(c_n + d_n)} + \frac{b_n}{d_n},$$

we obtain $d_n(c_n + d_n) \to 1$ and thus $d_n^2 \to 1$. Assume $\epsilon_n d_n \to 1$ by choosing suitable $\epsilon_n \in \{\pm 1\}$. Then we have $\epsilon_n a_n \to 1$, $b_n \to 0$, and $c_n \to 0$. This implies η_n converges to the identity and contradicts that Γ is a Kleinian group. □

From this theorem, we obtain the inverse of the claim of Theorem 5.1.10.

Theorem 5.1.12 *Let Γ be a Kleinian group and z and w in $\widehat{\mathbb{C}}$. If there is a sequence $\{\gamma_n\}_{n=1}^{\infty}$ of distinct elements of Γ such that $\gamma_n(w) \to z$, then z belongs to $\Lambda(\Gamma)$.*

Proof Assuming $z \in \Omega(\Gamma)$, we have a neighborhood $D \subset \Omega(\Gamma)$ of z on which Γ is a normal family. Fix N satisfying $\gamma_N(w) = w' \in D$ and set $\eta_n = \gamma_n \circ \gamma_N^{-1}$ $(n \geq N)$. Then $\eta_n(w') \to z$. Choose a domain $D_1 \subset D$ containing w' and z and satisfying $\overline{D_1} \subset D$. Since $\{\eta_n\}$ is a normal family on D, some subsequence converges uniformly on D_1. From Proposition 5.1.11, its limit function is a constant, which is equal to z. Again, by uniform convergence on D_1, there is m such that $\overline{\eta_m(D_1)} \subset D_1$. Hence we can easily see that η_m is a loxodromic transformation, one of whose fixed points is in D_1 and should be in $\Lambda(\Gamma)$. This contradicts $D \subset \Omega(\Gamma)$. □

Remark The claim of Theorem 5.1.12 is sometimes used as the definition of limit points.

Take a sequence $\{\gamma_n\}$ of Γ which converges uniformly on a compact set K of $\Omega(\Gamma)$; then we see that $\gamma_n(K)$ converges to a point in $\Lambda(\Gamma)$ by Proposition 5.1.11 and Theorem 5.1.12. Thus we have the following.

Theorem 5.1.13 *Let Γ be a Kleinian group and $z \in \widehat{\mathbb{C}}$. If there is a neighborhood U of z such that the number of elements γ of Γ satisfying*

$$\gamma(U) \cap U \neq \varnothing$$

is finite, then z belongs to $\Omega(\Gamma)$.

Definition For $z \in \widehat{\mathbb{C}}$, we say that Γ acts *properly discontinuously*, if there is a neighborhood U of z satisfying the condition in Theorem 5.1.13. Hence $\Omega(\Gamma)$ is also called the *region of discontinuity.*

Theorem 5.1.14 *For a non-elementary Kleinian group Γ, $\Lambda(\Gamma)$ is the smallest non-empty, Γ-invariant closed set.*

Proof We only show the minimality of $\Lambda(\Gamma)$ because $\Lambda(\Gamma)$ is a non-empty, Γ-invariant closed set by Corollary 1 to Proposition 5.1.2. Let A be a non-empty Γ-invariant closed set. Take $z \in \Lambda(\Gamma)$. For $w \in A$, there is a sequence $\{\gamma_n\}$ of Γ such that $\gamma_n(w) \to z$ by Theorem 5.1.10. Since A is a Γ-invariant closed set, z belongs to A. □

We consider the case where an ordinary set is not empty. We say a component Ω_1 of $\Omega(\Gamma)$ of a Kleinian group Γ is an *invariant component* of Γ if $\gamma(\Omega_1) = \Omega_1$ for all γ of Γ.

Proposition 5.1.15 *Let Ω_1 be an invariant component of a Kleinian group Γ. Then $\partial\Omega_1 = \Lambda(\Gamma)$.*

Proof We obtain the claim because $\Lambda(\Gamma) \subset \overline{\Omega_1}$ by Theorem 5.1.14 and $\partial\Omega_1 \subset \Lambda(\Gamma)$. □

Proposition 5.1.16 *The ordinary set of a Kleinian group has at most two invariant components.*

Proof Suppose that there were three invariant components Ω_1, Ω_2 and Ω_3. Since Γ is non-elementary in this case, we can find a loxodromic element γ of Γ and denote its fixed points by z_1 and z_2.

Since $\gamma(w) \in \Omega_1$ for $w \in \Omega_1$, we can take a suitable simple arc ℓ in Ω_1 connecting w and $\gamma(w)$ and satisfying $\ell \cap \gamma(\ell) = \{\gamma(w)\}$ so that $L_1 = \bigcup_{n=-\infty}^{\infty} \gamma^n(\ell)$ is a simple arc in Ω_1 connecting z_1 and z_2. Similarly, we take a simple arc L_2 in Ω_2 connecting z_1 and z_2 so that $L = L_1 \cup L_2 \cup \{z_1, z_2\}$ is a simple closed curve. We denote two complements of L by U_1 and U_2. The component Ω_3 is contained in only one of them, say $\Omega_3 \subset U_1$.

On the other hand, since Ω_1 and Ω_2 are distinct components, U_2 contains points in $\Lambda(\Gamma)$. Since there is a loxodromic fixed point z_3 in $\Lambda(\Gamma) \cap U_2$, we have an arc in Ω_3 connecting z_3 and a point in Ω_3 similarly as above. This is a contradiction. □

Proposition 5.1.17 *The number of components of* $\Omega(\Gamma)$ *for a Kleinian group* Γ *is either* 0, 1, 2 *or* ∞.

Proof Assume that the number of the components is finite and is equal to n (≥ 3) and set them as $\Omega_1, \Omega_2, \ldots, \Omega_n$. We denote the stabilizer of Ω_j by $\Gamma(\Omega_j)$. Since the number of the components is finite, $\Gamma(\Omega_j)$ is of finite index in Γ. Hence there is a subgroup Γ' of Γ which is of finite index in Γ such that $\gamma(\Omega_j) = \Omega_j$ for all $\gamma \in \Gamma'$ and for $j = 1, 2, \ldots, n$. It is easy to see that $\Lambda(\Gamma) = \Lambda(\Gamma')$. This contradicts Proposition 5.1.16. □

We end this section with an example of a Kleinian group whose limit set is $\widehat{\mathbb{C}}$.

Example 5.1.2 *The group* Γ *consisting of all transformations* γ *of the form*

$$\gamma(z) = \frac{az+b}{cz+d} \quad (ad - bc = 1,\ a, b, c, d \in \mathbb{Z} + i\mathbb{Z})$$

is a Kleinian group and $\Lambda(\Gamma) = \widehat{\mathbb{C}}$. *This Kleinian group is called the* Picard group.

Verification It is clear that Γ is a Kleinian group. To show $\Lambda(\Gamma) = \widehat{\mathbb{C}}$, we see that the set of all fixed points of the parabolic elements of Γ is dense in $\widehat{\mathbb{C}}$. Indeed, the set

$$\left\{ w = \frac{p+iq}{r} \;\middle|\; p, q, r \in \mathbb{Z},\ r \neq 0 \right\}$$

is dense in $\widehat{\mathbb{C}}$. On the other hand

$$\gamma(z) = \frac{(1 - r^2 w)z + r^2 w^2}{-r^2 z + (1 + r^2 w)}$$

is a parabolic element of Γ whose fixed point is w.

5.2 Sullivan's dictionary

Some results on Kleinian groups stated in §5.1 make readers recognize analogies with results in previous chapters. In this section, we explain some items of

Sullivan's dictionary, which enumerates analogies between iteration of holomorphic endomorphisms and action of Kleinian groups. In the following, we list terminology and results on complex dynamics in the left side of the table and those on Kleinian groups in the right. We start with fundamental definitions.

$f \in \mathrm{Rat} \cup \mathrm{Ent}$	non-elementary Γ
rational function with degree more than 2	non-elementary and finitely generated
Julia set $J(f)$	limit set $\Lambda(\Gamma)$
Fatou set $F(f)$	ordinary set $\Omega(\Gamma)$

Next we list analogies between fundamental properties of the Julia sets and those of the limit sets.

$J(f)$: smallest, closed, non-empty, completely invariant	$\Lambda(\Gamma)$: smallest, closed, non-empty, invariant
$\#J(f) \geq 3$ (Corollary 1 to Theorem 2.3.5)	$\#\Lambda(\Gamma) \geq 3$ (Theorem 5.1.14)
repelling periodic points are dense in $J(f)$ (first fundamental theorem)	loxodromic fixed points are dense in $\Lambda(\Gamma)$ (Corollary 2 to Theorem 5.1.10)

Further, we have the items as follows.

$\#\{\text{components of } F(f)\} = 0, 1, 2 \text{ or } \infty$ (Corollary to Theorem 4.2.16)	$\#\{\text{components of } \Omega(\Gamma)\} = 0, 1, 2 \text{ or } \infty$ (Proposition 5.1.17)
$\#\{\text{completely invariant components of } F(f)\} = 0, 1 \text{ or } 2$ (Corollary to theorem 4.2.16)	$\#\{\text{invariant components of } \Omega(\Gamma)\} = 0, 1 \text{ or } 2$ (Proposition 5.1.16)
$\partial(\text{completely invariant component}) = J(f)$ (Corollary 2 to Theorem 2.3.5)	$\partial(\text{invariant component}) = \Lambda(\Gamma)$ (Proposition 5.1.15)

The study of Kleinian groups was begun by Poincaré at the end of the nineteenth century. Thirty years later, the theory of iteration was started independently by Julia and Fatou, who already had recognized the analogy between Kleinian groups and iteration of holomorphic endomorphisms. More than fifty years later, Sullivan pushed forward the study by using such analogies. We state his results which refreshed the study of holomorphic dynamics. In what follows, we assume that $f \in \mathrm{Rat}$ and Γ is finitely generated,

Here, Ahlfors' finiteness theorem below is fundamental.

Theorem 5.2.1 *For a finitely generated Kleinian group* Γ, $\Omega(\Gamma)/\Gamma$ *is a finite union of analytically finite Riemann surfaces.*

Here, an *analytically finite* Riemann surface is a closed Riemann surface with finitely many points removed. Ahlfors proved this theorem by using the fact that the dimension of the Eichler cohomology group, which is regarded as the tangent space of the representation space by $\mathrm{PSL}(2, \mathbb{C})$, is finite. On the other hand, Sullivan paid attention to the analogies between iteration of rational functions and action of Kleinian groups, and proved both the finiteness theorem and the no wandering domain theorem by the same method. For the proof see §4.2 of Matsuzaki and Taniguchi (1998). As was stated in §4.2, the gist of this proof is to use quasiconformal mappings and deformation spaces, theories of which had not been established in the ages of Fatou and Julia. In his paper, Sullivan listed analogies between iteration of rational functions and action of Kleinian groups, which is the origin of Sullivan's dictionary. We list this item.

No wandering domains theorem (second fundamental theorem)	Ahlfors' finiteness theorem (Theorem 5.2.1)

The crucial facts about deformation spaces are the following.

degree k gives $2k - 2$ complex parameters	n generators give $3n - 3$ complex parameters

Further, we have a correspondence between Shishikura's theorem, which uses the number of critical points to estimate the number of cyclic components, and the following, Bers' area theorem, about the number of generators (see §4.1.2 of Matsuzaki and Taniguchi (1998)).

Theorem 5.2.2 *For a finitely generated Kleinian group without parabolic elements, the hyperbolic area of* $\Omega(\Gamma)/\Gamma$ *is not greater than*

$$4\pi(r(\Gamma) - 1),$$

where $r(\Gamma)$ *is the infimum of the number of generators.*

Shishikura's theorem (Theorem 4.3.1)	Bers' area theorem (Theorem 5.2.2)

Here we include the following two items on general structures of chaotic and non-chaotic parts.

#{completely invariant components of $F(f)$} = 2 ⇒ #{components of $F(f)$} = 2 (Theorem 4.2.16)	#{invariant components of $\Omega(\Gamma)$} = 2 ⇒ #{components of $\Omega(\Gamma)$} = 2 (Marden–Maskit)

Remark If a finitely generated Kleinian group has two invariant components of the ordinary set, then its limit set is a quasicircle. On the other hand, even if a rational function has two completely invariant components of the Fatou set, its Julia set may not be a quasicircle. For example, in the case of $P_{1/4}(z) = z^2 + 1/4$, $J(P_{1/4})$ is not a quasicircle but a Jordan curve.

We have an analogy between disconnected Julia sets and disconnected limit sets.

$J(f)$: uniformly perfect (Pommerenke 1984)	$\Lambda(\Gamma)$: uniformly perfect (Hinkkanen 1993, Mañé–Rocha 1992)

Actually, we can give proofs of these facts by the same argument.

Proof Let $X = J(f)$ or $\Lambda(\Gamma)$. And suppose that there were a sequence of ring domains $\{A_j\}$ in $\widehat{\mathbb{C}} - X$ whose moduli $\log m_j$ tend to ∞. Let E_j be the component of $\widehat{\mathbb{C}} - A_j$ with smaller spherical diameter; then the spherical diameter $\mathrm{diam}_0(E_j)$ tends to 0. So fix a sufficiently small $\delta > 0$ and assume that $\mathrm{diam}_0(E_j) < \delta$.

For every j, we define g_j as follows. In the case of a rational function f, define $k(j)$ by the condition $\mathrm{diam}_0(f^{k(j)}(E_j)) \geq \delta$ but $\mathrm{diam}_0(f^{k(j)-1}(E_j)) < \delta$, and set $g_j = f^{k(j)}$. In the case of a finitely generated Γ with generators $\{\gamma_i\}$, take an element γ such that, when γ can be presented as a word of generators, i.e. $\gamma = \gamma_{i(1)} \circ \cdots \circ \gamma_{i(N)}$, then $\mathrm{diam}_0(\gamma(E_j)) \geq \delta$ but $\mathrm{diam}_0((\gamma_{i(1)})^{-1}(\gamma(E_i))) < \delta$, and set $g_j = \gamma$. Here note that, letting C

be the Lipschitz constant for f or the maximum of those for γ_i, we have $g_j(E_j) < C\delta$.

Now, fix four points $\{z_l\}_{l=1}^4$ such that the $D(z_l, C\delta/2)$ are pairwise disjoint, and take a Riemann map ϕ_j of $A_j \cup E_j$ such that $0 \in K_j$, where $K_j = \phi_j(E_j)$. Since the moduli of $\phi_j(A_j)$ equal $\log m_j$, $\mathrm{diam}_0(K_j)$ tends to 0. And the three-point condition implies that $\{g_j \circ \phi_j^{-1}\}$ is a normal family. Hence we may assume that $g_j \circ \phi_j^{-1}$ converges, and hence the spherical diameters of $g_j \circ \phi_j^{-1}(K_j) = g_j(E_j)$ should tend to 0. This is a contradiction. □

Next, we classify Kleinian groups more closely. We say that a Kleinian group Γ is *Fuchsian* if there is a circle or a line S such that it is invariant under Γ and each complementary component of S in $\widehat{\mathbb{C}}$ is also invariant under Γ. Minimality of the limit set implies that $\Lambda(\Gamma) \subset S$, and we say that Γ is a Fuchsian group of the *first kind* if $\Lambda(\Gamma) = S$, and otherwise of the *second kind*. Note that $(\widehat{\mathbb{C}} - S)/\Gamma$ consists of two hyperbolic Riemann surfaces. Conversely, if we regard the universal covering surface of a hyperbolic Riemann surface as a disk, then its covering transformation group is a Fuchsian group.

A finitely generated Kleinian group with an invariant component is called a *function group*. Moreover, a function group whose invariant component is simply connected is called a *B-group*. Hence a finitely generated Fuchsian group is a function group and further if it is of the first kind, then it is a B-group.

A finitely generated Kleinian group is said to be a *quasi-Fuchsian group* if it is a B-group whose ordinary set consists of precisely two components. The limit set of a quasi-Fuchsian group is a quasicircle. A finitely generated Kleinian group which is not a function group but whose ordinary set consists of precisely two components contains a quasi-Fuchsian group as a subgroup of index 2.

Blaschke product	Fuchsian
hyperbolic rational with two completely invariant components	quasi-Fuchsian

Fix an analytically finite Riemann surface R and denote by Γ a Fuchsian group corresponding to R. The Kleinian group corresponding to the point in the quasiconformal deformation space $Q(\Gamma)$ of Γ is a quasi-Fuchsian group. Since the deformation space of complex structures on the Riemann surface R is the Teichmüller space $T(R)$, we can identify $Q(\Gamma)$ with $T(R) \times T(R)$. Denoting the point corresponding to Γ in $T(R)$ by 0, we call $Q_0(\Gamma) = T(R) \times \{0\}$ a *Bers slice* of $Q(\Gamma)$. In the case of rational functions with degree 2, the main cardioid of the Mandelbrot set corresponds to $Q_0(\Gamma)$. The boundary of $Q_0(\Gamma)$ is very complicated and many unsolved problem concerning it are left. A finitely generated Kleinian group corresponding to a boundary point of the Bers slice $Q_0(\Gamma)$ is called a boundary group and it has been proved to be a B-group. The converse is still unsolved. For the details, see §4.3.3 of Matsuzaki and Taniguchi (1998).

A B-group whose ordinary set is connected is called a *totally degenerate group*. An arbitrary B-group is either a quasi-Fuchsian group, a totally degenerate group or a Kleinian group which contain parabolic elements called accidental parabolic transformations (a theorem of Maskit, see Corollary 4.35 in Matsuzaki and Taniguchi (1998)).

dendrite	locally connected limit set of totally degenerate group

Let $\{C_j, C'_j\}_{j=1}^n$ be a set, each of whose elements is a simple closed curve with a common exterior D. If there are Möbius transformations g_j such that $g_j(C_j) = C'_j$ and $g_j(D) \cap D = \varnothing$, then the Kleinian group Γ generated by $g_1, g_2, \ldots, g_n$ is called a *Schottky group*. Every element but the identity of a Schottky group is loxodromic. The ordinary set consists of only one component and the limit set is a Cantor set. Hence a Schottky group is not a B-group but a function group.

hyperbolic rational with connected $F(f)$	Schottky

A Möbius transformation γ can be expressed as a composite of an even number of reflections with respect to a circle or a line. Since we can extend a reflection in $\widehat{\mathbb{C}}$ to a reflection in an upper half space $\mathbb{H}^3 = \{(z, t) \mid z \in \mathbb{C}, t > 0\}$, by extending a circle or a line to a hemisphere or a half-plane in $\mathbb{H}^3$, the action of γ extends naturally to the action on $\mathbb{H}^3$. Applying the hyperbolic metric $ds^2_{\mathbb{H}^3} = (|dz|^2 + dt^2)/t^2$ to $\mathbb{H}^3$, we see that the action of γ on $\mathbb{H}^3$ is an isometry with respect to this metric. And Γ is a Kleinian group if and only if Γ acts on $\mathbb{H}^3$ properly discontinuously. We say that Γ is *geometrically finite* if Γ has a fundamental polyhedron with finite sides in $\mathbb{H}^3$. A geometrically finite Kleinian group is finitely generated, but the converse is not necessarily true. For example, Schottky groups and quasi-Fuchsian groups are geometrically finite, but totally degenerate groups are not.

A Kleinian group is said to be *expanding* on the limit set $\Lambda(\Gamma)$, if, for $z \in \Lambda(\Gamma)$, there is $\gamma \in \Gamma$ such that $|\gamma'(z)| > 1$. A geometrically finite Kleinian group without parabolic transformations is expanding on the limit set.

hyperbolic rational	geometrically finite without parabolic elements
geometrically finite	geometrically finite

On size of the limit set of a Kleinian group, the following, Ahlfors' 0–1 theorem, is fundamental.

Theorem 5.2.3 *If the limit set of a geometrically finite Kleinian group does not coincide with* $\widehat{\mathbb{C}}$*, then its two-dimensional Lebesgue measure is* 0.

Here, the following, Ahlfors' 0–1 conjecture, still remains open.

Ahlfors' 0–1 conjecture If the limit set of a finitely generated Kleinian group does not coincide with $\widehat{\mathbb{C}}$, then its two-dimensional Lebesgue measure is 0.

Compare Ahlfors' 0–1 theorem with Theorem 4.4.3. Further refinements of these theorems have been obtained, some of which will be discussed in §5.4. Also see §6.3.4 of Matsuzaki and Taniguchi (1998).

As the study of holomorphic dynamics is advancing, items of the dictionary are still increasing. See also McMullen (1994b, 1996 and preprint). We end this section by referring the reader to Figure 5.1.

5.3 Stability

Sullivan's dictionary also provides a lot of conjectures. The typical examples are the score-cards due to Sullivan, which we will explain in this section.

Definition We say that a finitely generated, torsion free non-solvable subgroup Γ of $\mathrm{PSL}(2, \mathbb{C})$ is *structurally stable* if every sufficiently near $\mathrm{PSL}(2, \mathbb{C})$-representation of Γ is isomorphic, and that Γ is *rigid* if there are no representations in a neighborhood which are not conjugate in $\mathrm{PSL}(2, \mathbb{C})$.

Theorem 5.3.1 (Sullivan) *Let* Γ *be structurally stable. Then* Γ *is either rigid or a geometrically finite Kleinian group with no parabolic transformations.*

For the proof, see, for instance, §5.4.2 of Matsuzaki and Taniguchi (1998).

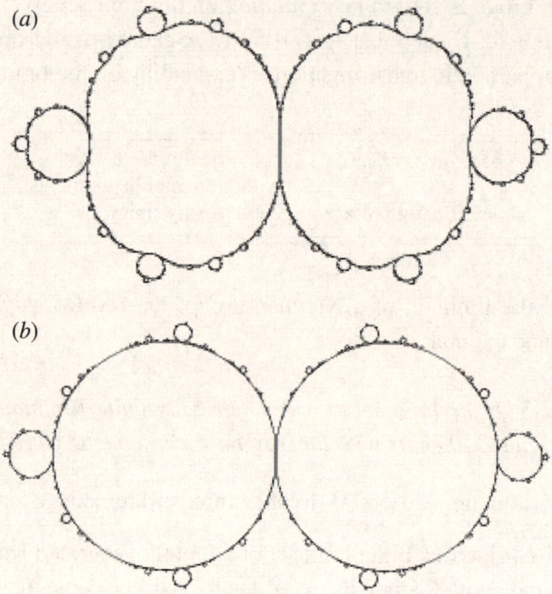

Fig. 5.1. (*a*) A Julia set and (*b*) a limit set

Conjecture I	Structural stability implies geometrical finiteness without parabolic elements

The first score-card is motivated by this result, and affords a central conjecture in the theory of complex dynamics.

Conjecture I Structurally stable rational function are hyperbolic.

To explain structural stability of rational functions, we begin with the following.

Definition Let X be a complex manifold. Then we say that $\mathcal{X} = \{f_\lambda(z)\}$ of rational functions is a *holomorphic family* over X if the map $(\lambda, z) \to f_\lambda(z)$ is a holomorphic map of $X \times \widehat{\mathbb{C}}$ to $\widehat{\mathbb{C}}$.

We say that f_λ is *J-stable* if there is a neighborhood U of λ in X such that every periodic point of f_α is either attracting, or repelling, or persistently indifferent for every $\alpha \in U$. Here we say that a periodic point z of f_α of period

p is *persistently indifferent* if there is a holomorphic map $w : V \to \widehat{\mathbb{C}}$ such that

$$f_\lambda^p(w(\lambda)) = w(\lambda), \quad |(f_\lambda^p)'(w(\lambda))| = 1$$

for some neighborhood V of α. We denote by X^J the set of all J-stable points in X.

Proposition 5.3.2 *A point $f_\alpha \in \mathcal{X}$ is J-stable if and only if any one of the following conditions holds.*

(i) *There is a neighborhood U of α such that the Julia set admits a holomorphic motion on U, compatible with dynamics, i.e., there is a continuous family of homeomorphisms $h_\lambda : J(f_\alpha) \to J(f_\lambda)$ such that $h_\lambda \circ f_\alpha = f_\lambda \circ h_\lambda$ for every $f_\lambda \in U$.*

(ii) *The total number of attracting and superattracting cycles of f_λ is constant in a neighborhood of α.*

(iii) *The maximum period of any (super)attracting cycle is locally bounded near α*

Here, a *holomorphic motion* of a set E on U with center α is a family of injections $\phi_\lambda : E \to \widehat{\mathbb{C}}$ $(\lambda \in U)$ such that $\phi_\alpha(z) = z$ and $\phi_\lambda(z)$ is holomorphic on U for every fixed $z \in E$.

Remark The name of J-stability comes from condition (i) in the proposition above, which is further equivalent to the condition that $J(f_\lambda)$ varies continuously in U with respect to Hausdorff distance between closed sets and compatible with dynamics.

In particular, at a point not J-stable, the Julia set varies discontinuously.

For the proof of Proposition 5.3.2, the λ lemma is crucial, for the proof of which we cite the paper by Mañé, Sad and Sullivan (1983).

Lemma 5.3.3 (The λ lemma) *A holomorphic motion of E on U has a unique extension to a holomorphic motion of $\overline{E}$ on U, which gives a continuous map $\phi : U \times \overline{E} \to \widehat{\mathbb{C}}$.*

Proof of Proposition 5.3.2 First, suppose that f_α is J-stable, and take a neighborhood U of α such that every indifferent periodic point of every f_λ with $\lambda \in U$ is persistently indifferent on U. Then attracting periodic points cannot change to be repelling on U, for otherwise they would be non-persistently indifferent at some point of U. Hence condition (ii) holds.

It is clear that condition (ii) implies (iii), for every attracting cycle remains attracting and of the same period.

Next assume that f_α satisfies condition (iii). Then we can take a neighborhood U of α and positive N such that every periodic point of any f_λ with $\lambda \in U$ of period greater than N is repelling. Further, two such repelling periodic points cannot collide on U, for otherwise they should be indifferent at the point. Thus repelling periodic points of order greater than N vary holomorphically without collision on U. Since such points are always dense in the Julia set, the λ lemma implies condition (i).

Finally, assume that f_α satisfies condition (i). To show that f_α is J-stable, it suffices to show that any indifferent periodic point z of f_λ with $\lambda \in U$ is persistently indifferent. Let n be the period of z.

If $z \notin J(f_\lambda)$, then $(f_\lambda^n)'(z) \neq 1$, for it is not parabolic. Hence we can find an analytic function $w(\lambda)$ such that $f_\lambda^p(w(\lambda)) = w(\lambda)$ on U and $w(\alpha) = z$. The assumption also implies that $w(\lambda)$ remains outside of $J(\lambda)$ near α, and hence $|(f_\lambda^p)'(w(\lambda))| \leq 1$. The maximal principle gives that z is persistently indifferent. And if $z \in J(f_\lambda)$, then the holomorphic motion of the Julia set implies that $w(\lambda)$ as above remains in the Julia set. Thus by the same argument, we conclude that z is persistently indifferent. □

Theorem 5.3.4 (Mañé, Sad and Sullivan) *J-stable elements are open and dense in $\mathcal{X}$.*

Proof Since each superattracting or attracting cycle of f_λ persists in a small neighborhood, the total number $N(\lambda)$ of superattracting and attracting cycles of f_λ is lower semi-continuous. Thus the set of points where $N(\lambda)$ takes its local maximum is open. This set coincides with X^J by Proposition 5.3.2, and is dense in X, for $N(\lambda)$ is bounded. □

Now structural stability for rational functions is defined as follows.

Definition We say that $f_\lambda \in \mathcal{X}$ is *topologically stable* or *structurally stable* if there is a neighborhood U of λ in X such that f_λ and f_α are topologically conjugate for every $\alpha \in U$. We denote by X^{top} the set of all topologically stable points in X.

Similarly, we can define *quasiconformal stability* and denote by X^{QC} the set of all quasiconformally stable points in X.

Theorem 5.3.5 (Mañé, Sad, Sullivan and McMullen (cf. McMullen 1994a, McMullen and Sullivan preprint)) *In any holomorphic family of rational* *maps, topologically stable maps are open and dense.*

Structural stability is open and dense	Conjecture II

Conjecture II Structurally stable subgroups of PSL(2, $\mathbb{C}$) are open and dense.

To prove Theorem 5.3.5, let a holomorphic family $\mathcal{X}$ be given. We say that a $f_\lambda \in \mathcal{X}$ is *postcritically stable* if there is a neighborhood U of λ in X such that any coincidence between the forward orbits of critical points persists on U. We denote by X^{PC} the set of all postcritically stable points in X.

Proposition 5.3.6

$$X^{QC} = X^{top} = X^{PC}.$$

Proof It is clear that $X^{QC} \subset X^{top} \subset X^{PC}$. On the other hand, by Lemma 5.3.7 below, we conclude that X^{QC} contains X^{PC}. □

Lemma 5.3.7 *Fix $\alpha \in X^{PC}$. Then there is a holomorphic motion of $\widehat{\mathbb{C}}$ on a neighborhood U of α compatible with dynamics.*

Proof Fix $\alpha \in X^{PC}$ and take a sufficiently small neighborhood U of α so that critical points $\{c_1(\lambda), \dots, c_n(\lambda)\}$ of f_λ define a holomorphic motion on U. Further, consider points

$$f_\lambda^m(c_j(\lambda))$$

for all non-negative integers m and j. Then since the f_λ are postcritically stable, they define a holomorphic motion of $C^+(f)$. Moreover, we can extend this motion, say $h_\lambda(z)$, to that of the grand orbits $[C^+(f)]$ compatible with dynamics.

Then by the λ lemma, we have a holomorphic motion of the closure $\overline{[C^+(f_\alpha)]}$ on U. If $\overline{[C^+(f_\alpha)]}$ contains at most two points, then f_α is conjugate to $z \mapsto z^k$, and the assertion easily follows. So we assume that $\overline{[C^+(f_\alpha)]}$ contains at least three points.

Then $\overline{[C^+(f_\alpha)]}$ contains the Julia set and the complement consists of hyperbolic domains. Then by the harmonic λ lemma below due to Bers and Royden, we can find a smaller neighborhood $V \subset U$ and a unique holomorphic motion $\phi_\lambda(z)$ of $\widehat{\mathbb{C}}$ on V such that

$$\phi_\lambda(z) = h_\lambda(z) \quad (\lambda \in V, z \in \overline{[C^+(f_\alpha)]})$$

and the Beltrami coefficient μ_λ of ϕ_λ is harmonic on $\widehat{\mathbb{C}} - \overline{[C^+(f_\alpha)]}$. This motion respects the dynamics. Indeed, since $f_\lambda : \widehat{\mathbb{C}} - \overline{[C^+(f_\alpha)]} \to \widehat{\mathbb{C}} - \overline{[C^+(f_\alpha)]}$ is a smooth covering, we can define another holomorphic motion by setting

$$\psi_\lambda(z) = f_\lambda^{-1} \circ \phi_\lambda \circ f_\alpha(z),$$

where we take the continuous branch of f_λ^{-1} such that $\psi_\alpha(z) = z$. Then since the Beltrami coefficient of ψ_λ is again harmonic, and hence coincides with μ_λ, $\psi_\lambda = \phi_\lambda$, which implies the assertion. □

Lemma 5.3.8 (Harmonic λ lemma) *Let $\phi : \{|z| < 1\} \times A \to \widehat{\mathbb{C}}$ be a holomorphic motion of A on the unit disk Δ. Suppose that A contains at least three points. Then there is a unique holomorphic motion*

$$\hat{\phi} : \{|z| < 1/3\} \times \widehat{\mathbb{C}} \to \widehat{\mathbb{C}}$$

such that $\hat{\phi} = \phi$ on $\{|z| < 1/3\} \times A$ and the Beltrami coefficient of $\hat{\phi}_\lambda$ is harmonic on $\widehat{\mathbb{C}} - \overline{A}$ for every $\lambda \in \{|z| < 1/3\}$.

Here as before, we say that a Beltrami coefficient on a domain D is harmonic if it can be written as $\overline{\varphi(z)}\lambda_D(z)^{-2}$ with a holomorphic function $\varphi(z)$ and the density $\lambda_D(z)$ of the hyperbolic metric on D. For the proof, see Bers and Royden (1986).

Lemma 5.3.7 further implies that $X^{top} = X^{PC}$ is an open subset of X^J. Thus, Theorem 5.3.5 follows from the lemma below.

Lemma 5.3.9 *X^{PC} is dense in X^J.*

Proof We consider the subset $\mathcal{X}'$ of $\mathcal{X}^J$ consisting of such f_λ that every superattracting cycle is persistently superattracting and that the number of critical points is locally maximum. Then $\mathcal{X}'$ is the complement of a proper complex analytic subvariety of $\mathcal{X}^J$, and hence open and dense.

Fix an open ball U in $X' = \{\lambda \mid f_\lambda \in \mathcal{X}'\}$ and label the critical points as $\{c_1(\lambda), \ldots, c_n(\lambda)\}$. Then it suffices to show that there is an open dense subset of U where the relations between the critical orbits of any c_i and c_j are locally constant.

Since $U \subset X^J$, any critical point in the Julia set remains in the Julia sets as was seen before, and hence the assertion follows for those in the Julia set. So assume that c_i and c_j are in the Fatou set. Then we will show that all relations between forward orbits of c_i and c_j are locally constant.

Consider the relation

$$f_\lambda^a(c_i(\lambda)) = f_\lambda^b(c_j(\lambda)) \tag{5.1}$$

with non-negative integers a and b, and denote by U_{ij} the set where no such relations with some a and b hold. If $U_{ij} = U$, then we have finished. If U_{ij} is empty, then some relation holds on an open set of U, and hence on the whole of U. Then since there remain a finite number of cases, the assertion follows. Hence, there remains the case that there are $a \neq b$, and $\lambda_0 \in U$, such that (5.1) holds, but (5.1) does not hold on the whole U. Hence (5.1) holds on a proper complex analytic subvariety of U. In this case, we will show that λ_0 is in the closure of the interior of U_{ij}, which implies the assertion.

First, assume that $i = j$. Then by Lemma 5.3.7 and the assumption on X', the Julia set moves continuously and there is the natural correspondence between the Fatou components which preserves the types. Further, (5.1) can hold only in a (super)attracting basin or in a Siegel disk. In these cases, there is a λ arbitrarily near λ_0 which has an infinite number of points in the forward orbit of c_i. Hence such λ satisfy no relation like (5.1) and hence belong to U_{ij}.

Next assume that $i \neq j$. From above, we may suppose that c_i and c_j have constant self-relations on U. If c_i has a finite forward orbit, then so does c_j by (5.1) and they land on the same cycle at λ_0. Hence (5.1) holds on the whole of U. So we assume that c_i and c_j have infinite forward orbits on U. Further, we may assume that

$$p = f_\lambda^a(c_i(\lambda)) = f_\lambda^b(c_i(\lambda))$$

is in a periodic Fatou component D.

When D is in an immediate attracting or parabolic basin, p can be supposed to be sufficiently near to the cycle. Then we can take a disk B containing p such that the elements of the set $\{f_\lambda^n(B)\}_{n\in\mathbb{Z}^+}$ are disjoint and the f_λ^n are univalent on B for every λ sufficiently near to λ_0. Then such λ belongs to U_{ij}, which shows the assertion.

When D is in an immediate superattracting basin, p can be supposed again to be sufficiently near to the cycle. Then we can find a holomorphic family $Z_\lambda(z)$ near (λ_0, p) of conformal maps such that

$$Z_\lambda(f_\lambda^m(z)) = Z_\lambda(z)^d$$

with some d, where m is the period of D. By assumption, there is a neighborhood V of λ_0 such that

$$Z_\lambda(f_\lambda^a(c_i(\lambda))) \neq Z_\lambda(f_\lambda^b(c_j(\lambda)))$$

on $V - \{\lambda_0\}$, and both quantities are non-zero on V. Hence we may assume that

$$|Z_\lambda(f_\lambda^a(c_i(\lambda)))|^d < |Z_\lambda(f_\lambda^b(c_j(\lambda)))| < |Z_\lambda(f_\lambda^a(c_i(\lambda)))|^{1/d}$$

on V. Then the complement of the proper real analytic subset determined by

$$|Z_\lambda(f_\lambda^a(c_i(\lambda)))| = |Z_\lambda(f_\lambda^b(c_j(\lambda)))|$$

in V contained in U_{ij}.

Finally, when D is a Siegel disk or a Herman ring, the forward orbit of $c_i(\lambda)$ forms a dense subset of a union $C_i(\lambda)$ of real analytic loops. This set moves injectively as λ varies, the λ lemma gives a holomorphic motion

$$\phi : U \times C_i(\lambda_0) \to \widehat{\mathbb{C}}$$

compatible with dynamics. By assumption, $f_\lambda^b(c_j(\lambda_0))$ belongs to $C_i(\lambda_0)$. If this is true on an open subset V of U, then $g(\lambda) = \phi_\lambda^{-1}(f_\lambda^b(c_j(\lambda)))$ is holomorphic on V, and hence is a constant. Thus (5.1) holds on V, and hence on U. If not, $c_i(\lambda)$ and $c_j(\lambda)$ have disjoint forward orbits for all λ outside a proper real analytic subset, which implies the assertion. □

Now as variants of Conjectures I and II, we recall the following.

HD conjecture Hyperbolic rational functions are open and dense in rational functions.

Conjecture of Thurston Geometrically finite Kleinian groups are open and dense in finitely generated Kleinian groups.

Finally, Sullivan also showed the following rigidity theorem for Kleinian groups.

Theorem 5.3.10 *There are no invariant Beltrami coefficients on the limit sets of any finitely generated Kleinian groups.*

For the proof, see for instance §5.2 of Matsuzaki and Taniguchi (1998). This result can be restated as that there are no invariant line fields on the limit set for any finitely generated Kleinian group.

NILF conjecture A rational function either is doubly covered by an integral torus endomorphism or admits no invariant line fields on the Julia set.

NILF conjecture	Rigidity theorem 5.3.10

Here the following fact is remarkable.

Theorem 5.3.11 *If the NILF conjecture is true, then so is the HD conjecture.*

Proof Let X be the space of rational functions of the fixed degree k (> 1). Then the subset X^{QC} of quasiconformally stable functions is open and dense. Let $f \in X^{QC}$; then we know that

$$\dim_{\mathbb{C}} \mathrm{Teich}(\widehat{\mathbb{C}}, f) = \dim_{\mathbb{C}} X - \dim_{\mathbb{C}} \mathrm{Aut}(\widehat{\mathbb{C}}) = 2k - 2.$$

Since $X^{QC} \subset X^J$, f has no indifferent cycles, for otherwise every element of X^{QC} would have one. Similarly f has no periodic critical points, and a theorem in Mañé (1985) implies that f has no Herman rings. Hence every periodic Fatou component is attracting. Also f is not doubly covered by an integral torus endomorphism, for they form a proper subvariety.

On the other hand, Theorem 2.4.15 gives that

$$\dim_{\mathbb{C}} \mathrm{Teich}(\widehat{\mathbb{C}}, f) = \dim_{\mathbb{C}} M(J(f), f) + N_{AC}.$$

Thus, if the NILF conjecture is true, then $N_{AC} = 2k - 2$, which implies that f is hyperbolic. □

Remark In the case of quadratic polynomials, the NILF conjecture is equivalent to the HD conjecture, which is true if so is the MCL conjecture stated in Chapter 1 (cf. Lyubich 1997).

5.4 Conformal density and Hausdorff dimension

In this section, we construct two typical probability measures on Julia sets. Further, using these measures, we show some results on Hausdorff dimensions of Julia sets.

First, we extend the Brolin measure stated in §1.4.3. Let f be a rational function with degree k and fix $\zeta \in J(f)$. For each natural number n, we define

$$\mu_n^{\zeta} : \text{each of } k^n \text{ points including multiplicity in } (f^n)^{-1} \text{ has}$$
$$\text{a point mass with weight } \frac{1}{k^n}.$$

Then we have the following theorem.

Theorem 5.4.1 *The sequence of probability measures $\{\mu_n^{\zeta}\}_{n=1}^{\infty}$ converges weakly to the unique probability measure μ independent of the choice of ζ. The support of it is $J(f)$ and it has positive mass at no points. Moreover, $\mu(E) = \mu(f^{-1}(E))$ for every Borel set with $\mu(\partial E) = 0$.*

We call this measure the *Lyubich measure* of f.

Proof We set

$$C(J) = \{\text{real valued continuous functions defined on } J(f)\}$$

with the maximum norm $\|\varphi\| = \max_{z \in J(f)} |\varphi(z)|$. Then $C(J)$ is the Banach space. Let $\zeta_1, \zeta_2, \ldots, \zeta_k$ be the solutions of $f(z) = \zeta$ including multiplicity, and set

$$(T\varphi)(\zeta) = \int_{J(f)} \varphi(z) d\mu_1^\zeta(z) = \frac{1}{k} \sum_{j=1}^{k} \varphi(\zeta_j).$$

Then T is a bounded linear operator. Further, $\|T^n \varphi\| \leq \|\varphi\|$ and we can show that one of the following two occurs.

(i) $\|T^n \varphi\| < \|\varphi\|$ for some n, or
(ii) $\|T^n \varphi\| = \|\varphi\|$ for all n and φ is a constant.

For each $\varphi \in C(J)$, since $\{T^n \varphi \mid n \in \mathbb{N}\}$ is equicontinuous, we can choose a subsequence $\{T^{n(l)} \varphi\}_{l=1}^{\infty}$ converging uniformly to the limit function, say ϕ. If necessary, we again choose a subsequence, so we may assume that $m(l) = n(l+1) - n(l) \to \infty$ as $l \to \infty$ and that $T^{m(l)} \phi$ converges. Setting $\omega = \lim_{l \to \infty} T^{m(l)} \phi$, we have $\phi = \omega$, that is, $T^{m(l)} \phi \to \phi$, because

$$\|T^{n(l+1)} \varphi - T^{m(l)} \phi\| = \|T^{m(l)}(T^{n(l)} \varphi - \phi)\| \leq \|T^{n(l)} \varphi - \phi\|.$$

Since

$$\|\phi\| = \lim_{l \to \infty} \|T^{m(l)-n}(T^n \phi)\| \leq \|T^n \phi\| \leq \|\phi\|$$

for all natural numbers n, ϕ is a constant function, which we denote by a_φ. Since

$$\|T^p \varphi - a_\varphi\| = \|T^{p-n(l)}(T^{n(l)}(\varphi - a_\varphi))\| \leq \|T^{n(l)}(\varphi - a_\varphi)\|$$

for $p > n(l)$, $T^m \varphi \to a_\varphi$. For each $\tilde{\varphi} \in C(J)$, we denote its constant limit by $a_{\tilde{\varphi}}$ as above. Then we have

$$|a_\varphi - a_{\tilde{\varphi}}| = \lim_{n \to \infty} \|T^n(\varphi - \tilde{\varphi})\| \leq \|\varphi - \tilde{\varphi}\|.$$

Hence the mapping which maps φ to a_φ is a non-negative bounded linear functional on $C(J)$. Thus we have

$$a_\varphi = \int_{J(f)} \varphi(z) d\mu(z)$$

by the Riesz representation theorem. Since the norm of this functional is 1, μ is a probability measure. We can also see that μ_n^ζ converges weakly to μ independently of the choice of ζ.

By the definition, it is clear that $\operatorname{supp}\mu \subset J(f)$, where $\operatorname{supp}\mu$ is the support of μ. Let ζ be in $J(f)$ and U its neighborhood. There is an m such that $f^m(U \cap J(f)) \supset J(f)$. Hence $f^{n+m}(z) = f^n(f^m(z)) = \zeta$ has at least k^n solutions in U. This means that $\mu_{n+m}^\zeta(U) \geq 1/k^m$. Thus we have $\mu(U) \geq 1/k^m$ and hence $\zeta \in \operatorname{supp}\mu$.

Since $J(f)$ does not contains any superattracting cycles, $\zeta \in J(f)$ is a critical point of f^m for any m with multiplicity at most $2^{2(k-1)} - 1$. Hence f^m takes each value at most $2^{2(k-1)}$ times in some neighborhood U of ζ and $f^{m+n}(z) = f^n(f^m(z)) = \zeta$ has at most $2^{2(k-1)}k^n$ solutions in U. This shows $\mu_{n+m}^\zeta(U) \leq 2^{2(k-1)}/k^m$ and we have $\mu(\{\zeta\}) = 0$

Since $\mu_{n+1}^\zeta(f^{-1}(E)) = \mu_n^\zeta(E)$ and μ_n^ζ converges to μ weakly, it is clear that $\mu(f^{-1}(E)) = \mu(E)$. □

Theorem 5.4.2 *The Lyubich measure is strongly mixing, i.e.,*

$$\lim_{n\to\infty} \int_{J(f)} \varphi(f^n(\eta))\phi(\eta)d\mu(\eta) = \int_{J(f)} \varphi(\eta)d\mu(\eta) \int_{J(f)} \phi(\eta)d\mu(\eta)$$

for all $\varphi, \phi \in L^2(\mu)$.

Proof We may assume that $\infty \in F(f)$. Cover $J(f)$ by a finite number of squares $Q_1, Q_2, \ldots, Q_l$ satisfying $\mu(\partial Q_i) = 0$ for each i. Since μ_n^ζ converges to μ weakly for every $\zeta \in J(f)$, for any $\epsilon > 0$, there is N such that

$$|\mu_n^\zeta(Q_j) - \mu(Q_j)| < \epsilon$$

for all $n \geq N$ and for $j = 1, 2, \ldots, l$. Hence, if φ and ϕ are step functions, we have the formula we need. For general measurable functions in $L^2(\mu)$, we obtain the claim by approximating them by step functions. □

We now estimate the Hausdorff dimensions of Julia sets by using the Lyubich measures.

Theorem 5.4.3 (Garber 1978, Beardon 1991) *Let f be a rational function with degree k and assume $F(f)$ contains ∞. Set $M = \max_{z\in J(f)} |f'(z)|$. Then*

$$\dim J(f) \geq \frac{\log k}{\log M}$$

holds. In particular, the Hausdorff dimensions of the Julia sets of rational functions are positive.

Proof By virtue of Lemma 3.2.6, it is enough to show that, for any fixed t with $0 < t < (\log k)/(\log M)$, there are positive constants c and r_0 such that $\mu(D) \le cr^t$ for all $D = D(z, r)$ with $z \in J(f)$ and $r < r_0$, where μ is the Lyubich measure in Theorem 5.4.1.

Choose a with $t < a < (\log k)/(\log M)$ and take an ϵ-neighborhood $J(f)_\epsilon$ of $J(f)$ so that $|f'(z)| < k^{1/a}$ for all $z \in J(f)_\epsilon$. Denoting the set of all the critical points of f by $C = C(f)$ and setting $C' = C \cap J(f)$, we may assume $J(f)_\epsilon \cap C = C'$. Fix a natural number p; then there is η with $0 < \eta < \epsilon$ such that $|w - f^l(w)| > \eta$ for all l with $0 < l < p$ and for all $w \in C'$ because each $w \in C'$ is not periodic. Choose ρ with $\rho < \eta$ satisfying $|(f^l)'(z)| < 1$ for every $z \in D(w, \rho)$, l with $0 < l < p$, and $w \in C'$. Denoting the ρ-neighborhood of C' by C'_ρ, we take a positive constant δ with $\delta < \rho/2$ so that f is univalent on $D(z, \delta)$ for all $z \in J(f) - C'_\rho$.

Fix a natural number p satisfying $a(1 - k/p) > t$. Choose $\zeta \in J(f)$ and set $D = D(\zeta, r)$. If we choose n so that $\delta(k^{1/a})^{-n-1} < r \le \delta(k^{1/a})^{-n}$, then $\Delta_j = f^j(D)$ is contained in $D(f^j(\zeta), \delta(k^{1/a})^{j-n})$. Since f^n takes each value at most $k^{kn/p}$ times in D,

$$\mu^\zeta_{n+m}(D) \le k^{kn/p} k^{-n} \mu^\zeta_m(\Delta_n)$$

for each m. Hence we have

$$\mu(D) \le k^{-n(1-\frac{k}{p})} \mu(\Delta_n)$$

by $m \to \infty$, and thus

$$\mu(D) \le (k^{\frac{1}{a}})^{-nt} < \left(\frac{k^{1/a}}{\delta}\right)^t r^t,$$

which shows our claim. □

Since every non-elementary Kleinian group contains a Schottky subgroup, whose limit set has a positive dimension, we have the following.

$f \in \mathrm{Rat} \Rightarrow$ $\dim J(f) > 0$ (Theorem 5.4.3)	Γ non-elementary $\Rightarrow$ $\dim \Lambda(\Gamma) > 0$

Next we construct another kind of probability measures on the Julia sets by a method similar to one used to construct probability measures, so-called Patterson–Sullivan measures, on limit sets of Kleinian groups.

As was stated in §5.1, Kleinian groups are regarded as groups acting properly discontinuously on the upper half space $\mathbb{H}^3$. Since $\mathbb{H}^3$ is conformally equivalent to the three-dimensional ball $\mathbb{B}^3$, hereafter we regard Kleinian groups as discontinuous groups consisting of Möbius transformations which leave $\mathbb{B}^3$ invariant. In this case, the limit set $\Lambda(\Gamma)$ of a Kleinian group Γ is a subset of $S^2 = \partial\mathbb{B}^3$.

Definition A positive finite measure μ on $\Lambda(\Gamma)$ is a δ-*conformal measure* for Γ if

$$\mu(\gamma(E)) = \int_E |\gamma'(\zeta)|^\delta d\mu(\zeta)$$

holds for all Borel sets E in S^2 and all $\gamma \in \Gamma$.

Now we give a rough sketch for construction of conformal densities. Denote the hyperbolic metric in $\mathbb{B}^3$ by λ. For an s-dimensional Poincaré series

$$S(s, x) = \sum_{\gamma\in\Gamma} e^{-s\lambda(x,\gamma(x))} \qquad (x \in \mathbb{B}^3),$$

we set

$$\delta = \delta(\Gamma) = \inf\{s \mid S(s, x) < \infty\},$$

which is called the *critical exponent* of the Poincaré series for Γ. The critical exponent δ is independent of the choice of $x \in \mathbb{B}^3$. If $s > \delta$ then $S(s, x)$ converges and if $s < \delta$ then $S(s, x)$ diverges for every x in $\mathbb{B}^3$. Whether $S(\delta, x)$ converges or not depends on Γ.

Though it is easy to see that $\delta \leq 2$, Beardon also showed the following.

Theorem 5.4.4 (Beardon 1968) *The critical exponent of the Poincaré series for a non-elementary Kleinian group is positive.*

We define a probability measure μ_s on $\widehat{\mathbb{C}}$ by

$$\mu_s = \frac{1}{S(s, x)} \sum_{\gamma\in\Gamma} e^{-s\lambda(x,\gamma(x))}\delta_{\gamma(x)},$$

where $\delta_{\gamma(x)}$ is the Dirac measure on $\gamma(x)$. If $S(\delta, x) = \infty$, then we can see that the weak limit μ of μ_s as $s \searrow \delta$ is a probability measure whose support is in $\Lambda(\Gamma)$, because the derived set of $\{\Gamma(x)\}$ is $\Lambda(\Gamma)$. If $S(\delta, x) < \infty$, then we use the continuous non-decreasing Patterson auxiliary function $h : \mathbb{R}^+ \to \mathbb{R}^+$ satisfying the following conditions:

(i) $\sum_{\gamma\in\Gamma} e^{-s\lambda(x,\gamma(x))}h(e^{\lambda(x,\gamma(x))})$ converges for $s > \delta$ and diverges for $s \leq \delta$;

(ii) for fixed $\epsilon > 0$, there is r_0 such that $h(rt) \leq t^{\epsilon}h(r)$ for all $r > r_0$ and all $t > 0$.

Since we have $h(e^{r+\epsilon})/h(e^r) \to 1$ as $r \to \infty$ from (ii), using $S^*(s,x) = \sum_{\gamma\in\Gamma} e^{-s\lambda(x,\gamma(x))}h(e^{\lambda(x,\gamma(x))})$ instead of $S(s,x)$ and defining

$$\frac{1}{S^*(s,x)}\sum_{\gamma\in\Gamma} e^{-s\lambda(x,\gamma(x))}h(e^{\lambda(x,\gamma(x))})\delta_{\gamma(x)},$$

we can show that the weak limit function μ of them is a probability measure whose support is in $\Lambda(\Gamma)$.

This measure constructed above is called a *Patterson–Sullivan measure* for Γ.

Theorem 5.4.5 *Let δ be the critical exponent of a Kleinian group Γ. Then the Patterson–Sullivan measure for Γ is a δ-conformal density.*

Further, Sullivan showed the following theorems on the connections between δ-conformal densities and Hausdorff measures.

Theorem 5.4.6 *Let δ be the critical exponent of a geometrically finite Kleinian group Γ and μ the Patterson–Sullivan measure for Γ. Then:*

(i) *There is a constant a such that*

$$H_\delta \leq a\mu, \quad \textit{and hence} \quad \dim(\Lambda(\Gamma)) \leq \delta.$$

(ii) *If Γ does not contains any parabolic elements, there is a constant a such that*

$$H_\delta = a\mu, \quad \textit{and hence} \quad \dim(\Lambda(\Gamma)) = \delta.$$

Combining Ahlfors' 0–1 theorem and Theorem 5.4.6, we obtain the following result.

Corollary *Let Γ be a geometrically finite Kleinian group without parabolic elements and suppose that $\Omega(\Gamma)$ is not empty. Then*

$$\dim(\Lambda(\Gamma)) < 2.$$

Furthermore, Bishop and Jones (1997) have shown the following.

Theorem 5.4.7 (i) *Let δ be the critical exponent of a non-elementary finitely generated Kleinian group Γ and assume that the area of $\Lambda(\Gamma)$ is equal to* 0. *Then*

$$\dim(\Lambda(\Gamma)) = \delta.$$

(ii) *The limit set $\Lambda(\Gamma)$ either is totally disconnected or a circle, or has Hausdorff dimension greater than* 1.

(iii) *Γ is geometrically finite and has non-empty $\Omega(\Gamma)$, if and only if* $\dim \Lambda(\Gamma) < 2$.

Now we apply the idea of conformal densities and the Patterson–Sullivan measures for Kleinian groups to the case of iteration of rational functions and show some of Sullivan's result.

Definition A positive finite measure μ on the Julia set $J(f)$ of a rational function f is a *δ-conformal density* if

$$\mu(f(A)) = \int_A |f'(\zeta)|^\delta d\mu(\zeta)$$

holds for all Borel sets $A \subset J(f)$ on which f is univalent.

Theorem 5.4.8 *Let f be a rational function and assume that $F(f)$ is not empty. Then there is a δ-conformal measure for f for some positive δ. Let $\delta(f)$ be the infimum of δ for which a δ-conformal density exists. Then* $0 < \delta(f) \leq 2$.

Proof We may assume $F(f) \ni \infty$. Choose $\zeta \in F(f)$ which belongs to neither Siegel disks, Herman rings, $C^+(f)$ nor $\{f^n(\infty)\}_{n=0}^\infty$ and is not a periodic point. Further, we choose a neighborhood U of ζ satisfying $U \cap C^+(f) = \varnothing$ and $\overline{U} \cap (\bigcup_{n=0}^\infty f^n(\infty)) = \varnothing$. Then each branch of f^{-n} on U for all n is single valued. Denoting generally by S any branch of any f^{-n} on U, we have

$$\sum \int_U |S'(z)|^2 dm < \infty,$$

where summation is over all single valued branches of f^{-n} with $n = 1, 2, \ldots$ and m denotes the two-dimensional Lebesgue measure.

Set $I(x) = \bigcup_{n=0}^\infty f^{-n}(x)$ for $x \in U$ and $d(y) = |(f^n)'(y)|^{-1}$ for $y \in I(x)$ contained in $f^{-n}(x)$. From the inequality above and the Lebesgue monotone convergence theorem, we have

$$\sum_{y \in I(x)} d(y)^2 < \infty$$

for almost all $x \in U$. Fix $x \in U$ satisfying this inequality and set

$$\delta = \inf \left\{ s \,\middle|\, \sum_{y \in I(x)} d(y)^s < \infty \right\}.$$

If the degree of f is k, then $f^{-n}(x)$ consists of k^n points. By the assumption $\infty \in F(f)$, $|f'(z)|$ is bounded on some neighborhood of $J(f)$. Hence we have $\delta > 0$.

First, we assume $\sum_{y \in I(x)} d(y)^\delta = \infty$. For $s > \delta$, the measure μ_s has a point mass on each $y \in I(x)$ with weight $d(y)^s / \sum_{w \in I(x)} d(w)^s$. Note that $\mu_s(\widehat{\mathbb{C}}) = 1$. Take a weak limit μ of $\{\mu_s\}$ as $s \searrow \delta$. Then the support of μ is in $J(f)$, because the series diverges at δ. Take $\zeta \in J(f)$ which is not a critical point and choose a neighborhood U of ζ on which f is univalent. Then f is a bijection from $I(x) \cap U$ onto $I(x) \cap f(U)$. Set $|f'(\zeta)| = \lambda$ and, with a given $\epsilon > 0$, replace U so that

$$\lambda(1 - \epsilon) < |f'(z)| < \lambda(1 + \epsilon)$$

for all $z \in U$. It follows that

$$\lambda^s \mu_s(U)(1 - \epsilon) \leq \mu_s(f(U)) \leq \lambda^s \mu_s(U)(1 + \epsilon).$$

We see that μ is a δ-conformal density by letting $s \searrow \delta$ and $\epsilon \searrow 0$.

Next, in the case of $\sum_{y \in I(x)} d(y)^\delta < \infty$, using the Patterson auxiliary function h introduced in the case of Kleinian groups and considering the series $\sum_{y \in I(x)} h(d(y)^{-1}) d(y)^s$, we obtain a δ-conformal density as a weak limit of them.

Since the conformal densities of f are closed in the space of probability measures with respect to the weak topology, we can find the minimum $\delta(f)$ of δ. Suppose that $\delta(f) = 0$ and let μ be a 0-conformal density. If μ has a point mass on ζ, then

$$\mu(\{f(\zeta)\}) = \mu(\{\zeta\}).$$

The backward orbit of ζ is an infinite set, which contradicts that μ is a probability measure. Defining $f^*\mu$ by $f^*\mu(A) = \mu(f^{-1}(A))$, we have

$$f^*\mu(J(f)) = \mu(J(f)) = 1.$$

On the other hand, we should have $f^*\mu = k\mu$ where a single valued branch of f^{-1} is defined.

Thus we have obtained $\delta(f) > 0$. □

Theorem 5.4.9 *Let f be a hyperbolic rational function. If there is a δ-conformal density, then $\delta = \delta(f)$. Furthermore,*

$$\dim J(f) = \delta(f), \qquad H_{\delta(f)}(J(f)) > 0.$$

Corollary *For the Julia set $J(f)$ of a hyperbolic rational function f,*

$$0 < \dim(J(f)) < 2.$$

Proof If $\delta(f) = 2$, then $H_2(J(f)) > 0$, which contradicts Theorem 4.4.3. □

f: hyperbolic rational $\Rightarrow \dim J(f) = \delta(f) < 2$ (Corollary to Theorem 5.4.9)	Γ: geometrically finite without parabolic elements and $\Omega(\Gamma) \neq \varnothing$ $\Rightarrow \dim \Lambda(\Gamma) = \delta(\Gamma) < 2$ (Corollary to Theorem 5.4.6)

Proof of Theorem 5.4.9 It is enough to show that if there is a δ-conformal measure then $0 < H_\delta(J(f)) < \infty$. Assume that $\infty \in F(f)$ and choose a positive constant λ such that $|f'(z)| \geq \lambda$ on $J(f)$. Since f is hyperbolic, for each $\zeta \in J(f)$, there is $r > 0$ such that each branch of f^{-n} $(n = 1, 2, \ldots)$ on $D(\zeta, r)$ is single valued. We denote a single valued branch of f^{-1} on $D(\zeta, r)$ by S. Koebe's distortion theorem shows that there is a constant c_0 so that

$$\sup\left\{|S'(z)| \,\middle|\, z \in D(\zeta, r/2)\right\} \leq c_0 \inf\{|S'(z)| \,|\, z \in D(\zeta, r/2)\}.$$

Fix $z_0 \in J(f)$ and denote the single valued branch of f^{-n} which maps $z_n = f^n(z_0)$ to z_0 by S_n. Setting $l = r/c_0$ and

$$r_n = |(S_n)'(z_n)|l,$$

we have $r_n \to 0$ as $n \to \infty$ and

$$S_n(D(z_n, c_0^{-1}l)) \subset D(z_0, r_n) \subset S_n(D(z_n, c_0 l)).$$

Since μ is a conformal density, for a fixed positive constant a, there is a constant $M(a)$ such that

$$\mu(D(\zeta, a)) > M(a)$$

for all $\zeta \in J(f)$. From this estimate, Koebe's distortion theorem and the fact that μ is a δ-conformal density, we obtain

$$\begin{aligned}\mu(D(z_0, r_n)) &\geq c_0^{-\delta}|(S_n)'(z_n)|^\delta \mu(D(z_n, c_0^{-1}l)) \\ &\geq c_0^{-\delta}l^{-\delta}r_n^\delta M(c_0^{-1}l)\end{aligned}$$

and

$$\begin{aligned}\mu(D(z_0, r_n)) &\leq c_0^\delta|(S_n)'(z_n)|^\delta \mu(D(z_n, c_0 l)) \\ &\leq c_0^\delta l^{-\delta} r_n^\delta.\end{aligned}$$

Thus there is a constant $c_1 > 1$ such that

$$c_1^{-1} \leq \frac{\mu(D(z_0, r_n))}{r_n^\delta} \leq c_1.$$

Set $c_2 = \max_{z \in J(f)} |f'(z)|/\lambda$, and we can find an r_n such that

$$c_2^{-1} r_n \leq t \leq c_2 r_n$$

for all $t < r_1$. Since

$$(c_1 c_2^\delta)t^\delta \geq c_1 r_n^\delta \geq \mu(D(z_0, r_n)) \geq c_1^{-1} r_n^\delta \geq (c_1 c_2^\delta)^{-1} t^\delta$$

for all $t < r_1$, we have

$$c^{-1}t^\delta \leq \mu(D(z_0, t)) \leq ct^\delta$$

for all t with $0 < t < r_1$, where $c = c_1 c_2^{2\delta}$. By Lemma 3.2.6, we obtain $\dim(J(f)) \geq \delta$ and $H_\delta(J(f)) > 0$.

Next we define a covering $\{B_i\}$ of $J(f)$ as follows. Each B_i is an open disk whose center is in $J(f)$ and out of $B_1 \cup B_2 \cup \cdots \cup B_{i-1}$, and whose radius l_i satisfies $\epsilon > l_1$ and $l_i > l_{i+1}$ for a sufficiently small given ϵ. If we take an open disk B_i' whose center is the same as B_i's and whose radius is $l_i/2$ for all i, then the B_i $(i = 1, 2, \ldots)$ are pairwise disjoint. Hence we have

$$\sum_{i=1}^\infty r_i^\delta = 2^\delta \sum_{i=1}^\infty \left(\frac{1}{2}r_i\right)^\delta \leq 2^\delta c\mu\left(\bigcup_{i=1}^\infty B_i'\right) \leq 2^\delta c\mu(J(f)),$$

which implies that $H_\delta(J(f)) \leq 2^\delta c\mu(J(f))$. Thus we have $\dim(J(f)) \leq \delta$.
□

Remark As was stated in §1.4.3, the Brolin measure for a polynomial P is a harmonic measure. Hence its dimension is always equal to 1. Thus, in the case of a hyperbolic polynomial P, if the Hausdorff dimension of $J(P)$ is not equal to 1, then the conformal density for P is different from the Brolin measure.

Recently, McMullen has proved the following counterpart of the theorem of Bishop and Jones.

Theorem 5.4.10 *Let* $f \in \mathrm{Rat}$ *be geometrically finite; then*

$$\dim J(f) = \delta(f).$$

Moreover, if $J(f) \neq \widehat{\mathbb{C}}$, *then* $\dim J(f) < 2$.

f geometrically finite	Γ geometrically finite
$J(f) \neq \widehat{\mathbb{C}}$	$\Lambda(\Gamma) \neq \widehat{\mathbb{C}}$
$\Rightarrow \dim J(f) = \delta(f) < 2$	$\Rightarrow \dim \Lambda(\Gamma) = \delta < 2$
(Theorem 5.4.10)	(Theorem 5.4.7)

Note that there are various rational functions f with $\dim J(f) < 2$ and not geometrically finite. For the details, see McMullen (preprint (Hausdorff dimension and conformal dynamics II)).

6

Dynamics of holomorphic maps of several variables

In Chapters 6–9, we will treat holomorphic dynamics of several complex variables, in particular, dynamics of polynomial automorphisms of $\mathbb{C}^2$. For other directions of the study on (holomorphic and meromorphic) dynamics of several complex variables, see for instance Fornaess (1966), Fornaess and Sibony (1994, 1995) and Ueda (1993, 1994). In this chapter, we will mainly investigate the behavior of holomorphic maps of two variables in a neighborhood of a fixed point.

In §6.1, we shall summarize the fundamental results concerning holomorphic functions, maps, and 1-forms of several complex variables, used in Chapters 6 and 7. We will need only fundamental properties, which will be regarded as a natural generalization of the calculus of several real variables and the theory of complex functions of one variable.

In §6.2 we shall give the normal forms for holomorphic maps in a neighborhood of an attracting fixed point for holomorphic maps of two complex variables. This can be considered as the two-dimensional version of the solution to the Schröder equation of the one variable case. In §6.3, we apply the result of the previous section to holomorphic automorphisms of $\mathbb{C}^2$ and construct Fatou–Bieberbach domains, i.e., proper subdomains of $\mathbb{C}^2$ that are biholomorphically equivalent to $\mathbb{C}^2$. In §6.4, we treat fixed points of saddle type and their stable and unstable manifolds.

We will mainly discuss the case of two variables for the sake of simplicity. Many of the results can be generalized to the case of three or more variables.

6.1 Preparation from complex analysis of several variables

The set consisting of all N-tuples $z = (z_1, \ldots, z_N)$ of complex numbers is said to be the complex number space of dimension N and denoted by $\mathbb{C}^N$. If we write $z_k = x_k + iy_k \ \ (k = 1, \ldots, N)$, the points $(z_1, \ldots, z_N) \in \mathbb{C}^N$ and

$(x_1, \ldots, x_N, y_1, \ldots, y_N) \in \mathbb{R}^{2N}$ correspond bijectively. With this correspondence, $\mathbb{C}^N$ and $\mathbb{R}^{2N}$ can be identified and the topology of $\mathbb{C}^N$ can be defined. Further any (real or complex valued) function defined on a subset of $\mathbb{C}^N$ can be regarded as a function of $2N$ real variables.

We may define holomorphic functions of several variables in various ways (which will turn out to be equivalent). Here we choose the following definition.

Definition A complex valued function f defined in an open set $U \subset \mathbb{C}^N$ is said to be holomorphic if the following two conditions are satisfied:

(i) f is continuous in U;

(ii) for each k $(1 \le k \le N)$, the function $f(z_1, \ldots, z_k, \ldots, z_N)$ is holomorphic with respect to the single variable z_k.

We denote by $\frac{\partial f}{\partial z_k}$ the derivative of f as holomorphic function of the variable z_k with other variables fixed. It is clear from the definition that a polynomial in $z_1, \ldots, z_N$ is holomorphic on all of $\mathbb{C}^N$. Many of the results in the one variable theory can be immediately generalized to the several variable case. We mention two such theorems.

Theorem 6.1.1 *Let $\{f_n\}_{n=0}^{\infty}$ be a sequence of holomorphic functions that is uniformly convergent on compact sets in an open set U. Then the limit $\lim_{n\to\infty} f_n$ is also holomorphic in U.*

Theorem 6.1.2 (Theorem of coincidence) *Let f and g be holomorphic functions on a domain (connected open set) U. If they coincide on a non-empty open subset of U, then they coincide on the whole of U.*

Cauchy's integral formula is also generalized to the several variable case. Let $a = (a_1, \ldots, a_n)$ be a point in $\mathbb{C}^N$ and $r_1, \ldots, r_N$ be positive numbers. A domain of the form

$$\Delta = \Delta(a; r_1, \ldots, r_N) = \left\{ z \in \mathbb{C}^N \;\middle|\; |z_k - a_k| < r_k \ (1 \le k \le N) \right\}$$

is said to be a polydisk with center a.

Theorem 6.1.3 (Cauchy's integral formula) *Let f be a holomorphic function on an open set $U \subset \mathbb{C}^N$ and let $\Delta = \Delta(a; r_1, \ldots, r_N)$ be a polydisk such that the closure $\bar{\Delta}$ is contained in U. Then $f(z)$, $z \in \Delta$, is represented by the N-tuple integral*

$$f(z) = \frac{1}{(2\pi i)^N} \int \cdots \int_{\Gamma} \frac{f(\zeta_1, \ldots, \zeta_N)}{(\zeta_1 - z_1) \cdots (\zeta_N - z_N)} d\zeta_1 \cdots d\zeta_N \qquad (6.1)$$

where $\Gamma = \{(\zeta_1, \ldots, \zeta_N) \mid |\zeta_k - a_k| = r_k \quad (k = 1, \ldots, N)\}$.

Further f *is infinitely many times differentiable and*

$$\frac{\partial^{\nu_1+\cdots+\nu_N} f}{\partial z_1^{\nu_1} \cdots \partial z_N^{\nu_N}}(z) = \frac{\nu_1! \cdots \nu_N!}{(2\pi i)^N} \int \cdots \int_\Gamma \frac{f(\zeta_1, \ldots, \zeta_N)}{(\zeta_1 - z_1)^{\nu_1+1} \cdots (\zeta_N - z_N)^{\nu_N+1}} d\zeta_1 \cdots d\zeta_N \tag{6.2}$$

Proof We note that the integral on the right means the N-tuple integral on the set $\{(\theta_1, \ldots, \theta_N) \mid 0 \le \theta_k \le 2\pi\} \subset \mathbb{R}^N$, with

$$\zeta_k = a_k + r_k e^{i\theta_k}, \quad d\zeta_k = i r_k e^{i\theta_k} d\theta_k.$$

This N-tuple integral is deformed into an iterated integral. Hence the repeated use of the Cauchy integral formula for one variable yields (6.1). This integral can be differentiated with respect to z_k so that we obtain (6.2). □

As in the case of a single variable, the essential property of holomorphic functions lies in the fact that they are expressed by power series. Let $a = (a_1, \ldots, a_N) \in \mathbb{C}^N$. An expression

$$\sum_{\nu_1, \ldots, \nu_N = 0}^{\infty} c_{\nu_1 \cdots \nu_N} (z_1 - a_1)^{\nu_1} \cdots (z_N - a_N)^{\nu_N} \tag{6.3}$$

is said to be a power series with center a. It is said to be a convergent power series if it is absolutely uniformly convergent on a neighborhood of a.

Theorem 6.1.4 *Under the condition of the previous theorem,* $f(z)$, $z \in \Delta$, *is represented as the sum of the convergent power series of the form (6.3). Here*

$$c_{\nu_1 \cdots \nu_N} = \frac{1}{\nu_1! \cdots \nu_N!} \frac{\partial^{\nu_1+\cdots+\nu_N} f}{\partial z_1^{\nu_1} \cdots \partial z_N^{\nu_N}}(a). \tag{6.4}$$

Proof We suppose $a = (a_1, \ldots, a_N) = (0, \ldots, 0)$ for simplicity. In the integral formula (6.1), we have $|z_k| < r_k = |\zeta_k|$ for each k. Hence

$$\frac{1}{\zeta_k - z_k} = \frac{1}{\zeta_k} \frac{1}{1 - z_k/\zeta_k} = \sum_{\nu=0}^{\infty} \frac{z_k^\nu}{\zeta_k^{\nu+1}}.$$

Therefore

$$\frac{1}{(\zeta_1 - z_1) \cdots (\zeta_N - z_N)} = \sum_{\nu_1, \ldots, \nu_N = 0}^{\infty} \frac{z_1^{\nu_1} \cdots z_N^{\nu_N}}{\zeta_1^{\nu_1+1} \cdots \zeta_N^{\nu_N+1}}.$$

We substitute this into (6.1) and change the order of integration and summation. Then we obtain (6.3). □

Now, if we consider a complex valued function as a pair of real valued functions of real variables, we can apply the results of the calculus of several real variables. Let $f = u+iv$ be a complex valued function of N complex variables that is of C^1 class with respect to the $2N$ real variables. The function f is holomorphic with respect to z_k if and only if it satisfies the Cauchy–Riemann equations

$$\frac{\partial u}{\partial x_k} = \frac{\partial v}{\partial y_k}, \qquad \frac{\partial v}{\partial x_k} = -\frac{\partial u}{\partial y_k} \quad (k = 1, \dots, N). \tag{6.5}$$

By combining the real and imaginary parts, this can be expressed as

$$\frac{\partial f}{\partial x_k} = -i\frac{\partial f}{\partial y_k} \quad (k = 1, \dots, N). \tag{6.6}$$

The partial derivatives with respect to z_k of f as holomorphic function are given by

$$\frac{\partial f}{\partial z_k} = \frac{\partial f}{\partial x_k} = \frac{\partial u}{\partial x_k} + i\frac{\partial v}{\partial x_k}.$$

From these observations we conclude as follows.

Theorem 6.1.5 *Let f be a complex valued function defined on an open set U. The following three conditions are equivalent to one another:*

(A) *f is holomorphic;*
(B) *for every point $a \in U$, there is a neighborhood U of a such that f can be represented as the sum of a convergent power series in U;*
(C) *f is of C^1 class and satisfies the Cauchy–Riemann equation.*

We could have adopted the condition (B) or (C) as the definition of holomorphic function. In any case the equivalence of these conditions is an important fact.

Definition A map

$$F : U \ni (z_1, \dots, z_N) \mapsto (w_1, \dots, w_M) = (f_1(z), \dots, f_M(z)) \in \mathbb{C}^M$$

defined by an M-tuple of holomorphic functions $f_1, \dots, f_M$ on an open set $U \subset \mathbb{C}^N$ is said to be a *holomorphic map*.

The (M, N) matrix $F'(z) = \left(\partial w_j/\partial z_k\right)$ is said to be the (complex) *Jacobian matrix* of the holomorphic map F. In general, a map (not necessarily holomorphic) F from $U \subset \mathbb{C}^N$ to $\mathbb{C}^M$ can be regarded as a map from $U \subset \mathbb{R}^{2N}$ to

$\mathbb{R}^{2M}$. Suppose that this map is of class C^1. Then the Jacobian matrix is a $(2M, 2N)$ matrix

$$\begin{pmatrix} \left(\frac{\partial u_j}{\partial x_k}\right) & \left(\frac{\partial u_j}{\partial y_k}\right) \\ \left(\frac{\partial v_j}{\partial x_k}\right) & \left(\frac{\partial v_j}{\partial y_k}\right) \end{pmatrix}. \tag{6.7}$$

We will call this the real Jacobian matrix, to distinguish it from the (complex) Jacobian matrix of a holomorphic map. A map F of C^1 class is holomorphic if and only if all f_j satisfy the Cauchy–Riemann equations (6.5). This is equivalent to saying that the real Jacobian matrix (6.7) for F is of the form

$$\begin{pmatrix} A & -B \\ B & A \end{pmatrix}$$

with

$$A = \left(\frac{\partial u_j}{\partial x_k}\right) = \left(\frac{\partial v_j}{\partial y_k}\right), \quad B = \left(\frac{\partial v_j}{\partial x_k}\right) = -\left(\frac{\partial u_j}{\partial y_k}\right).$$

The Jacobian matrix of a holomorphic map F is given by

$$F'(z) = \left(\frac{\partial w_j}{\partial z_k}\right) = A + iB.$$

With these observations, we can derive the theorems for composition and the inversion of holomorphic maps.

Theorem 6.1.6 *Let $U \subset \mathbb{C}^N$ and $V \subset \mathbb{C}^M$ be open sets, and $F : U \to V$, $G : V \to \mathbb{C}^L$ be holomorphic maps. Then the composed map $G \circ F : U \to \mathbb{C}^L$ is also a holomorphic map and its complex Jacobian matrix is given by $(G \circ F)'(z) = G'(F(z))F'(z)$. In other words, the chain rule holds for holomorphic maps.*

Proof Let the real Jacobian matrices of F and G be $\begin{pmatrix} A & -B \\ B & A \end{pmatrix}$ and $\begin{pmatrix} C & -D \\ D & C \end{pmatrix}$. Then, by the chain rule, the real Jacobian matrix of the composed map $G \circ F$ is given by

$$\begin{pmatrix} C & -D \\ D & C \end{pmatrix} \begin{pmatrix} A & -B \\ B & A \end{pmatrix} = \begin{pmatrix} CA - DB & -(DA + CB) \\ DA + CB & CA - DB \end{pmatrix}.$$

Hence $G \circ F$ is a holomorphic map, whose complex Jacobian matrix is

$$(G \circ F)' = (CA - DB) + i(DA + CB) = (C + iD)(A + iB) = G'F'.$$

□

Now we let $M = N$ and consider the inverse of a holomorphic map. Let F be a holomorphic map from an open set $U \subset \mathbb{C}^N$ to $\mathbb{C}^N$. The determinant of the complex (or real) Jacobian matrix of F is called the complex (or real) *Jacobian determinant* of F. The complex Jacobian determinant of F is denoted by $J_F(z) = \det F'(z)$, which is a holomorphic function on U.

Theorem 6.1.7 *Let F be an injective holomorphic map from an open set $U \subset \mathbb{C}^N$ into $\mathbb{C}^N$. If the complex Jacobian determinant $J_F(z)$ does not vanish on U, then the image $F(U)$ is an open set, and the inverse $F^{-1} : F(U) \to U$ is a holomorphic map.*

Proof The real and complex Jacobian determinants of F are related as follows:

$$\det\begin{pmatrix} A & -B \\ B & A \end{pmatrix} = \det\begin{pmatrix} A+iB & O \\ O & A-iB \end{pmatrix}$$
$$= |\det(A+iB)|^2 = |\det F'|^2 = |J_F(z)|^2.$$

Since this does not vanish, we know from the inverse map theorem for the real variable case that $F(U)$ is an open set and that F^{-1} is of C^1 class. Further, if we denote the inverse matrix of $A + iB$ by $X + iY$, the inverse matrix of $\begin{pmatrix} A & -B \\ B & A \end{pmatrix}$ is given by $\begin{pmatrix} X & -Y \\ Y & X \end{pmatrix}$, as is easily verified by straightforward calculation. Therefore F^{-1} is a holomorphic map. □

As the matter of fact, we have the following stronger theorem.

Theorem 6.1.8 *If F is an injective holomorphic map from an open set $U \subset \mathbb{C}^N$ into $\mathbb{C}^N$, then the complex Jacobian determinant $J_F(z)$ does not vanish on U (hence the inverse $F^{-1} : F(U) \to U$ is holomorphic).*

For the proof this theorem, we refer to Lemma 6.13 in Imayoshi and Taniguchi, for example.

A bijective holomorphic map from an open set $U \subset \mathbb{C}^N$ onto $V \subset \mathbb{C}^N$ is called *biholomorphic*. Two open sets that correspond by a biholomorphic map are said to be biholomorphically equivalent to each other. In general, a map F from a topological space X to Y is said to be an open map if the image of any open subset of X is an open subset of Y. A holomorphic function of one variable is an open map if it is non-constant. In the several variable case, a non-constant holomorphic map from an open set $U \subset \mathbb{C}^N$ into $\mathbb{C}^N$ is not necessarily an open map.

Example 6.1.1 *Consider a holomorphic map from* $\mathbb{C}^2$ *into* $\mathbb{C}^2$

$$F : (z_1, z_2) \mapsto (w_1, w_2) = (z_1, z_1 z_2).$$

By F, *the* z_2*-axis* $z_1 = 0$ *is mapped to the single point* $(0, 0)$; *the other part* $\{z_1 \neq 0\}$ *is mapped bijectively onto* $\{w_1 \neq 0\}$. *Therefore the image* $F(\mathbb{C}^2)$ *is not an open set.*

We remark that, in the following situation, we have an analogue to the one variable case. Recall that a map is proper if the inverse image of any compact set is compact.

Theorem 6.1.9 *Let* F *be a holomorphic map from an open set* $U \subset \mathbb{C}^N$ *into* $\mathbb{C}^N$ *with* $a \in U, b = F(a)$. *Suppose that* a *is an isolated point in the inverse image* $F^{-1}(b)$ *of* b. *Then there exist neighborhoods* V *of* a *and* W *of* b *such that* $F|V : V \to W$ *is a proper and open map.*

Now we summarize the facts concerning differential 1-forms that are necessary for later purposes. For simplicity, we suppose that the functions are differentiable as many times as we need. We will denote the points in $\mathbb{R}^N$ by $x = (x_1, \ldots, x_N)$. Let $\varphi_1(x), \ldots, \varphi_N(x)$ be complex valued functions in an open set $U \subset \mathbb{R}^N$. An expression

$$\varphi = \sum_{k=1}^{N} \varphi_k dx_k$$

is said to be a (differential) *1-form*. For a function $g(x)$ on U, the 1-form

$$dg = \sum_{k=1}^{N} \frac{\partial g}{\partial x_k} dx_k$$

is said to be the differential of the function g. A 1-form that is a differential of a function is said to be *exact*. A 1-form φ is said to be *closed* if

$$\frac{\partial \varphi_k}{\partial x_j} = \frac{\partial \varphi_j}{\partial x_k} \qquad (j, k = 1, \ldots, N). \tag{6.8}$$

It is clear from the commutativity of partial differentiations, that an exact 1-form is closed.

A (C^1 class) map $\gamma(t) = (\gamma_1(t), \ldots, \gamma_N(t))$ from an interval $[a, b] \subset \mathbb{R}$ into U is called a curve (or a path) in U. It is said to be a closed curve if the initial point $\gamma(a)$ and the end point $\gamma(b)$ coincide. Two curves γ_1, γ_2 with common initial and end points in U are said to be homotopic in U if γ_1 can be continuously deformed into γ_2 in U with the initial and the end points fixed. In particular, a closed curve is said to be *zero-homotopic* in U if it can be continuously deformed in U to a single point. Let γ be a curve in U and φ a 1-form on U. The line integral of the 1-form φ along the curve γ is defined by

$$\int_\gamma \varphi = \int_a^b \left(\sum_{k=1}^N \varphi_k(\gamma(t)) \frac{d\gamma_k}{dt} \right) dt.$$

Suppose that φ is a closed 1-form on U. If curves γ_1 and γ_2 are homotopic in U, then we have $\int_{\gamma_1} \varphi = \int_{\gamma_2} \varphi$. In particular, if U is a simply connected domain, then $\int_\gamma \varphi = 0$ for any closed curve in U. Now suppose that $\int_\gamma \varphi = 0$ for any closed curve γ in U. We fix a point $a \in U$ and define a function $g(x) = \int_a^x \varphi$, where the integral is taken along any curve in U that connects a and x. Then we have $dg = \varphi$. Hence we conclude that, if U is simply connected, then any closed 1-form is exact in U. Thus any closed 1-form is locally an exact 1-form. That is, for each point in U, there exist a (simply connected) neighborhood W and a function g such that $\varphi = dg$ on W. Now suppose that

$$F : V \ni \xi = (\xi_1, \ldots, \xi_M) \mapsto x = (x_1, \ldots, x_N) = (f_1(\xi), \ldots, f_N(\xi)) \in U$$

is a map (of C^1 class) from an open set $V \subset \mathbb{R}^M$ into U. For a function g on U, the pullback F^*g of g by F is defined to be the composed function $g \circ F$. For a 1-form $\varphi = \sum_{k=1}^N \varphi_k dx_k$ on U, the pullback $F^*\varphi$ of φ by F is defined by

$$F^*\varphi = \sum_{k=1}^N \varphi_k(F(\xi)) df_k = \sum_{k=1}^N \sum_{j=1}^M \varphi_k(F(\xi)) \frac{\partial f_k}{\partial \xi_j} d\xi_j.$$

The operations of pullback by F and differentiation commute: $d(F^*g) = F^*dg$. It follows that the pullback of an exact 1-form is also exact. Further, the pullback of a closed 1-form is also closed, since it is locally exact. If γ is a curve in V, then its image $F(\gamma)$ under F is a curve in U. Then we have

$$\int_\gamma F^*\varphi = \int_{F(\gamma)} \varphi$$

for a 1-form φ on U. This follows immediately from the definitions of the pullback and the line integral of 1-forms.

If we identify the complex number space $\mathbb{C}^N$ with coordinates $(z_1, \ldots, z_N)$ and the real number space $\mathbb{R}^{2N}$ with coordinates $(x_1, \ldots, x_N, y_1, \ldots, y_N)$, any complex valued 1-form on an open set $U \subset \mathbb{C}^N$ is expressed as $\varphi = \sum_{k=1}^N (\varphi_k dx_k + \psi_k dy_k)$. This can be expressed as

$$\varphi = \frac{1}{2} \sum_{k=1}^N [(\varphi_k - i\psi_k) dz_k + (\varphi_k + i\psi_k) d\bar{z}_k]$$

using $dz_k = dx_k + i dy_k$ and $d\bar{z}_k = dx_k - i dy_k$.

We say that a 1-form φ is a $(1,0)$-form if the coefficients of $d\bar{z}_k$ are 0, i.e., if $\psi_k = i\varphi_k$:

$$\varphi = \sum_{k=1}^{N} \varphi_k dz_k = \sum_{k=1}^{N} (\varphi_k dx_k + i\varphi_k dy_k).$$

A $(1,0)$-form with holomorphic φ_k is said to be a *holomorphic 1-form.* The differential of a function g on U is given by

$$dg = \sum_{k=1}^{N} \left(\frac{\partial g}{\partial x_k} dx_k + \frac{\partial g}{\partial y_k} dy_k \right).$$

By the Cauchy–Riemann equation, g is holomorphic if and only if its differential dg is a $(1,0)$-form. Then dg is a holomorphic form written as

$$dg = \sum_{k=1}^{N} \frac{\partial g}{\partial z_k} dz_k$$

using the partial derivatives of g as a holomorphic function.

A $(1,0)$-form $\varphi = \sum_{k=1}^{N} \varphi_k dz_k$ is closed if and only if φ_k are holomorphic and satisfy the relation

$$\frac{\partial \varphi_j}{\partial z_k} = \frac{\partial \varphi_k}{\partial z_j} \quad (j,k = 1,\ldots,N).$$

This can be easily verified from the condition (6.8). A closed $(1,0)$-form will also be called a holomorphic closed 1-form. A closed 1-form φ is locally a differential of a holomorphic function. The concepts of line integral and pullback of a form on a complex number space are naturally defined, viewing them as the objects in the real number space.

Let f be a holomorphic function on an open set $U \subset \mathbb{C}^N$. By the Cauchy–Riemann equation, the real part u of f satisfies the equation

$$\frac{\partial^2 u}{\partial x_j \partial x_k} + \frac{\partial^2 u}{\partial y_j \partial y_k} = 0, \quad \frac{\partial^2 u}{\partial x_j \partial y_k} - \frac{\partial^2 u}{\partial y_j \partial x_k} = 0 \quad (j,k = 1,\ldots,N). \tag{6.9}$$

Any function that satisfies this equation is called *pluriharmonic.* A real valued pluriharmonic function is locally the real part of a holomorphic function. In fact, if u satisfies the equation (6.9), the 1-form

$$\varphi = \sum_{k=1}^{N} \left(-\frac{\partial u}{\partial y_k} dx_k + \frac{\partial u}{\partial x_k} dy_k \right)$$

is a closed form. Hence there exists, locally, a real valued function v such that $dv = \varphi$. Then $f = u + iv$ is a holomorphic function with real part u, since $\partial v/\partial x_k = -\partial u/\partial y_k$, $\partial v/\partial y_k = \partial u/\partial x_k$. The concept of subharmonic functions can also be generalized to the case of several variables.

Definition A function $h : U \to \mathbb{R} \cup \{-\infty\}$ defined on an open set $U \subset \mathbb{C}^n$ is *plurisubharmonic* if the following two conditions are satisfied:

(i) h is upper semi-continuous;
(ii) for any domain $\Omega \subset \mathbb{C}$ and any holomorphic map $f : \Omega \to U$, the composed function $h \circ f$ is subharmonic.

Remark For later purposes, we will assume that a plurisubharmonic function is not identically $-\infty$ on any connected component.

To conclude this section, we will mention a remarkable phenomenon that we encounter only in the case of several variables. There exists a pair $U \subset V$ of domains such that any holomorphic function on U can always be analytically continued to V.

Example 6.1.2 *We define the domains* Δ *and* $H \subset \mathbb{C}^2$ *by*

$$\begin{aligned} \Delta &= \left\{ z = (z_1, z_2) \in \mathbb{C}^2 \,\middle|\, |z_k| < R_k \ (k = 1, 2) \right\}, \\ H &= \{ z \in \Delta \mid \text{either } |z_1| < r_1 \text{ or } |z_2| > r_2 \} \end{aligned}$$

where $0 < r_k < R_k$ $(k = 1, 2)$. *(See Figure 6.1.)*

Then any holomorphic function on H *can be extended to a holomorphic function on* Δ.

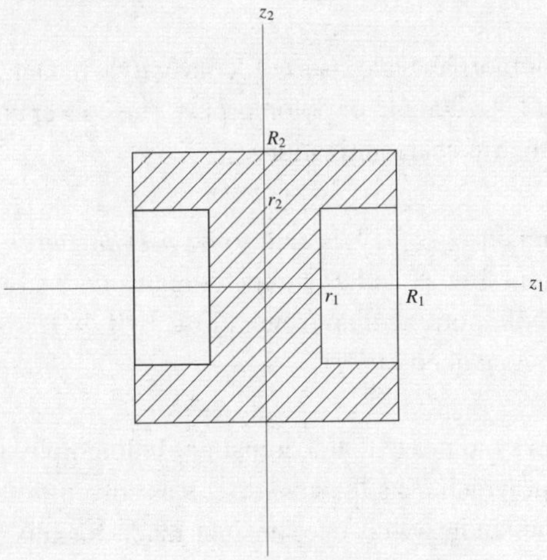

Fig. 6.1. Slices of Δ and H by $\mathbb{R}^2$

To see this, we define, for a holomorphic function $f(z)$ on H,

$$\tilde{f}(z) = \frac{1}{2\pi i}\int \frac{f(z_1, \zeta)}{\zeta - z_2} d\zeta, \quad z = (z_1, z_2) \in \Delta.$$

Here the integral is taken over the circle $|\zeta| = \rho$ $(r_2 < \rho < R_2)$. Then $\tilde{f}$ is holomorphic on Δ and coincides with f on H.

This phenomenon leads to the concept of domain of holomorphy. A domain $U \subset \mathbb{C}^N$ is said to be a domain of holomorphy if there exists a holomorphic function on U that cannot be analytically continued beyond the boundary of U. The precise definition is as follows.

Definition A domain $U \subset \mathbb{C}^N$ is called a *domain of holomorphy* if there exists a holomorphic function f on U with the following property.

If V is a domain such that $U \cap V \neq \varnothing$ and g is a holomorphic function on V such that f and g coincide on a non-empty open subset of $U \cap V$, then V is contained in U.

The following are some examples of domains of holomorphy.

(a) A connected component of an open set given by $|f(z)| < R$ or $|f(z)| > R$, where f is a holomorphic function on $\mathbb{C}^N$ and R is a positive constant.

(b) The intersection of a finite number of domains of holomorphy.

(c) The union of the increasing sequence of domains of holomorphy.

Domains of holomorphy have a sort of 'convexity'. In fact, a convex domain (in the usual sense) is a domain of holomorphy. The concept of pseudoconvexity gives this geometric characterization.

Definition A domain $U \subset \mathbb{C}^N$ is said to be *pseudoconvex* if it satisfies the following condition. Let Δ and H be the domains defined in Example 6.1.2 and let Φ be any holomorphic map from Δ to $\mathbb{C}^N$. If $\Phi(H)$ is contained in U, then $\Phi(\Delta)$ is also contained in U.

It is relatively easy to prove that a domain of holomorphy is pseudoconvex. The Levi problem, which asks the converse, was solved in the affirmative by Kiyoshi Oka. The studies with deep insight made for this solution laid the foundation of the modern theory of several complex variables. But we will not go into this subject further.

6.2 Local theory at attracting or repelling fixed points

Let F be a holomorphic map from an open set $U \subset \mathbb{C}^N$ to $\mathbb{C}^N$. We say that $a \in U$ is a fixed point of F if $F(a) = a$. In this section we will study the behavior of the points in a neighborhood of the fixed point a under the iteration F^n $(n = 1, 2, \ldots)$ of F.

We will be concerned with the behavior of points sufficiently close to the fixed point. In other words, we are interested in the germ of the map F. Therefore the domain of definition U of F is supposed to be a sufficiently small neighborhood and will be replaced by a smaller neighborhood whenever it is necessary. We will call F a holomorphic map with a fixed point $a \in \mathbb{C}^N$ without mentioning the domain of definition.

A holomorphic map F with a fixed point a has an inverse F^{-1} (in a sufficiently small neighborhood of a) if and only if the Jacobian determinant $J_F(a) = \det F'(a)$ is non-zero. Such an F will be called *invertible*.

Holomorphic maps F and G with fixed points a and $b \in \mathbb{C}^N$ are said to be *conjugate* if there is a biholomorphic map Φ from a neighborhood of a to one of b such that $\Phi(a) = b$ and that $G \circ \Phi = \Phi \circ F$ in a neighborhood of a, in other words $G = \Phi \circ F \circ \Phi^{-1}$ in a neighborhood of b. If we regard $\Phi : z = (z_1, \ldots, z_N) \to (w_1, \ldots, w_N) = (\varphi_1(z), \ldots, \varphi_N(z))$ as a change of local coordinates, G can be regarded as the expression of F with respect to the local coordinates $(w_1, \ldots, w_N)$. We will sometimes express this fact as that F is reduced to G by the change of local coordinates Φ, or by regarding $(\varphi_1(z), \ldots, \varphi_N(z))$ as local coordinates.

If F and G are conjugate by Φ, then we have $G = \Phi \circ F \circ \Phi^{-1}$ and hence, $G'(b) = \Phi'(a)F'(a)\Phi'(a)^{-1}$ for the Jacobian matrices. This shows that the eigenpolynomial and the set of eigenvalues of $F'(a)$ are invariant under conjugation.

Definition Let F be a holomorphic map with a fixed point $a \in \mathbb{C}^N$. We say that a is an *attracting* (a *repelling*) fixed point of F if all the eigenvalues of the Jacobian matrix $F'(a)$ are less (greater) than 1 in absolute value.

A holomorphic map with a repelling fixed point is invertible, and the inverse has an an attracting fixed point. So we will study the case of an attracting fixed point.

Proposition 6.2.1 *A holomorphic map F is attracting at a fixed point $a \in \mathbb{C}^N$ if and only if there exists a neighborhood U of a with the following property.*

$F(U) \subset U$ and the sequence $F^n : U \to U$ $(n = 1, 2, \ldots)$ is uniformly convergent to the constant map a.

Lemma 6.2.2 *Suppose that A is a square matrix of order N with eigenvalues $\lambda_1, \ldots, \lambda_N$. Let s be any real number greater than $\max_{1 \le i \le N} |\lambda_i|$. Then there exists an invertible matrix P such that, for any column vector $v = {}^t(v_1, \ldots, v_N)$, the inequality $\|PAP^{-1}v\| \le s\|v\|$ holds, where $\|v\| = (|v_1|^2 + \cdots + |v_N|^2)^{1/2}$.*

Proof First we note that the lemma is trivial if A is a diagonal matrix with diagonal entries $\lambda_1, \ldots, \lambda_N$. In the general case, we can assume that $A = (a_{ij})_{1 \le i,j \le N}$ is a lower triangular matrix:

$$a_{ii} = \lambda_i \ (1 \le i \le N), \quad a_{ij} = 0 \ (1 \le i < j \le N).$$

We choose, as P, a diagonal matrix with diagonal entries $\varepsilon^{-1}, \varepsilon^{-2}, \ldots, \varepsilon^{-N}$ ($\varepsilon \neq 0$) so that $PAP^{-1} = (\varepsilon^{i-j} a_{ij})$. The matrix PAP^{-1} tends to the diagonal matrix with diagonal components $\lambda_1, \ldots, \lambda_N$ when $\varepsilon \to 0$. This shows the lemma. □

Proof of Proposition 6.2.1 Suppose that a is an attracting fixed point of F and let $\lambda_1, \ldots, \lambda_N$ be the eigenvalues of $F'(a)$. We choose real numbers s, s' so that $\max_{1 \le i \le N} |\lambda_i| < s' < s < 1$. By the lemma, shifting to a conjugate by a suitable linear change of coordinates of $\mathbb{C}^N$, we have $\|F'(a)v\| \le s'\|v\|$ for all $v \in \mathbb{C}^N$. Therefore, if r is sufficiently small we have $\|F(z) - a\| \le s\|z - a\|$ for $\|z - a\| < r$. Thus the neighborhood $U = \{z \in \mathbb{C}^N \mid \|z - a\| < r\}$ satisfies the requirement of the theorem.

Conversely, suppose that there exists a U satisfying the condition. Then the Jacobian matrix $(F^n)'(a) = (F'(a))^n$ of F^n at a converges to the zero matrix. Hence the eigenvalues of $F'(a)$ are all smaller than 1 in absolute value. □

In the case of two variables, any invertible holomorphic map with attracting fixed point can be reduced to some normal form by conjugation (in other words by change of local coordinates). In what follows, we will denote the points in $\mathbb{C}^2$ by $z = (x, y)$.

Definition We denote by $L_{\lambda,\mu}$ and $E_{\lambda,k}$ the following holomorphic maps from $\mathbb{C}^2$ to $\mathbb{C}^2$.

$$\begin{aligned} L_{\lambda,\mu} &: (x, y) \mapsto (\lambda x, \mu y), \quad (\lambda, \mu \neq 0), \\ E_{\lambda,k} &: (x, y) \mapsto (\lambda x, \lambda^k(y + x^k)), \quad (\lambda \neq 0, k = 1, 2, \ldots). \end{aligned}$$

These are biholomorphic maps of $\mathbb{C}^2$ that have the origin O as fixed point. In the following theorem, however, we are concerned only with their germs at the origin.

Theorem 6.2.3 (Lattès) *Let F be an invertible holomorphic map with an attracting fixed point $a \in \mathbb{C}^2$. Suppose that the eigenvalues λ, μ of $F'(a)$ satisfy the condition $0 < |\mu| \le |\lambda| < 1$.*

(1) *If $\lambda^k \neq \mu$ for any positive integer k, then F is conjugate to $L_{\lambda,\mu}$.*

(2) *If $\lambda^k = \mu$ for some positive integer k, then F is conjugate either to $L_{\lambda,\mu}$ or to $E_{\lambda,k}$.*

Proof We assume that the fixed point is the origin O. The proof is divided into three steps. First we will show that F can be reduced to a map of the form

$$(x, y) \mapsto (\lambda x, g(x, y)). \tag{6.10}$$

Next we show that a map of this form can be reduced to the form

$$(x, y) \mapsto (\lambda x, \mu y + h(x)) \tag{6.11}$$

where $h(x)$ is a holomorphic function defined in a neighborhood of $x = 0$ such that $h(0) = 0$. Finally we show that a map of the form (6.11) can be reduced to either $L_{\lambda,\mu}$ or $E_{\lambda,k}$.

(Step 1) This step can be proved in the same manner as the proof of Theorem 1.2.4. First we choose a real number s so that $s^2 < |\lambda| < s < 1$. By a linear change of coordinates, we can suppose that F has the form

$$z = (x, y) \mapsto (f(z), g(z)) = (\lambda x + \cdots, \nu x + \mu y + \cdots),$$

where $\cdots$ denotes the terms of order ≥ 2 in x, y. Further, by choosing a sufficiently small neighborhood U of the origin, we can suppose that, if $z \in U$, then $F(z) \in U$ and $\|F(z)\| \le s\|z\|$. We denote the nth iterates of F by $F^n(z) = (f_n(z), g_n(z))$ with holomorphic functions $f_n(z), g_n(z)$ defined in U. We assert that the sequence of functions $\frac{1}{\lambda^n} f_n(z)$, $n = 1, 2, \ldots$, converges uniformly on U. To show this, note that $f(z)$ is expressed as $f(z) = \lambda x + A(z)$. Here $A(z)$ is a holomorphic function of $z = (x, y)$ consisting of terms of order ≥ 2. Hence there exists a constant K such that $|A(z)| \le K\|z\|^2$. If we replace z by $F^n(z)$ for each $n = 0, 1, 2, \ldots$ and remark that $f(F^n(z)) = f_{n+1}(z)$, we obtain $f_{n+1}(z) = \lambda f_n(z) + A(F^n(z))$, and hence

$$\frac{1}{\lambda^{n+1}} f_{n+1}(z) - \frac{1}{\lambda^n} f_n(z) = \frac{1}{\lambda^{n+1}} A(F^n(z)). \tag{6.12}$$

Since $\|F^n(z)\| \le s^n\|z\|$ for $z \in U$ ($n = 0, 1, 2, \ldots$), we have

$$\left| \frac{1}{\lambda^{n+1}} A(F^n(z)) \right| \le \frac{K}{|\lambda|^{n+1}} \|F^n(z)\|^2 \le \frac{K}{|\lambda|^{n+1}} (s^n\|z\|)^2$$
$$= \frac{K\|z\|^2}{|\lambda|} \left(\frac{s^2}{|\lambda|} \right)^n.$$

Since we have assumed that $s^2/|\lambda| < 1$, it follows that

$$\varphi(z) = \lim_{n\to\infty} \frac{1}{\lambda^n} f_n(z) = x + \sum_{n=0}^{\infty} \frac{1}{\lambda^{n+1}} A(F^n(z))$$

is uniformly convergent on U. The series on the right consist of terms of order ≥ 2. If we let $n \to \infty$ in the equation

$$\frac{1}{\lambda^n} f_n(F(z)) = \lambda \frac{1}{\lambda^{n+1}} f_{n+1}(z)$$

we obtain $\varphi(F(z)) = \lambda\varphi(z)$. Thus, by regarding $(\varphi(z), y)$ as local coordinates, the map F is reduced to the form (6.10).

(Step 2) Now suppose that the holomorphic map $F : (x, y) \to (f(z), g(z))$ is of the form (6.10), i.e., $f(z) \equiv \lambda x$. Then $g(z)$ is of the form $g(z) = \nu x + \mu y +$ (terms of order ≥ 2) and its partial derivative with respect to y can be written as $\frac{\partial g}{\partial y}(z) = \mu(1 + B(z))$, where $B(z)$ has no constant term, and $|B(z)| \leq L\|z\|$ for some $L > 0$. By the chain rule and the fact that $\frac{\partial f}{\partial y} \equiv 0$ we have

$$\frac{\partial g_n}{\partial y}(z) = \frac{\partial g_{n-1}}{\partial y}(F(z))\frac{\partial g}{\partial y}(z). \tag{6.13}$$

Hence

$$\frac{\partial g_n}{\partial y}(z) = \prod_{j=0}^{n-1} \frac{\partial g}{\partial y}(F^j(z)) = \mu^n \prod_{j=0}^{n-1}(1 + B(F^j(z)).$$

Now, since we have the estimate

$$|B(F^j(z))| \leq L\|F^j(z)\| \leq Ls^j\|z\| \quad (0 < s < 1),$$

the infinite product

$$\gamma(z) = \lim_{n\to\infty} \frac{1}{\mu^n} \frac{\partial g_n}{\partial y}(z) = \prod_{n=0}^{\infty}(1 + B(F^n(z))$$

is uniformly convergent on U. Hence $\gamma(z)$ is a holomorphic function on U with $\gamma(O) = 1$. By dividing both sides of (6.13) by μ^{n-1} and letting $n \to \infty$, we see that $\gamma(z)$ satisfies the equation

$$\gamma(F(z))\frac{\partial g}{\partial y}(z) = \mu\gamma(z).$$

Now we choose a holomorphic function $\psi(z)$ on a neighborhood of O such that $\psi(0) = 0$ and $\frac{\partial \psi}{\partial y}(z) = \gamma(z)$.

Then

$$\frac{\partial(\psi \circ F)}{\partial y}(z) = \frac{\partial \psi}{\partial y}(F(z))\frac{\partial g}{\partial y}(z) = \gamma(F(z))\frac{\partial g}{\partial y}(z) = \mu\gamma(z) = \mu\frac{\partial \psi}{\partial y}(z).$$

Therefore there exists a holomorphic function on a neighborhood of $x = 0$ such that $h(0) = 0$ and that $\psi(F(z)) = \mu\psi(z) + h(x)$. Thus, by regarding $(x, \psi(z))$ as local coordinates, F is reduced to the form (6.11).

(Step 3) Finally suppose that F is of the form (6.11). Let $q(x)$ be a holomorphic function in a neighborhood of $x = 0$ with $q(0) = 0$, and put $\eta(z) = y + q(x)$. Then $\eta(F(z)) = \mu y + h(x) + q(\lambda x)$. Therefore we have

$$\eta(F(z)) - \mu\eta(z) = h(x) + q(\lambda x) - \mu q(x). \tag{6.14}$$

Now let

$$h(x) = h_1 x + h_2 x^2 + \cdots, \quad q(x) = q_1 x + q_2 x^2 + \cdots$$

be the power series expansions. Then the right hand side of (6.14) takes the form

$$[h_1 + (\lambda - \mu)q_1]x + [h_2 + (\lambda^2 - \mu)q_2]x^2 + \cdots + [h_j + (\lambda^j - \mu)q_j]x^j + \cdots.$$

Suppose that $\lambda^j \neq \mu$ for any integer $j \geq 1$. Then we define $q(x)$ by $q_j = \frac{h_j}{\mu - \lambda^j}$. Since $\lambda^j \to 0$ as $j \to \infty$, the series thus determined has the same radius of convergence as that of $h(x)$. Now the right hand side of (6.14) is equal to 0, so that we have $\eta(F(z)) = \mu\eta(z)$. Regarding $(x, \eta(z))$ as local coordinates, we can reduce F to $L_{\lambda,\mu}$. Suppose that $\lambda^k = \mu$ for some integer $k \geq 1$. We define $q(x)$ by $q_j = \frac{h_j}{\mu - \lambda^j}$ $(j \neq k)$, whereas q_k is an arbitrary number. Then the right hand side of (6.14) is equal to $h_k x^k$, so that we have $\eta(F(z)) = \lambda^k \eta(z) + h_k x^k$. If $h_k \neq 0$, then we replace $\frac{\lambda^k}{h_k}\eta(z)$ by $\eta(z)$, so that we obtain $\eta(F(z)) = \lambda^k(\eta(z) + x^k)$. Regarding $(x, \eta(z))$ as local coordinates, we can reduce F to $L_{\lambda,\mu}$ when $h_k = 0$ and to $E_{\lambda,k}$ when $h_k \neq 0$. □

Remarks (1) We could have chosen the map $(x, y) \to (\lambda x, \lambda^k y + x^k))$ as the canonical form, since it is conjugate to $E_{\lambda,k}$. The reason why we have chosen $E_{\lambda,k}$ is the fact that the nth iterates $E^n_{\lambda,k}$ take the simple form $(x, y) \to (\lambda^n x, \lambda^{nk}(y + nx^k))$.

(2) The first step of the proof is valid also in the case $\mu = 0$. Therefore an attracting fixed point can be reduced to the form (6.10) if one of the eigenvalues is non-zero.

Proposition 6.2.4 *The maps L_{λ,λ^k} and $E_{\lambda,k}$ are not conjugate to each other.*

Proof Suppose that $E_{\lambda,k}$ is conjugate to $L_{\lambda,\lambda^k} : (u, v) \mapsto (\lambda u, \lambda^k v)$ by a map $\Phi : (x, y) \mapsto (u, v) = (\varphi(x, y), \psi(x, y))$. Then φ and ψ satisfy the functional equations

$$\varphi(\lambda x, \lambda^k(y+x^k)) = \lambda\varphi(x, y), \ \psi(\lambda x, \lambda^k(y+x^k)) = \lambda^k\psi(x, y).$$

We substitute $\psi(x, y) = \sum_{i,j>0} c_{ij}x^i y^j$ in the second equation. By equating the coefficients of x and x^k, we obtain

$$\lambda c_{10} = \lambda^k c_{10}, \quad \lambda^k c_{01} + \lambda^k c_{k0} = \lambda^k c_{k0}.$$

If $k \geq 2$, we have $c_{10} = c_{01} = 0$, which contradicts that Φ is invertible at the origin. If $k = 1$, we have $c_{01} = 0$. Since φ satisfies the same equation as ψ, the coefficient of y in the power series expansion vanishes. This again contradicts the invertibility of Φ. □

6.3 Holomorphic automorphisms of $\mathbb{C}^2$ and Fatou–Bieberbach domains

A biholomorphic map of $\mathbb{C}^N$ onto itself is said to be a holomorphic automorphism of $\mathbb{C}^N$. The set of all holomorphic automorphisms of $\mathbb{C}^N$ is a group with respect to the composition of maps. In the case $N \geq 2$, this constitutes a huge group and it seems to be a difficult problem to describe the structure (to give a set of generators for example) of this group. This is in contrast to the fact that, in the one-dimensional case, holomorphic automorphisms of $\mathbb{C}$ are only linear transformations. Another striking difference of the case of higher dimension is the following phenomenon.

For $N \geq 2$, there exist proper subdomains of $\mathbb{C}^N$ that are biholomorphically equivalent to $\mathbb{C}^N$.

Such domains are called *Fatou–Bieberbach domains.* In the one-dimensional case, any simply connected proper subdomain of $\mathbb{C}$ is conformally equivalent to the unit disk by the Riemann mapping theorem and is not conformally equivalent to $\mathbb{C}$.

These phenomena are closely related to each other. In fact, we can construct Fatou–Bieberbach domains using holomorphic automorphism of $\mathbb{C}^N$.

In what follows, we restrict our considerations to the case of dimension 2. Let $F : \mathbb{C}^2 \to \mathbb{C}^2$ be a holomorphic automorphism of $\mathbb{C}^2$ that has an attracting fixed point $a \in \mathbb{C}^2$. The *basin of attraction* of a is defined by

$$\Omega = \{z \in \mathbb{C}^2 \mid F^n(z) \to a \ (n \to \infty)\}.$$

This set Ω is a non-empty open set. It is connected and simply connected. In fact, let U be a connected and simply connected neighborhood of a such that $F(U) \subset U$. Then Ω is the union of the increasing sequence $F^{-n}(U), n = 1, 2, \ldots$. The basin of attraction Ω may coincide with the whole of $\mathbb{C}^2$ or it may be a proper subset of $\mathbb{C}^2$.

Example 6.3.1 *The maps $L_{\lambda,\mu}$, $E_{\lambda,k}$ defined in the previous section can be regarded as holomorphic automorphism of $\mathbb{C}^2$. If $|\lambda|, |\mu| < 1$, then the origin $O = (0,0)$ is an attracting fixed point. In this case the basin of attraction is the whole of $\mathbb{C}^2$. In fact, for $z = (x, y) \in \mathbb{C}^2$, we have $L^n_{\lambda,\mu}(z) = (\lambda^n x, \mu^n y)$ and $E^n_{\lambda,k}(z) = (\lambda^n x, \lambda^{kn}(y + nx^k))$, $n = 1, 2, \ldots$. When $n \to \infty$, we have $\lambda^n, \mu^n, n\lambda^{kn} \to 0$. Hence $L^n_{\lambda,\mu}(z)$ and $E^n_{\lambda,k}(z)$ converge to O.*

Theorem 6.3.1 *Suppose that a holomorphic automorphism F of $\mathbb{C}^2$ has an attracting fixed point a. Let λ, μ $(0 < |\mu| \le |\lambda| < 1)$ be the eigenvalues of $F'(a)$. Then there exists a biholomorphic map Φ from the basin of attraction Ω onto $\mathbb{C}^2$ such that $\Phi \circ (F|\Omega) = G \circ \Phi$. Here G is either $L_{\lambda,\mu}$ or $E_{\lambda,k}$.*

Proof By Theorem 6.2.3, there exist neighborhoods U of the attracting fixed point a and V of the origin $O \in \mathbb{C}^2$ and a biholomorphic map Φ_0 from U onto V such that $\Phi_0(a) = 0$ and $\Phi_0 \circ F(z) = G \circ \Phi_0(z)$ for every $z \in U$. Here either $G = L_{\lambda,\mu}$ or $G = E_{\lambda,k}$. We can assume that $F(U) \subset U$, by shrinking U if it is necessary. Then we have also $G(V) \subset V$. Hence we have the increasing sequences of domains

$$U \subset F^{-1}(U) \subset \cdots \subset F^{-n}(U) \subset \cdots,$$
$$V \subset G^{-1}(V) \subset \cdots \subset G^{-n}(V) \subset \cdots.$$

The union of the first sequence is the basin of attraction Ω of F to a, and that of the second is the basin of attraction of G to O:

$$\bigcup_{n=0}^{\infty} F^{-n}(U) = \Omega, \quad \bigcup_{n=0}^{\infty} G^{-n}(V) = \mathbb{C}^2.$$

The map $\Phi_0 : U \to V$ is biholomorphic and satisfies the equation $G^{-1} \circ \Phi_0 \circ F(z) = \Phi_0(z)$ for $z \in U$. Using this equation, we will extend the map Φ_0 to a biholomorphic map of Ω onto $\mathbb{C}^2$. To do this we define holomorphic maps $\Phi_n : F^{-n}(U) \to F^{-n}(V)$ for $n = 1, 2, \ldots$, by

$$\Phi_n(z) = G^{-n} \circ \Phi_0 \circ F^n(z).$$

Since the maps $F^n : F^{-n}(U) \to U$, $\Phi_0 : U \to V$ and $G^{-n} : V \to G^{-n}(V)$ are biholomorphic, the composition Φ_n: $F^{-n}(U) \to G^{-n}(V)$ is also biholomorphic. Further, if $z \in F^{-n}(U)$, then $F^n(z) \in U$ and

$$\Phi_{n+1}(z) = G^{-n} \circ G^{-1} \circ \Phi_0 \circ F \circ F^n(z) = G^{-n} \circ \Phi_0 \circ F^n(z) = \Phi_n(z).$$

This shows that the restriction of Φ_{n+1} to $F^{-n}(U)$ coincides with Φ_n. For any point $z \in \Omega$, we choose n so that $z \in F^{-n}(U)$ and define $\Phi(z) = \Phi_n(z)$. The map Φ is determined independently of the choice of n. It is now clear that the map Φ thus defined is a biholomorphic map from Ω onto $\mathbb{C}^2$ satisfying the equation $G^{-1} \circ \Phi \circ F(z) = \Phi(z)$. □

Corollary *If the basin of attraction Ω is not the whole of $\mathbb{C}^2$, then it is a Fatou–Bieberbach domain. The basin of attraction Ω coincides with the whole of $\mathbb{C}^2$ if and only if F is conjugate to $L_{\lambda,\mu}$ or $E_{\lambda,k}$ by a holomorphic automorphism of $\mathbb{C}^2$.*

Example 6.3.2 *A holomorphic automorphism of $\mathbb{C}^2$ of the form*

$$F : (x, y) \mapsto (y, y^2 + c - \delta x), \quad \delta, c \in \mathbb{C}, \delta \neq 0,$$

is said to be an Hénon map. (We will study Hénon maps in detail in the next chapters.) The fixed points (x, y) of F are given as the solutions of the equations $x = y$, $y = y^2 + c - \delta x$. If α is a root of the quadratic equation $\zeta^2 - (1 + \delta)\zeta + c = 0$, then the point (α, α) is a fixed point of F. Therefore F has exactly two fixed points, if the discriminant $(\delta + 1)^2 - 4c \neq 0$.

Proposition 6.3.2 *If $|\delta| < 1$ and $|c| < \frac{1}{4}(1 - |\delta|)^2$, then the Hénon map F has exactly two fixed points. One of them is attracting and its basin of attraction is a Fatou–Bieberbach domain. (See Figure 6.2.)*

Fig. 6.2. The slice of a Fatou–Bieberbach domain for $F_{\delta,c}$ by $\{x = 0\}$; $\delta = -0.25$, $c = -0.1$

Proof It is easy to verify that the discriminant $\neq 0$ so that there exist exactly two fixed points. To show that one of them is attracting, we might explicitly calculate the eigenvalues. Here we will apply Theorem 6.3.5, given below at the end of this section. Let $D = \{(x, y) \mid |x| < r, |y| < s\}$ for positive numbers r, s. Under the assumption of the proposition, we will show that r, s can be chosen so that $\overline{F(D)} \subset D$. Let $z = (x, y) \in D$ and $F(z) = (x_1, y_1)$. Then we have

$$|x_1| = |y| < s, \quad |y_1| = |y^2 + c - \delta x| < s^2 + |c| + |\delta| r.$$

This shows that, if $0 < s < r$ and $s^2 + |c| + |\delta| r < s$, then $\overline{F(D)} \subset D$. Such r and s exist if there exists an $s > 0$ with $s^2 - (1 - |\delta|)s + |c| < 0$. But this is equivalent to the condition given in the proposition. The basin of attraction is a proper subdomain of $\mathbb{C}^2$, since it does not contain the other fixed point. □

Remark The other fixed point is not attracting (see §7.1).

Example 6.3.3 *Let h be an entire function of one complex variable and δ a non-zero constant. We consider the holomorphic automorphism $F : (x, y) \mapsto (y, h(y) - \delta x)$. The fixed points are given by (α, α), with zeros α of the function $h(\zeta) - (1 + \delta)\zeta$.*

In particular, let $h(\zeta) = (1 + \delta)\zeta + c \sin \zeta$ $(c \neq 0)$. Then F has infinitely many fixed points $a_m = (m\pi, m\pi)$, $m \in \mathbb{Z}$. The Jacobian matrix at a_m with even m is $\begin{pmatrix} 0 & 1 \\ -\delta & 1 + \delta + c \end{pmatrix}$ and the eigenpolynomial is $t^2 - (1 + \delta + c)t + \delta = 0$. Hence, if δ, c are so chosen that the absolute values of the roots of this quadratic equation are both less than 1, then all a_m are attracting fixed points. The basins of attraction to a_m are disjoint from one another. Thus we have constructed infinitely many Fatou–Bieberbach domains that are disjoint from one another.

An example with cubic polynomial h and two Fatou–Bieberbach domains is shown in Figure 6.3.

Example 6.3.4 *Let $g(\zeta)$ and $h(\zeta)$ be non-constant and non-vanishing entire functions of a single variable ζ such that $|g(0)|, |h(0)| < 1$. We define the automorphism F of $\mathbb{C}^2$ by*

$$F : (x, y) \mapsto (h(g(x)y)x, g(x)y).$$

The map F is the composition of the automorphisms $(x, y) \mapsto (x, g(x)y)$ and $(x, y) \mapsto (h(y)x, y)$. Let $\alpha_1, \alpha_2, \ldots$ be the solutions of $g(\zeta) = 1$ and $\beta_1, \beta_2, \ldots$ the solutions of $h(\zeta) = 1$. We note that these equations have infinitely many solutions by the big Picard theorem. The fixed points of F are $O = (0, 0)$ and (α_i, β_j) $(i, j = 1, 2, \ldots)$. The origin O is an attracting fixed

Fig. 6.3. Slices of (white and gray) Fatou–Bieberbach domains for $(x, y) \mapsto (y, h(y) - \delta x)$ by $\mathbb{R}^2$; $\delta = 0.64$, $h(y) = 0.02y^3 + 1.66y$

point, since $F'(O) = \begin{pmatrix} h(0) & 0 \\ 0 & g(0) \end{pmatrix}$. *The x- and y-axes are contained in the basin of attraction* Ω *to* O, *since* $F(x, 0) = (h(0)x, 0)$, $F(0, y) = (0, g(0)y)$. *Other fixed points* (α_i, β_j) *are not contained in* Ω. *Thus we have constructed a Fatou–Bieberbach domain that contains two complex lines in* $\mathbb{C}^2$.

Proposition 6.3.3 *Let* $\Omega \subset \mathbb{C}^2$ *be a basin of attraction to an attracting fixed point of a holomorphic automorphism of* $\mathbb{C}^2$. *Let* $H : \mathbb{C} \to \mathbb{C}^2$ *be a holomorphic map. Then the set* $H^{-1}(\Omega)$ *is a simply connected open subset of* $\mathbb{C}$. *(This set may be empty, or disconnected.)*

Proof Let $F^n(z) = (f_n(z), g_n(z))$, $n = 1, 2, \ldots$. Then $f_n(H(\zeta)), g_n(H(\zeta))$ are holomorphic functions on $\mathbb{C}$, the set $H^{-1}(\Omega)$ consists of the points $\zeta \in \mathbb{C}$ such that both of $f_n \circ H$, $g_n \circ H$ converge uniformly on some neighborhood of ζ. Let E be any compact subset of $\mathbb{C}$. If the boundary of E is contained in $H^{-1}(\Omega)$, then E is contained in $H^{-1}(\Omega)$, by the principle of maximum modulus. Hence $H^{-1}(\Omega)$ is simply connected. □

Remark It can be proved that the basin of attraction Ω is a Runge domain. Proposition 6.3.3 can be considered as a consequence of this fact. Here, a domain $U \subset \mathbb{C}^N$ is said to be a *Runge domain* if U is a domain of holomorphy and if any holomorphic function on U can be approximated uniformly on compact sets in U by polynomials. Now this Proposition 6.3.3 has the following interesting consequence.

Theorem 6.3.4 *The unit disk Δ can be embedded into $\mathbb{C}^2$ as a closed complex submanifold.*

Proof Let H be a holomorphic map that embeds $\mathbb{C}$ into $\mathbb{C}^2$ (non-constant linear map, for example) such that the image $H(\mathbb{C})$ intersects both Ω and the complement of Ω. Let D be a connected component of $H^{-1}(\Omega)$. Then D is a simply connected proper subdomain of $\mathbb{C}$, and hence conformally equivalent to the unit disk Δ by the Riemann mapping theorem. Then D is embedded into Ω by the map $H|D$ as a complex submanifold. Hence D is embedded into $\mathbb{C}^2$ by the composition $\Phi \circ H|D$, where Φ is as in Theorem 6.3.1. □

We will conclude this section by giving a sufficient condition for the existence of an attracting fixed point.

Theorem 6.3.5 (Contraction principle) *Let F be a holomorphic map from a domain $U \subset \mathbb{C}^N$ to $\mathbb{C}^N$. Suppose that the closure $\overline{F(U)}$ of the image $F(U)$ is a compact set contained in U. Then the map F has a single fixed point $a \in U$. The fixed point a is attracting and the sequence $\{F^n\}_{n=1}^{\infty}$ converges to the constant map a uniformly on U.*

First we prepare a lemma, which is considered as a generalization of the Schwarz lemma.

Lemma 6.3.6 *For a domain $V \subset \mathbb{C}^M$ and a compact set $L \subset V$, there exists a constant $\rho < 1$ with the following property: if f is a bounded holomorphic function on V that vanishes at a point in L, then the inequality $\sup_{\zeta \in L} |f(\zeta)| \le \rho \sup_{\zeta \in V} |f(\zeta)|$ holds.*

Proof Suppose that there is no such ρ. Then we can choose a sequence $\{f_n\}$ of holomorphic functions on V and sequences $\{a_n\}$, $\{b_n\}$ of points in L such that

$$\sup_{\zeta \in V} |f_n(\zeta)| \le 1, \quad |f_n(a_n)| \ge 1 - 1/n, \quad f_n(b_n) = 0.$$

This sequence $\{f_n\}$ is a normal family. By shifting to a subsequence, we can assume that $\{f_n\}$ converges to a holomorphic function f_0 uniformly on compact sets in V. Further shifting to a subsequence, we can assume that $\{a_n\}$, $\{b_n\}$ converge to points $a_0, b_0 \in L$ respectively. Then f_0 satisfies the conditions

$$\sup_{\zeta \in V} |f_0(\zeta)| \le 1, \quad |f_0(a_0)| = 1, \quad f_0(b_0) = 0.$$

From the first two conditions, we conclude that $f_0(\zeta)$ is a constant of absolute value 1, by the principle of maximum modulus. But this contradicts the third condition. □

Lemma 6.3.7 *Let $K_1 \supset K_2 \supset \cdots \supset K_n \supset \cdots$ be a decreasing sequence of compact sets in $\mathbb{C}^N$. Suppose that there exist a domain $V \subset \mathbb{C}^M$, a compact set $L \subset V$ and a sequence of holomorphic maps $\Phi_n : V \to \mathbb{C}^N$ $(n = 1, 2, \ldots)$ such that*

$$K_n \supset \Phi_n(V), \quad \Phi_n(L) \supset K_{n+1}$$

for all n. Then the intersection $\bigcap_{n=1}^{\infty} K_n$ consists of a single point.

Proof Since $\bigcap_{n=1}^{\infty} K_n$ is non-empty, we assume that this set contains the origin $O = (0, \ldots, 0)$, by translation of the coordinates. We write $|z| = \max_{k=1,\ldots,N} |z_k|$ for $z = (z_1, \ldots, z_N) \in \mathbb{C}^N$. Let $r_n = \sup_{z \in K_n} |z|$. It suffices to show that $r_n \to 0$ as $n \to \infty$. We write $\Phi_n(\zeta) = (\varphi_{n1}(\zeta), \ldots, \varphi_{nN}(\zeta))$. Since $\Phi_n(L) \supset K_{n+1} \ni O$, every $\varphi_{nk}(\zeta)$ has a zero in L. By Lemma 6.3.6, we have

$$\sup_{\zeta \in L} |\varphi_{nk}(\zeta)| \leq \rho \sup_{\zeta \in V} |\varphi_{nk}(\zeta)|.$$

Hence

$$r_{n+1} \leq \sup_{z \in \Phi_n(L)} |z| = \sup_{\zeta \in L} |\Phi_n(\zeta)| \leq \rho \sup_{\zeta \in V} |\Phi_n(\zeta)| = \rho \sup_{z \in \Phi_n(V)} |z| \leq \rho r_n.$$

This shows that $r_n \to 0$ $(n \to \infty)$. □

Proof of Theorem 6.3.5 Let $K_n = \overline{F^n(U)}$, $V = U$, $L = \overline{F(U)}$, $\Phi_n = F^n$. Then we have $K_n \supset F^n(U)$ and $F^n(L) = K_{n+1}$. By Lemma 6.3.7, the set $\bigcap_{n=1}^{\infty} \overline{F^n(U)}$ consists of a single point. The sequence $\{F^n\}_{n=1}^{\infty}$ converges to this point uniformly on U. □

6.4 Behavior at a fixed point of saddle type

Now we will consider fixed points that are neither attracting nor repelling. First we define fixed points of saddle type, which can be relatively easily treated as far as the local behavior is concerned.

Definition Let F be a holomorphic map with a fixed point $a \in \mathbb{C}^2$. We will say that F is *of saddle type* if one of the eigenvalues of $F'(a)$ is less than 1 and the other is greater than 1, in absolute value.

Theorem 6.4.1 *Let F be a holomorphic map with a fixed point $a \in \mathbb{C}^2$. Suppose that F is of saddle type and let λ, μ ($0 < |\lambda| < 1 < |\mu|$) be the eigenvalues of $F'(a)$. Then F is conjugate to a holomorphic map of the form*

$$G : (x, y) \mapsto (\lambda x + xy\alpha(x, y), \mu y + xy\beta(x, y))$$

with the fixed point $O = (0, 0)$. Here $\alpha(x, y)$ and $\beta(x, y)$ are holomorphic functions in a neighborhood of the origin O.

Before the proof of this theorem, we will mention some of its consequences. Let G be a holomorphic map as in the theorem. Let

$$\Delta = \{z = (x, y) \mid |x|, |y| < r\}, \quad \Gamma^+ = \Gamma_G^+ = \{z = (x, 0) \mid |x| < r\}$$

for some positive number r.

Proposition 6.4.2 *If r is sufficiently small, then the following statements hold:*

(1) *if $z \in \Gamma^+$, then $G^n(z) \in \Gamma^+$ for all $n = 1, 2, \ldots$ and the sequence $G^n(z)$ converges to O as $n \to \infty$;*
(2) *if $z \in \Delta - \Gamma^+$, then $G^n(z) \notin \Delta$ for some n.*

Proof First we note that assertion (1) holds since $G^n(x, 0) = (\lambda^n x, 0)$. To show assertion (2), we choose r sufficiently small so that

$$L = \inf_{z \in \Delta} |\mu + x\beta(z)| > 1.$$

Then we have $|\mu y + xy\beta(x, y)| \geq L|y|$. Suppose that $G^n(x, y) = (x_n, y_n) \in \Delta$ for all $n \geq 1$. Then $r > |y_n| \geq L^n|y|$. Hence we obtain $y = 0$. This shows (2). □

We will call the set $\Gamma^+ = \Gamma_G^+$ the *(local) stable manifold* of the map G. For a fixed point of saddle type of a holomorphic map F, we define the (local) stable manifold by $\Gamma_F^+ = \Phi^{-1}(\Gamma_G^+)$, where Φ is a biholomorphic map which gives the conjugation of F to G in Theorem 6.4.1.

We will prove the following theorem, which slightly generalizes Theorem 6.4.1.

Theorem 6.4.3 *Let F be a holomorphic map with fixed point $a \in \mathbb{C}^2$. Suppose that the eigenvalues λ, μ of $F'(a)$ satisfy either of the conditions (i) and (ii):*

(i) $0 < |\lambda| < 1$ *and* $|\mu| \geq 1$;
(ii) $|\lambda| > 1$ *and* $|\mu| \leq 1$.

Then there exists a holomorphic map H from a disk $\Delta_\rho = \{|t| < \rho\} \subset \mathbb{C}$ into U such that $H(0) = a$ and that

$$F(H(t)) = H(\lambda t) \quad (t, \lambda t \in \Delta_\rho).$$

Proof First we note that case (i) can be reduced to case (ii), by considering the inverse map of F. So we will consider case (ii). We can assume that the fixed point a is the origin O. Further, taking conjugate, by a linear transformation, we can assume that F is of the form $F(x, y) = (f(x, y), g(x, y)) = (\lambda x + \cdots, \mu y + \cdots)$ where $\cdots$ stands for the terms of degree ≥ 2 in x, y. We choose and fix a positive number s such that $|\lambda| < s < |\lambda|^2$. As the domain of definition of F we choose a ball with sufficiently small radius r. Then for $z, z' \in U$, we have

$$\|F(z) - F(z')\| \leq s\|z - z'\|. \tag{6.15}$$

We define $H_0 : \Delta_r \to U$ by $H_0(t) = (t, 0)$. Further we define H_1 by

$$H_1(t) = F\left(H_0\left(\frac{t}{\lambda}\right)\right) = F\left(\frac{t}{\lambda}, 0\right) = (t + \cdots, \ldots).$$

Then there exists a constant K such that

$$\|H_1(t) - H_0(t)\| \leq K|t|^2 \quad (t \in \Delta_r).$$

Now we choose ρ so that

$$0 < \rho < r/2 \quad \text{and} \quad \sum_{n=0}^{\infty}\left(\frac{s}{|\lambda|^2}\right)^n K\rho^2 < r/2.$$

We will prove the following assertions (C_n) for $n = 1, 2, \ldots$:
(C_n) The map $H_n : \Delta_\rho \to \mathbb{C}^2$ is well defined by

$$H_n(t) = F\left(H_{n-1}\left(\frac{t}{\lambda}\right)\right) \tag{6.16}$$

and the inequality

$$\|H_n(t) - H_{n-1}(t)\| \leq \left(\frac{s}{|\lambda|^2}\right)^{n-1} K|t|^2 \tag{6.17}$$

holds.

The assertion (C_1) is already proved above. Suppose that $(C_1), \ldots, (C_n)$ have been already proved. Then, by (6.17),

$$\|H_n(t)\| \leq \|H_0(t)\| + \sum_{j=1}^{n} \|H_j(t) - H_{j-1}(t)\| < r.$$

It follows that $H_n(\Delta_\rho) \subset U$, so that $H_{n+1}(t) = F(H_n(t/\lambda))$ is well defined. By (6.15) and (6.17), we have

$$\begin{aligned}
\|H_{n+1}(t) - H_n(t)\| &= \left\| F\left(H_n\left(\frac{t}{\lambda}\right)\right) - F\left(H_{n-1}\left(\frac{t}{\lambda}\right)\right)\right\| \\
&\leq s\left\| H_n\left(\frac{t}{\lambda}\right) - H_{n-1}\left(\frac{t}{\lambda}\right)\right\| \\
&\leq s\left(\frac{s}{|\lambda|^2}\right)^{n-1} K\left|\frac{t}{\lambda}\right|^2 \\
&= \left(\frac{s}{|\lambda|^2}\right)^{n} K|t|^2.
\end{aligned}$$

This implies (C_{n+1}). Thus the sequence $\{H_n\}_{n=0}$ is uniformly convergent on Δ_ρ. By letting $n \to \infty$ in (6.16), we obtain $H(t) = F(H(\frac{t}{\lambda}))$. □

Proof of Theorem 6.4.1 By Theorem 6.4.3, we have the maps H, K satisfying $F(H(t)) = H(\lambda t), F(K(u)) = K(\mu u)$. We can choose the coordinates (x, y) in such a way that $H : t \to (t, 0)$, $K : u \to (0, u)$. Then $F(x, 0) = (\lambda x, 0)$, $F(0, y) = (0, \mu y)$. This F has the required form. □

Now we will apply the above results to holomorphic automorphisms of $\mathbb{C}^2$. Suppose that a holomorphic automorphism F of $\mathbb{C}^2$ has a fixed point of saddle type a. The set of all points that converge to a by the iteration of F,

$$W^s(a) = \{z \in \mathbb{C}^2 \mid F^n(z) \to a \ (n \to \infty)\},$$

will be called the *(global) stable manifold* of F at a.

Theorem 6.4.4 *There exists an injective holomorphic map $H : \mathbb{C} \to \mathbb{C}^2$ such that $H(\mathbb{C}) = W^s(a)$ and that $F \circ H(t) = H(\lambda t)$.*

Proof We choose a neighborhood D and a coordinate system (x, y) around the fixed point a as in the previous lemma. Then Δ_r is mapped bijectively onto Γ^+ by $H_0 : t \to (x, y) = (t, 0)$. Now it is clear that $W^s(a) = \bigcup_{n=0}^{\infty} F^{-n}(\Gamma^+)$. For any $t \in \mathbb{C}$, we define $H(t) = F^{-n} \circ H_0(\lambda^n t)$ where n is sufficiently large so that $\lambda^n t \in \Delta$. We can prove that $H(t)$ is well defined and has the required property, in the same way as in the proof of Theorem 6.4.3. □

Remark The stable manifold $H(\mathbb{C}) = W^s(a)$ is not necessarily a closed submanifold of $\mathbb{C}^2$. Suppose that $|h(0)| < 1 < |g(0)|$ in Example 6.3.4. In this case the origin is a fixed point of saddle type and the stable manifold coincides with the x-axis. As to an example of the non-closed case, see the Hénon maps studied in the next chapter.

We conclude this chapter by mentioning those fixed points that are neither attracting, repelling nor of saddle type, i.e., the case where at least one of the eigenvalues λ, μ is equal to 1 in absolute value. See Ueda.

In general the structure of these fixed points is quite complicated. Here we will briefly describe some of the known results for the particular case $\lambda = 1, 0 < |\mu| < 1$. Such a map can be expressed as

$$(x, y) \mapsto \left(x + \sum_{j+k \geq 2} a_{jk} x^j y^k, \mu y + \sum_{j+k \geq 2} b_{jk} x^j y^k \right)$$

with respect to an appropriate coordinate system (x, y) with center a. We will restrict our consideration to the case $a_{20} \neq 0$. In this situation, we can construct an analogue of the Leau–Fatou petals for parabolic fixed points of one variable.

Theorem 6.4.5 (Ueda) *We can construct a domain D whose boundary contains the fixed point a satisfying the following conditions:*

(i) $F(D) \subset D$;
(ii) *the sequence $\{F^n\}$ converges to the constant map a uniformly on D;*
(iii) *if $\{F^n\}$ is uniformly convergent on some neighborhood of a point $z \in U$, then $F^m(z) \in D$ for a sufficiently large m.*

Now consider the case where F is a holomorphic automorphism of $\mathbb{C}^2$ and a is a fixed point satisfying the above conditions. Then if we let

$$\Omega = \left\{ z \in \mathbb{C}^2 \;\middle|\; \begin{array}{l} \text{the sequence } F^n \text{ converges uniformly on a} \\ \text{neighborhood of } z \text{ to the constant map} = a \end{array} \right\},$$

we have $\Omega = \bigcup_{n=0}^{\infty} F^{-n}(D)$. Further the following facts can be proved.

There exists a biholomorphic map $\Phi : \Omega \to \mathbb{C}^2$ such that $\Phi \circ F \circ \Phi^{-1}$ is conjugate to the translation $(x, y) \mapsto (x + 1, y)$ in $\mathbb{C}^2$. The domain Ω is a Fatou–Bieberbach domain since it does not contain the fixed point a.

In the Hénon map given in Example 6.3.2, consider the case where the discriminant $(\delta + 1)^2 - 4c = 0$. In this case, F has a unique fixed point $a = (\frac{\delta+1}{2}, \frac{\delta+1}{2})$ and the eigenvalues of $F'(a)$ are 1 and δ.

7

Dynamics of generalized Hénon maps

In this chapter we will study holomorphic automorphisms of $\mathbb{C}^2$ defined by polynomials – in particular, generalized Hénon maps.

In §7.1, we introduce some fundamental results concerning polynomial automorphisms of $\mathbb{C}^2$ and give a classification of the polynomial automorphisms of degree 2. From the discussion of this section, we will see that the class of generalized Hénon maps is the most fundamental and essential among all polynomial automorphisms of $\mathbb{C}^2$. To understand the arguments in the next section, we will only need the definition of generalized Hénon maps.

In §§7.2, 7.3 and 7.4, we will study the dynamics of generalized Hénon maps. As an analogy to polynomial maps of a single variable, we define the set I^+ of escaping points, and the set K^+ of non-escaping points. In §7.3, we study the structure of the set I^+. In particular we will prove that the universal covering manifold of I^+ is biholomorphic to the product of the unit disk Δ and $\mathbb{C}$. In §7.4 we study structure of the set K^+ under the horseshoe condition.

7.1 Polynomial automorphisms of $\mathbb{C}^2$ of degree 2

A holomorphic automorphism $F : (x, y) \mapsto (u, v) = (f(x, y), g(x, y))$ of $\mathbb{C}^2$ is called a *polynomial automorphism* if $f(x, y), g(x, y)$ are polynomials in the variables x, y. The degree of F is defined by $\deg F = \max\{\deg f, \deg g\}$.

Since the Jacobian determinant $J_F(x, y)$ of a polynomial map is a polynomial, this must be a constant if it has no zeros. Hence we have the following.

Proposition 7.1.1 *The Jacobian determinant $J_F(x, y)$ of a polynomial automorphism F of $\mathbb{C}^2$ is a non-zero constant.*

Remark Suppose that $F : \mathbb{C}^2 \to \mathbb{C}^2$ is a polynomial map whose Jacobian determinant is a non-zero constant. Is F an automorphism of $\mathbb{C}^2$? This is a well-known open problem, called the *Jacobian conjecture*.

Suppose that F and G are polynomial automorphisms of $\mathbb{C}^2$. Then it is clear that the composition $G \circ F$ is also a polynomial automorphism of $\mathbb{C}^2$ with $\deg G \circ F \leq \deg G \cdot \deg F$. Further, from the following proposition, we conclude that the set of all polynomial automorphisms of $\mathbb{C}^2$ constitutes a subgroup of the group consisting of all holomorphic automorphisms.

Proposition 7.1.2 *If F is a polynomial automorphism of $\mathbb{C}^2$, then its inverse map F^{-1} is also a polynomial automorphism with* $\deg F^{-1} = \deg F$.

Lemma 7.1.3 *Let $\varphi(u, v)$ be a holomorphic function on $\mathbb{C}^2$. Suppose that, for any $\alpha, \beta \in \mathbb{C}$, the function $\varphi(\alpha t, \beta t)$ is a polynomial of degree $\leq k$ in t. Then $\varphi(u, v)$ is a polynomial of degree $\leq k$ in the two variables u, v.*

Proof We develop $\varphi(u, v)$ into the series

$$\varphi(u, v) = P_0 + P_1(u, v) + \cdots + P_j(u, v) + \cdots$$

where $P_j(u, v)$ are homogeneous polynomials of degree j in u, v.

Suppose that $P_j(u, v)$ is non-zero for some $j \geq k + 1$. Then we can choose α, β so that $P_j(\alpha, \beta) \neq 0$. Then

$$\varphi(\alpha t, \beta t) = P_0 + P_1(\alpha, \beta)t + \cdots + P_j(\alpha, \beta)t^j + \cdots$$

is either a polynomial of degree $\geq k + 1$ or an entire function. □

Proof of Proposition 7.1.2 The inverse of F is a holomorphic automorphism of $\mathbb{C}^2$ represented by a pair of holomorphic functions.

$$F^{-1} : (u, v) \mapsto (x, y) = (\varphi(u, v), \psi(u, v)).$$

We will show that, if F is of degree k, then the polynomials $\varphi(u, v)$ and $\psi(u, v)$ are of degree $\leq k$. We define complex lines $L_c = \{(c, y) \mid y \in \mathbb{C}\}$ and $L_{\alpha,\beta} = \{(\alpha t, \beta t) \mid t \in \mathbb{C}\}$ and consider the sets

$$\begin{aligned} L_c \cap F^{-1}(L_{\alpha,\beta}) &= \{(c, y) \mid \beta f(c, y) - \alpha g(c, y) = 0\}, \\ F(L_c) \cap L_{\alpha,\beta} &= \{(\alpha t, \beta t) \mid \varphi(\alpha t, \beta t) = c\}. \end{aligned}$$

These two sets are in one-to-one correspondence under F. Since f, g are polynomials of degree $\leq k$, the set $L_c \cap F^{-1}(L_{\alpha,\beta})$ consists of at most k points (or else coincides with L_c). Hence for any c, the number of the solutions of $\varphi(\alpha t, \beta t) = c$ does not exceed k, or $\varphi(\alpha t, \beta t) \equiv c$ identically. Therefore $\varphi(\alpha t, \beta t)$ is a polynomial of degree $\leq k$ for any α, β. By Lemma 7.1.3 we conclude that $\varphi(u, v)$ is a polynomial of degree $\leq k$. The same argument shows that $\psi(u, v)$ is also a polynomial of degree $\leq k$.

Thus we have shown that F^{-1} is a polynomial map with $\deg F^{-1} \le \deg F$. If we exchange the roles of F and F^{-1}, we conclude that $\deg F^{-1} = \deg F$. □

Definition We define the following three fundamental types of polynomial automorphisms of $\mathbb{C}^2$:

(a) *affine map* (linear map)

$$(x, y) \mapsto (ax + by + c, a'x + b'y + c') \quad (ab' - a'b \neq 0);$$

(b) *elementary map*

$$(x, y) \mapsto (ax + b, sy + P(x)) \quad (as \neq 0),$$

where $P(x)$ is a polynomial in the single variable x;

(c) *generalized Hénon map*

$$(x, y) \mapsto (y, P(y) - \delta x) \quad (\delta \neq 0),$$

where $P(y)$ is a polynomial in the single variable y with $\deg P \ge 2$.

Two holomorphic automorphisms F, G of $\mathbb{C}^2$ are said to be *conjugate* if there exists a holomorphic automorphism Φ of $\mathbb{C}^2$ such that $G = \Phi \circ F \circ \Phi^{-1}$. Then G is considered to be an expression of F obtained by the change Φ of coordinates. From the viewpoint of complex dynamics, F and G are regarded as the same object.

It is known that the group of polynomial automorphisms of $\mathbb{C}^2$ is generated by the affine maps and the elementary maps – the Jung theorem. The conjugacy classes of the polynomial automorphisms are studied using this Jung theorem.

Theorem 7.1.4 (Friedland–Milnor) *Every polynomial automorphism of $\mathbb{C}^2$ is conjugate by a polynomial automorphism to one of the following maps:*

(a) *an affine map,*
(b) *an elementary map,*
(c) *a composition of some generalized Hénon maps.*

For an exact formulation and proof of this theorem, see Friedland and Milnor (1989). Here we will restrict our consideration to quadratic automorphisms, i.e., polynomial automorphisms of degree 2.

Definition (i) By an *Hénon map*, we mean a generalized Hénon map of degree 2 of the form

$$F_{\delta,c} : (x, y) \mapsto (y, y^2 + c - \delta x), \quad (\delta, c) \in \mathbb{C}^* \times \mathbb{C}.$$

(ii) The following maps are called *canonical quadratic elementary maps*:

$$\begin{aligned} E_a &: (x, y) \mapsto (ax, a^2(y + x^2)), \quad a \in \mathbb{C}^*; \\ E_+ &: (x, y) \mapsto (x, y + x^2 - 1); \\ E_- &: (x, y) \mapsto (-x, y + x^2 - 1). \end{aligned}$$

Theorem 7.1.5 *A quadratic polynomial automorphism of $\mathbb{C}^2$ is conjugate, by a polynomial automorphism, to one of the following maps:*

(i) *an Hénon map,*

(ii) *a canonical quadratic elementary map,*

(iii) *an affine map.*

Further there are no conjugacies by holomorphic automorphisms among the maps in (i), (ii). They are not conjugate to affine maps.

As the first step of the proof, we show the following.

Proposition 7.1.6 *A quadratic polynomial automorphism of $\mathbb{C}^2$ is conjugate to a generalized Hénon map or an elementary map by an affine map.*

Lemma 7.1.7 *Let F be a quadratic polynomial automorphism of $\mathbb{C}^2$. Then there exists a linear function $u(x, y) = \alpha x + \beta y$ such that $u(F(x, y)) = \alpha f(x, y) + \beta g(x, y)$ is again a linear function $\alpha' x + \beta' y + \gamma'$.*

Remark This implies geometrically that the family of parallel lines $\alpha' x + \beta' y = c$ (constant) is mapped by F to the family of parallel lines $\alpha x + \beta y = c + \gamma'$.

Proof of Lemma 7.1.7 Suppose that F is given by quadratic polynomials

$$\begin{aligned} f(x, y) &= a + px + qy + \frac{1}{2}(rx^2 + 2sxy + ty^2), \\ g(x, y) &= a' + p'x + q'y + \frac{1}{2}(r'x^2 + 2s'xy + t'y^2). \end{aligned}$$

Then the Jacobian determinant of F is of the form

$$\begin{aligned} J_F(x, y) &= \begin{vmatrix} p + rx + sy & q + sx + ty \\ p' + r'x + s'y & q' + s'x + t'y \end{vmatrix} \\ &= (pq' - qp') + (ps' + rq' - qr' - sp')x \\ &\quad + (pt' + sq' - qs' - tp')y + (rs' - sr')x^2 \\ &\quad + (rt' - tr')xy + (st' - ts')y^2. \end{aligned}$$

Since this is a constant, we have

$$rs' - sr' = rt' - tr' = st' - ts' = 0.$$

Hence (r, s, t) and (r', s', t') are linearly dependent. There exists a pair of complex numbers $(\alpha, \beta) \neq (0, 0)$ such that

$$\alpha(rx^2 + 2sxy + ty^2) + \beta(r'x^2 + 2s'xy + t'y^2) = 0.$$

Then $\alpha f(x, y) + \beta g(x, y)$ is a linear function. Thus $u(x, y) = \alpha x + \beta y$ satisfies the desired condition. □

Proof of Proposition 7.1.6 Let $u(x, y) = \alpha x + \beta y$ and $u(F(x, y)) = \alpha' x + \beta' y + \gamma'$. We will distinguish two cases.

(A) The case $\alpha\beta' - \beta\alpha' \neq 0$. We take the linear map

$$L : (x, y) \mapsto (\xi, \eta) = (\alpha x + \beta y, \alpha' x + \beta' y + \gamma').$$

Then

$$L \circ F \circ L^{-1} : (\xi, \eta) \mapsto (\xi_1, \eta_1) = (\eta, h(\xi, \eta)).$$

Since the Jacobian determinant of this map is constant ($= \delta$), the polynomial $h(\xi, \eta)$ must be of the form

$$h(\xi, \eta) = c_0\eta^2 + c_1\eta + c_2 - \delta\xi.$$

Thus $L \circ F \circ L^{-1}$ is a generalized Hénon map of degree 2.

(B) The case $\alpha\beta' - \beta\alpha' = 0$. In this case we have $u(F(x, y)) = au(x, y) + b$. Choose a linear function $v(x, y)$ so that

$$L : (x, y) \mapsto (\xi, \eta) = (\alpha x + \beta y, v(x, y))$$

is an automorphism of $\mathbb{C}^2$. Then

$$L \circ F \circ L^{-1} : (\xi, \eta) \mapsto (\xi_1, \eta_1) = (a\xi + b, h(\xi, \eta)).$$

Since the Jacobian determinant is constant, we have $a \neq 0$ and $h(\xi, \eta)$ must be of the form

$$h(\xi, \eta) = s\eta + c_0\xi^2 + c_1\xi + c_2.$$

Thus $L \circ F \circ L^{-1}$ is an elementary map of degree 2. □

Further, taking conjugates, we will reduce these maps to the canonical ones stated in Theorem 7.1.5. Let us begin with the generalized Hénon maps.

Proposition 7.1.8 *Any generalized Hénon map*

$$F : (x, y) \mapsto (y, c_0 y^2 + c_1 y + c_2 - \delta x) \quad (\delta \neq 0, c_0 \neq 0)$$

of degree 2 can be reduced to an Hénon map by conjugation by an affine map.

Proof By taking conjugates by the affine map

$$L : (x, y) \mapsto (\xi, \eta) = \left(c_0 x + \frac{c_1}{2}, c_0 y + \frac{c_1}{2}\right),$$

we obtain

$$L \circ F \circ L^{-1} : (\xi, \eta) \mapsto (\eta, \eta^2 + c - \delta\xi).$$

□

Now we look at the elementary maps. In this case we have to introduce conjugations by polynomial automorphisms of higher degree, in addition to conjugations by affine maps.

Proposition 7.1.9 *An elementary map of degree* 2

$$F : (x, y) \mapsto (ax + b, sy + P(x)) \quad \textit{with} \quad P(x) = c_0 x^2 + c_1 x + c_2,\ as \neq 0,$$

is conjugate by a polynomial automorphism either to a canonical elementary map of degree 2 *or to an affine map.*

Proof We distinguish three cases according to the linear transformation $l : x \mapsto ax + b$ of $\mathbb{C}$:

(A) $a \neq 1$; (B) $a = 1, b \neq 0$; (C) $a = 1, b = 0$.

(A) In this case l reduces to the map $x \mapsto ax$ by a translation of the coordinate x. Hence we may suppose that F is of the form $(x, y) \mapsto (ax, sy+P(x))$. Let $Q(\xi) = q_0\xi^2 + q_1\xi + q_2$ be a polynomial of degree 2 and consider the polynomial automorphism

$$G : (\xi, \eta) \mapsto (x, y) = (\xi, \eta + Q(\xi)).$$

Then $(\xi, \eta) \in \mathbb{C}^2$ is mapped by $G^{-1} \circ F \circ G$ as follows:

$$(\xi, \eta) \mapsto (\xi, \eta + Q(\xi)) \mapsto (a\xi, s(\eta + Q(\xi)) + P(\xi)) \mapsto (a\xi, s\eta + R(\xi))$$

where

$$\begin{aligned} R(\xi) &= P(\xi) + sQ(\xi) - Q(a\xi) \\ &= \{c_0 + (s - a^2)q_0\}\xi^2 + \{c_1 + (s - a)q_1\}\xi + \{c_2 + (s - 1)q_2\}. \end{aligned}$$

(A$_1$) When $s \neq a^2$, we let $q_0 = c_0/(a^2 - s)$. Then $R(\xi)$ is linear and $G^{-1} \circ F \circ G$ is an affine map.

(A$_2$) When $s = a^2$, we have $a - s = a(1 - a) \neq 0$. Hence by letting $q_1 = c_1/(a - s)$, we can assume that the term of degree 1 of $R(\xi)$ vanishes. Further we distinguish the two cases.

If $a \neq -1$, then $s = a^2 \neq 1$. Hence by letting $q_0 = c_2/(1 - s)$, we can assume that the constant term in $R(\xi)$ vanishes. Then we have $G^{-1} \circ F \circ G : (\xi, \eta) \mapsto (a\xi, a^2\eta + c_0\xi^2)$. Further choosing $(\xi, (a^2/c_0)\eta)$ as a coordinate system, this reduces to the canonical quadratic elementary map E_a $(a \neq \pm 1)$. If $a = -1$, then $s = a^2 = 1$. We have $G^{-1} \circ F \circ G : (\xi, \eta) \mapsto (-\xi, \eta + c_0\xi^2 + c_2)$. If $c_2 = 0$, then $G^{-1} \circ F \circ G$ reduces to the canonical quadratic elementary map E_{-1}, by a similar argument as above. If $c_2 \neq 0$, we let α be one of the square roots of $-c_0/c_2$ and $\beta = -1/c_2$. Choosing $(\alpha\xi, \beta\eta)$ as a coordinate system we obtain the canonical quadratic elementary map E_-.

(B) In this case l is reduced to the form $x \mapsto x + 1$ by translation of the coordinate x. We may suppose that F is of the form $(x, y) \mapsto (x + 1, sy + P(x))$. Let $Q(\xi)$ be a polynomial and define G as above. Then $G^{-1} \circ F \circ G$ has the form

$$(\xi, \eta) \mapsto (\xi, \eta + Q(\xi)) \mapsto (\xi + 1, s(\eta + Q(\xi)) + P(\xi)) \mapsto (\xi + 1, s\eta + R(\xi))$$

where $R(\xi) = P(\xi) + sQ(\xi) - Q(\xi + 1)$. By Lemma 7.1.10 below, we can choose a polynomial $Q(\xi)$ of degree 2 or 3 for which $R(\xi) = 0$. Hence $G^{-1} \circ F \circ G$ is an affine map $(\xi, \eta) \mapsto (\xi + 1, s\eta)$.

(C) In this case F has the form $(x, y) \mapsto (x, sy + P(x))$.

(C$_1$) The case $s \neq 1$. We choose G as above. Then $G^{-1} \circ F \circ G$ takes the form

$$\begin{aligned}(\xi, \eta) \mapsto (\xi, \eta + Q(\xi)) \quad &\mapsto \quad (\xi, s(\eta + Q(\xi)) + P(\xi)) \\ &\mapsto \quad (\xi, s(\eta + Q(\xi)) - Q(\xi) + P(\xi)).\end{aligned}$$

Hence, if we let $Q(\xi) = \frac{1}{1-s} P(\xi)$, then $G^{-1} \circ F \circ G$ is an affine map $(\xi, \eta) \mapsto (\xi, s\eta)$.

(C$_2$) The case $s = 1$. The polynomial $P(x)$ of degree 2 has either a multiple zero or two distinct zeros. If $P(x)$ has a multiple zero, we may assume that $P(x) = c_0x^2$ by a translation of coordinate x. Then $F : (x, y) \mapsto (x, y + c_0x^2)$. This reduces to the canonical quadratic elementary map E_1. If $P(x)$ has two distinct zeros, we may assume that $P(x) = c_0(x^2 - 1)$ by a linear change of coordinate x. This reduces to the canonical quadratic elementary map E_+. □

Lemma 7.1.10 *Let $P(x)$ be a polynomial of degree k and s a complex number. If $s \neq 1$, then the functional equation $Q(x+1) - sQ(x) = P(x)$ has a unique polynomial solution $Q(x)$, whose degree is k. If $s = 1$, the equation has a unique solution up to the constant term, whose degree is $k+1$.*

The proof is left to the reader.

Remark Thus far we have proved the first half of Theorem 7.1.5. The proof shows the following fact: a quadratic polynomial automorphism of $\mathbb{C}^2$ is either (1) conjugate by an affine map to an Hénon map or a canonical elementary map, or (2) conjugate by a polynomial automorphism of degree ≤ 3 to an affine map.

Now in order to prove the latter half of Theorem 7.1.5, we have to take notice of a property invariant under conjugation by holomorphic automorphisms. Such are properties concerning fixed points. The following propositions for affine maps and elementary maps are easily proved.

Proposition 7.1.11 *The set of the fixed points of an affine map is either empty, a set consisting of a single point, a complex line, or the whole of $\mathbb{C}^2$.*

Proposition 7.1.12 (1) *E_a $(a \neq \pm 1)$ has the sole fixed point* $(0, 0)$*;*
(2) *the set of the fixed points of E_1, E_{-1} is the complex line $x = 0$;*
(3) *the set of the fixed points of E_+ consists of two complex lines $x = \pm 1$;*
(4) *E_- has no fixed points.*

Proposition 7.1.13 *The Hénon map $F_{\delta,c}$ has either one or two fixed points. Further the parameters δ and c can be expressed by the eigenvalues λ_1 and λ_2 at one of the fixed points.*

Proof The fixed points of $F_{\delta,c}$ are given by (α, α), where α is a root of the quadratic equation (*) $\zeta^2 - (\delta+1)\zeta + c = 0$. Hence the number of the fixed points is either 1 or 2. The eigenpolynomial at a fixed point (α, α) is $\begin{vmatrix} t & -1 \\ \delta & t - 2\alpha \end{vmatrix} = t^2 - 2\alpha t + \delta$. Therefore, if λ_1, λ_2 are eigenvalues, we have $\delta = \lambda_1\lambda_2,\ 2\alpha = \lambda_1 + \lambda_2$. Further we have $c = (\delta+1)\alpha - \alpha^2$ since α is the root of $(*)$. Thus δ and c are expressed by λ_1, λ_2. □

Proof of the latter part of Theorem 7.1.5 If holomorphic automorphisms F_1 and F_2 are conjugate by a holomorphic automorphism G, i.e., $F_2 \circ G = G \circ F_1$, then the fixed points of F_1 and those of F_2 correspond by G and the sets of eigenvalues at the corresponding fixed points must coincide. First we note that, among the maps given in (i), (ii), (iii) of Theorem 7.1.5, Hénon maps are characterized by the following properties: F has exactly two fixed points or F has a single fixed point with one of the eigenvalues 1. Hence an Hénon map is not conjugate to any map of another type. Further, if Hénon maps $F_{\delta,c}$ and $F_{\delta',c'}$ are conjugate, then we have $(\delta, c) = (\delta', c')$ by Proposition 7.1.13. We can prove that a canonical elementary map E_a is not conjugate to an affine map, in the same way as Proposition 6.2.4. Further, if E_a and $E_{a'}$ are conjugate, then $a = a'$. (For E_a and E_{a^2} where a is a primitive cube root of 1, the pairs of eigenvalues at the fixed point coincide. We can prove that they are not conjugate in the same way as in the proof of Proposition 6.2.4.) Finally, the set of fixed points for E_+ consists of two lines. The set of fixed points for E_- is empty and E_-^2 is conjugate to E_+. Hence E_+ and E_- are not conjugate to any maps of other types. □

We conclude this section by mentioning the index formula for fixed points of generalized Hénon maps. This formula can be regarded as an analogue of the index formula for rational functions of one variable in §4.2.

Let $F : (x, y) \mapsto (y, P(y) - \delta x)$ be a generalized Hénon map of degree k. Here $P(y)$ is a polynomial of degree k. The fixed points of F are of the form (α, α), where α are the roots of the equation $P(\zeta) - (\delta + 1)\zeta = 0$. Hence a generalized Hénon map has at least one and at most k fixed points.

The Jacobian matrix of F at a fixed point $a = (\alpha, \alpha)$ is given by $F'(a) = \begin{pmatrix} 0 & 1 \\ -\delta & P'(\alpha) \end{pmatrix}$. We note that $\lambda_1 + \lambda_2 = P'(\alpha)$, $\lambda_1\lambda_2 = \delta$, since the eigenvalues λ_1, λ_2 of $F'(a)$ are the zeros of the eigenpolynomial $t^2 - P'(\alpha)t + \delta$.

Theorem 7.1.14 *Let F be a generalized Hénon map of degree k that has k distinct fixed points $a_\nu = (\alpha_\nu, \alpha_\nu)$, $\nu = 1, \dots, k$. Let $\lambda_{\nu,1}, \lambda_{\nu,2}$ denote the eigenvalues of $F'(a_\nu)$. Then we have the following relations:*

$$\sum_{\nu=1}^{k} \frac{1}{(1 - \lambda_{\nu,1})(1 - \lambda_{\nu,2})} = 0,$$

$$\sum_{\nu=1}^{k} \left(\frac{1}{1 - \lambda_{\nu,1}} + \frac{1}{1 - \lambda_{\nu,2}} \right) = k.$$

Lemma 7.1.15 *Let $Q(\zeta)$ be a polynomial of degree k (≥ 2) that has k distinct* *zeros* $\alpha_1, \ldots, \alpha_k$. *Then*

$$\sum_{\nu=1}^{k} \frac{1}{Q'(\alpha_\nu)} = 0.$$

Proof We decompose $1/Q(\zeta)$ into partial fractions,

$$\frac{1}{Q(\zeta)} = \sum_{\nu=1}^{k} \frac{A_\nu}{\zeta - \alpha_\nu},$$

with

$$A_\nu = \lim_{\zeta \to \alpha_\nu} \frac{\zeta - \alpha_\nu}{Q(\zeta)} = \frac{1}{Q'(\alpha_\nu)}.$$

Consider the Laurent series expansion of $1/Q(z)$ with center the point at infinity. This begins with the term $1/\zeta^k$. On the other hand, the Laurent series of the partial fraction has the form $(\sum_{\nu=1}^{k} A_\nu)/\zeta + \cdots$. This shows that $\sum_{\nu=1}^{k} A_\nu = 0$ if $k \geq 2$. □

Proof of Theorem 7.1.14 By applying the preceding lemma to the polynomial $(\delta + 1)\zeta - P(\zeta)$, we obtain

$$\sum_{\nu=1}^{k} \frac{1}{1 + \delta - P'(\alpha_\nu)} = 0.$$

The denominator of each term is the value for $t = 1$ of the eigenpolynomial of $F'(a_\nu)$. Hence we have the first formula. Further, noting that $\lambda_{\nu,1}\lambda_{\nu,2} = \delta$, we have

$$\frac{1}{1 - \lambda_{\nu,1}} + \frac{1}{1 - \lambda_{\nu,2}} = 1 + \frac{1 - \delta}{(1 - \lambda_{\nu,1})(1 - \lambda_{\nu,2})}.$$

Summation for $\nu = 1, \ldots, k$ yields the second relation. □

Theorem 7.1.16 *Let F be a generalized Hénon map of degree k. Suppose that the absolute value of the Jacobian determinant δ is different from* 1, *and that F has k distinct fixed points. Then at least one of the fixed points is of saddle type.*

Proof We consider the case $|\delta| < 1$. Since $|\lambda_{\nu,1}\lambda_{\nu,2}| = |\delta| < 1$ for each ν, if a_ν is not of saddle type then $|\lambda_{\nu,1}|, |\lambda_{\nu,2}| \leq 1$ and at least one of them is less than 1. Therefore both Re $\frac{1}{1-\lambda_{\nu,1}}$ and Re $\frac{1}{1-\lambda_{\nu,2}}$ are not less than $\frac{1}{2}$ and at least one of them is greater than $\frac{1}{2}$. Thus we have

$$\mathrm{Re}\,\frac{1}{1-\lambda_{\nu,1}}+\mathrm{Re}\,\frac{1}{1-\lambda_{\nu,2}}>1.$$

If there is no fixed point of saddle type, the sum of the above expression for all the fixed points is greater than k, contradicting Theorem 7.1.14. The case $|\delta|>1$ can be proved in the same manner. □

7.2 Sets of escaping points and non-escaping points

We will study the behavior of the points in $\mathbb{C}^2$ under the iteration of a holomorphic automorphism $F:\mathbb{C}^2\to\mathbb{C}^2$. In what follows we will consider generalized Hénon maps $F(x,y)=(y,P(y)-\delta x)$. We begin with the classification of the points z in $\mathbb{C}^2$ into two classes, according to the behavior of the forward orbits:

$$F(z),\ F^2(z),\ldots,\ F^n(z),\ldots.$$

First we define the set I^+ of escaping points and the set K^+ of non-escaping points for F as follows:

$$\begin{aligned}I^+ &= \{z\in\mathbb{C}^2 \mid \|F^n(z)\|\to\infty\ (n\to\infty)\},\\ K^+ &= \{z\in\mathbb{C}^2 \mid \text{the sequence } \{F^n(z)\}_{n=1}^{\infty} \text{ is bounded}\}.\end{aligned}$$

The study of the Hénon maps in the complex domain started with the pursuit of analogy to polynomial maps of a single variable. The sets I^+ and K^+ defined above are considered as counterparts of the set of escaping points and the filled Julia set for polynomial maps of one variable.

Remark For a positive integer n, we mean by F^n the nth iterate of the automorphism F of $\mathbb{C}^2$. By a $(-n)$th iterate F^{-n} we mean the nth iterate $(F^{-1})^n=F^{-1}\circ\cdots\circ F^{-1}$ of the inverse map F^{-1}. The 0th iterate F^0 is defined to be the identity map. Thus we can speak of the nth iterate F^n of F for any integer n. The inverse of the generalized Hénon map F is given by $F^{-1}(x,y)=(\frac{1}{\delta}(P(x)-y),x)$. This is conjugate to a generalized Hénon map by the linear transformation that permutes the coordinates (x,y). So the following arguments can be applied to the backward orbit

$$F^{-1}(z),\ F^{-2}(z),\ldots,\ F^{-n}(z),\ldots.$$

The suffix $^+$ for K^+, I^+ indicates that they are determined by the behavior of the forward orbit. Objects that are determined by the backward iterations will be indicated with suffix $^-$.

It is clear that the sets I^+, K^+ are invariant under F, i.e., $F(I^+) = I^+, F(K^+) = K^+$. The following theorem is fundamental in our discussion.

Theorem 7.2.1 *For a generalized Hénon map F, we have*

$$I^+ \cup K^+ = \mathbb{C}^2, \quad I^+ \cap K^+ = \varnothing.$$

The set I^+ is a non-empty connected open set.

Let us first make some preparations for the proof. For a positive number R, we define the sets

$$\begin{aligned}
D_R &= \left\{(x, y) \in \mathbb{C}^2 \mid \max\{|x|, |y|\} < R\right\}, \\
V_R^+ &= \left\{(x, y) \in \mathbb{C}^2 \mid \max\{R, |x|\} < |y|\right\}, \\
V_R^- &= \left\{(x, y) \in \mathbb{C}^2 \mid \max\{R, |y|\} < |x|\right\}, \\
W_R^+ &= \mathbb{C}^2 - \overline{V_R^-} \quad (\text{the interior of } D_R \cup \overline{V_R^+}), \\
W_R^- &= \mathbb{C}^2 - \overline{V_R^+} \quad (\text{the interior of } D_R \cup \overline{V_R^-}).
\end{aligned}$$

In what follows, we drop the subscript R and write simply D, V^+, V^-. (See Figure 7.1.)

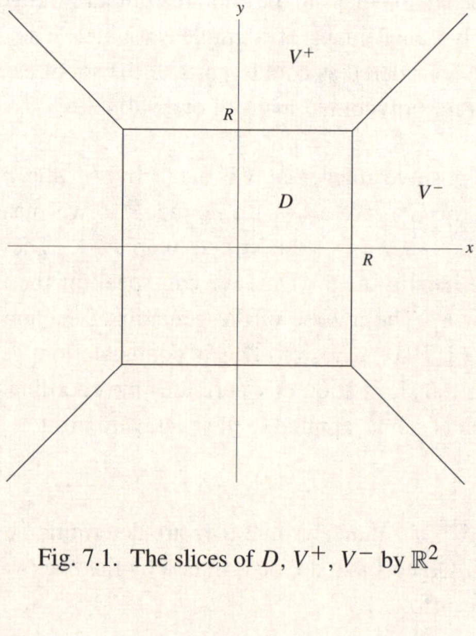

Fig. 7.1. The slices of D, V^+, V^- by $\mathbb{R}^2$

Definition We say that a holomorphic automorphism F of $\mathbb{C}^2$ satisfies *condition (A) for a positive number* R if there exists a real number $\rho > 1$, with the following properties:

(i) if $z = (x, y)$ is a point in $\overline{V^+}$ and $F(z) = (x_1, y_1)$ is its image under F, then $F(z) \in V^+$ and $|y_1| > \rho|y|$;
(ii) if $z = (x, y)$ is a point in $\overline{V^-}$ and $F^{-1}(z) = (x_{-1}, y_{-1})$ is its image under F^{-1}, then $F^{-1}(z) \in V^-$ and $|x_{-1}| > \rho|x|$.

Remark If two holomorphic automorphisms F_1 and F_2 satisfy condition (A) for a positive number R, then their composition $F_2 \circ F_1$ satisfies condition (A) for the same R. In particular, if F satisfies condition (A) for R, then its iterates F^n satisfy condition (A) for R.

Further the inverse F^{-1} of F satisfies condition (A) if the roles of x and y are permuted. It can be proved that a holomorphic automorphism satisfying condition (A) is necessarily a polynomial automorphism.

For the proof of Theorem 7.2.1, it suffices to show the following, Propositions 7.2.2 and 7.2.3.

Proposition 7.2.2 *A generalized Hénon map satisfies condition (A) for a sufficiently large R.*

Proposition 7.2.3 *For a holomorphic automorphism F of $\mathbb{C}^2$ satisfying condition (A), we have*

$$I^+ = \bigcup_{n=0}^{\infty} F^{-n}(V^+), \quad K^+ = \mathbb{C}^2 - I^+ = \bigcap_{n=0}^{\infty} F^{-n}(W^-).$$

Proof of Proposition 7.2.2 We choose and fix a real number $\rho > 0$. Since $P(\zeta)$ is a polynomial of degree ≥ 2, we have $|P(\zeta)/\zeta| \to \infty$ as $|\zeta| \to \infty$. Hence, we can choose a sufficiently large positive number R so that

$$|P(\zeta)| > \rho(1 + |\delta|)|\zeta| \quad \text{for} \quad |\zeta| \geq R.$$

(i) Suppose $z = (x, y) \in \overline{V^+}$, i.e., $|y| \geq R$, $|y| \geq |x|$. Then

$$|y_1| = |P(y) - \delta x| \geq |P(y)| - |\delta x| \geq \rho(1 + |\delta|)|y| - |\delta y| > \rho|y|.$$

Hence $|y_1| > \rho|y| > |y| > \max\{R, |x_1|\}$. This implies $F(z) \in V^+$.

(ii) Suppose $z = (x, y) \in \overline{V^-}$, i.e., $|x| \geq R$, $|x| \geq |y|$. Then

$$|x_{-1}| = \frac{1}{|\delta|}|P(x) - y| \geq \frac{1}{|\delta|}(|P(x)| - |y|) \geq \frac{1}{|\delta|}(\rho(1 + |\delta|)|x| - |x|) > \rho|x|.$$

Hence $|x_{-1}| > \rho|x| > |x| > \max\{R, |y_{-1}|\}$. This implies $F^{-1}(z) \in V^-$. □

Remark Let us estimate the number R for the Hénon map $F : (x, y) \mapsto (y, y^2+c-\delta x)$. From the above proof, it suffices to choose R so that $|\zeta^2+c| > (1+|\delta|)|\zeta|$ for $|\zeta| \geq R$ (with ρ sufficiently close to 1). Since

$$|\zeta^2 + c| - (1 + |\delta|)|\zeta| \geq |\zeta|^2 - (1 + |\delta|)|\zeta| - |c|,$$

we conclude that R has the required property if it is greater than the positive root of the quadratic equation $Q(t) = t^2 - (1 + |\delta|)t - |c| = 0$.

The proof of Proposition 7.2.3 is divided into several steps. In what follows, F is a holomorphic automorphism of $\mathbb{C}^2$ that satisfies condition (A) for some positive number R.

Lemma 7.2.4 *For the open set V^+, we have*

$$V^+ \subset \overline{V^+} \subset F^{-1}(V^+) \subset F^{-1}(\overline{V^+}) \subset \cdots \subset F^{-n}(V^+) \subset \cdots$$

and $\bigcup_{n=0}^{\infty} F^{-n}(V^+) = \bigcup_{n=0}^{\infty} F^{-n}(\overline{V^+}) \subset I^+$.

Proof By (i) of condition (A), we have $F(\overline{V^+}) \subset V^+$. If we map this relation by the automorphisms F^{-n} $(n = 1, 2, \ldots)$, we obtain the first assertion. Suppose that $z = (x, y) \in V^+$ and write $F^n(z) = (x_n, y_n)$. Then

$$|y_n| > \rho|y_{n-1}| > \rho^2|y_{n-2}| > \cdots > \rho^n|y| \geq \rho^n R.$$

Hence $\|F^n(z)\| \to \infty$ $(n \to \infty)$. This shows that $z \in I^+$. In general, if $z \in \bigcup_{n=0}^{\infty} F^{-n}(V^+)$ then $F^m(z) \in V^+$ for some $m \geq 0$. Since $\|F^{m+n}(z)\| \to \infty$ $(n \to \infty)$, we have $z \in I^+$. □

Lemma 7.2.5 *For the open set W^+, we have*

$$W^+ \subset F^{-1}(W^+) \subset \cdots \subset F^{-n}(W^+) \subset \cdots$$

and $\bigcup_{n=0}^{\infty} F^{-n}(W^+) = \mathbb{C}^2$.

Proof Shifting to the complements, we will show that

$$\overline{V^-} \supset F^{-1}(\overline{V^-}) \supset F^{-2}(\overline{V^-}) \supset \cdots \supset F^{-n}(\overline{V^-}) \supset \cdots$$

and that $\bigcap_{n=0}^{\infty} F^{-n}(\overline{V^-}) = \varnothing$. By (ii) of condition (A), we have $V^- \supset F^{-1}(\overline{V^-})$. By mapping this relation by F^{-n} $(n = 1, 2, \ldots)$, we obtain the first assertion. Further, for any $z = (x, y) \in \overline{V^-}$, write $F^{-n}(z) = (x_{-n}, y_{-n})$ ($n = 1, 2, \ldots$). Then

$$|x_{-n}| > \rho|x_{-(n-1)}| > \cdots > \rho^n|x| \geq \rho^n R.$$

Therefore $F^{-n}(\overline{V^-}) \subset \{(x, y) \,|\, |x| > \rho^n R\}$. This implies the second assertion. $\square$

Proof of Proposition 7.2.3 By Lemma 7.2.4, for any point $z \in \mathbb{C}^2$, there exists a number $n_0 \geq 1$ such that $F^{n_0}(z), F^{n_0+1}(z), \ldots$ are all contained in $W^+ \subset D \cup \overline{V^+}$. Hence one of the following two cases occurs.

(1) There exists an m such that $F^m(z) \in \overline{V^+}$.
(2) $F^n(z) \in D$ for every $n \geq n_0$.

In case (1), we have $F^{m+1}(z) \in V^+$ and hence z is contained in I^+. In case (2) we have $z \in K^+$ by definition. Thus we have $I^+ = \bigcup_{n=0}^{\infty} F^{-n}(V^+)$. This set I^+ is a non-empty connected open set since it is the union of an increasing sequence of such sets. $\square$

Now let us note some properties of the sets I^+ and K^+.

Proposition 7.2.6 (1) *For any compact subset E of I^+, there exists an $n \geq 0$ such that $F^n(E) \subset V^+$.*

(2) *For any compact subset E of K^+, there exists an $n \geq 0$ such that $F^n(E) \subset D$.*

Proof (1) Since $E \subset F^{-n}(V^+)$ for some $n \geq 0$, we have $F^n(E) \subset V^+$.

(2) By Lemma 7.2.5, we have $E \subset F^{-n}(W^+)$ for some $n \geq 0$. But since $E \cap F^{-n}(\overline{V^+}) = \varnothing$ we have $E \subset F^{-n}(D)$. $\square$

Theorem 7.2.7 *The set I^+ of escaping points is a domain of holomorphy. Further, if the interior of the set K^+ of non-escaping points is non-empty, then its connected components are domains of holomorphy.*

Proof Since V^+ is a domain of holomorphy, so are the sets $F^{-n}(V^+)$, $n = 1, 2, \ldots$. Therefore the union I^+ of the increasing sequence $\{F^{-n}(V^+)\}$ is also a domain of holomorphy. In general, the set of those points that have neighborhoods on which a family of holomorphic functions on $\mathbb{C}^2$ is uniformly bounded is a domain of holomorphy. This shows the latter part of the theorem. $\square$

The set K^+ is non-compact, unlike the filled Julia set of a polynomial in one complex variable.

Theorem 7.2.8 *The set K^+ has a non-empty intersection with every complex line parallel to the y-axis. Hence K^+ is an unbounded set.*

Proof Let $L_a = \{(a, y) \mid y \in \mathbb{C}\}$ be a complex line parallel to the y-axis. Consider the compact subset $E_a = \big\{(a, y) \,\big|\, |y| \le \max\{R, |a|\}\big\}$ of L_a. Since $L_a - E_a \subset V^+$, we have $F^n(L_a - E_a) \subset V^+$ for all $n \ge 0$. Suppose that $L_a \subset I^+$. Then, for a sufficiently large n, we have $F^n(E_a) \subset V^+$ and hence $F^n(L_a) \subset V^+$. Then, writing $F^n(x, y) = (f_n(x, y), g_n(x, y))$, we have $|g_n(a, y)| > R$. Hence $g_n(a, y)$ must be a constant. Further $f_n(a, y)$ is also a constant, since $|f_n(a, y)| < |g_n(a, y)|$. This implies that $F^n(L_a)$ is a point, which contradicts that F is a biholomorphic map. □

Remark It is known that any composition of generalized Hénon maps has a fixed point. This shows that K^+ is non-empty, since a fixed point is contained in K^+.

Theorem 7.2.9 *We denote by $m(K^+)$ the four-dimensional volume (Lebesgue measure) of K^+. Let $J_F = \delta$ (constant) be the Jacobian determinant of F. Then the following assertions hold:*

(1) *if $|\delta| > 1$, then $m(K^+) = 0$;*
(2) *if $|\delta| = 1$, then $m(K^+) < +\infty$;*
(3) *if $|\delta| < 1$, then $m(K^+) = 0$, or $+\infty$.*

Proof We let $E = D \cap K^+$. Then we have the sequence

$$E \subset F^{-1}(E) \subset F^{-2}(E) \subset \cdots \subset F^{-n}(E) \subset \cdots$$

with $K^+ = \bigcup_{n=0}^{\infty} F^{-n}(E)$. Therefore $m(F^{-n}(E))$, $n = 0, 1, 2, \ldots$, is a non-decreasing sequence and $m(K^+) = \lim_{n\to\infty} m(F^{-n}(E))$.

On the other hand, $m(F^{-n}(E)) = |\delta|^{-2n} m(E)$. If $m(E) = 0$, then $m(K^+) = 0$. If $m(E) > 0$, then $|\delta| \le 1$. Further, if $|\delta| < 1$, then $m(K^+) = \infty$. Finally, if $|\delta| = 1$, then $m(K^+) = m(E)$ is finite. □

Remark Another important property of K^+ is that it is connected (see Bedford and Smillie). See also §7.4 for a particular case.

Now we consider the set $K = K^+ \cap K^-$. This is the set of all points that have bounded grand orbit $\{F^n(z)\}_{n\in\mathbb{Z}}$. It is clear that K is a closed set that is invariant under F. Further, since $K^+ \subset W^-$, $K^- \subset W^+$, we have $K \subset W^- \cap W^+ = D$. Thus K is a compact set. If the Jacobian determinant of F is not equal to 1 in absolute value, then $m(K) = 0$ by Theorem 7.2.9, so that K has no interior points.

In a generalized Hénon map F, if $P(x)$ is a polynomial with real coefficients and δ is a real number, the restriction of F to $\mathbb{R}^2$ is an automorphism of $\mathbb{R}^2$. In this case we call F a real generalized Hénon map. The real plane $\mathbb{R}^2$ is divided into the sets of the escaping points $I^+ \cap \mathbb{R}^2$ and the non-escaping points $K^+ \cap \mathbb{R}^2$. The set $K^+ \cap \mathbb{R}^2$ may be empty.

Proposition 7.2.10 *Let F be a real generalized Hénon map. The set $K^+ \cap \mathbb{R}^2$ is non-empty if and only if F has a fixed point in $\mathbb{R}^2$.*

Proof If F has a fixed point in $\mathbb{R}^2$, then it belongs to $K^+ \cap \mathbb{R}^2$; hence $K^+ \cap \mathbb{R}^2$ is non-empty. Conversely suppose that F has no real fixed points. Then the equation $y = P(y) - \delta y$ has no real roots and the polynomial $P(y) - (\delta + 1)y$ is always positive or negative on $\mathbb{R}$. Suppose that it is always positive and let m be its minimum. Consider the linear function $u(x, y) = y - \delta x$. Then

$$u(F(z)) = y_1 - \delta x_1 = P(y) - \delta x - \delta y \geq u(z) + m.$$

This shows that $u(F^n(z))$ diverges to $+\infty$ as $n \to \infty$. □

7.3 Structure of the set of escaping points

In this section we will study the analytic structure of the set I^+ for generalized Hénon maps. We begin with the construction of a function that can be considered as an analogue of the Böttcher function. (See §1.2.)

Theorem 7.3.1 *For a generalized Hénon map F of degree k, there exists a holomorphic function $\varphi(z)$ on V^+ with the following properties. Such a function is uniquely determined up to a constant factor $(k-1)$th root of unity.*

(i) *The function $\varphi(z)$ satisfies the functional equation $\varphi(F(z)) = \{\varphi(z)\}^k$.*

(ii) *There exist positive numbers m, M such that $m \leq \frac{|\varphi(z)|}{|y|} \leq M$ on V^+. Here $z = (x, y)$.*

Proof For a point $z = (x, y)$ we write $F^n(z) = (f_n(z), g_n(z))$, $n = 1, 2, \ldots$. We will show that, for each n, we can choose a branch $\varphi_n(z)$ on V^+ of the k^nth root of $g_n(z)$ and the sequence $\varphi_n(z)$ converges uniformly to the required holomorphic function.

First, if $z = (x, y) \in V^+$ then $|g(z)| > R$ and hence we can choose a single valued branch of the logarithm $\beta(z)$ of

$$\frac{g(z)}{y^k} = \frac{P(y) - \delta x}{y^k} = c_0 + \frac{c_1}{y} + \cdots + \frac{c_k - \delta x}{y^k}$$

on V^+. Then

$$g(z) = P(y) - \delta x = y^k e^{\beta(z)}, \quad z = (x, y) \in V^+.$$

Such a $\beta(z)$ is determined modulo $2\pi i$. We choose and fix a choice of $\beta(z)$. Then $\beta(z)$ is bounded on V^+. In the above equation we replace z by $F^{n-1}(z) = (f_{n-1}(z), g_{n-1}(z))$. Then, noting that $g(F^{n-1}(z)) = g_n(z)$, we obtain

$$g_n(z) = g_{n-1}(z)^k \exp \beta(F^{n-1}(z)).$$

Using similar relations for $F^{n-2}(z), F^{n-3}(z), \ldots$, we have

$$g_n(z) = y^{k^n} \exp\{k^{n-1}\beta(z) + \cdots + k\beta(F^{n-2}(z)) + \beta(F^{n-1}(z))\}.$$

We define, as a k^nth root of $g_n(z)$,

$$\phi_n(z) = y \exp\left\{\frac{1}{k}\beta(z) + \cdots + \frac{1}{k^{n-1}}\beta(F^{n-2}(z)) + \frac{1}{k^n}\beta(F^{n-1}(z))\right\}.$$

Since $\beta(z)$ is bounded on V^+, the series

$$\gamma(z) = \frac{1}{k}\beta(z) + \frac{1}{k^2}\beta(F(z)) + \cdots$$

converges uniformly on V^+. Hence the limit $\varphi(z) = \lim_{n\to\infty} \varphi_n(z) = ye^{\gamma(z)}$ exists. By letting $n \to \infty$ in the relation $\varphi_n(F(z)) = \{\varphi_{n+1}(z)\}^k$, we obtain property (i). Further, (ii) holds since $\gamma(z)$ is bounded.

To prove the uniqueness, let $\tilde{\varphi}(z)$ be a function satisfying the condition. Let $\xi(z) = \tilde{\varphi}(z)/\varphi(z)$. There exist positive numbers m' and M' such that $m' \leq |\xi(z)| \leq M'$ ($z \in V^+$). Since $\xi(z)$ satisfies the relation $\xi(F(z)) = \{\xi(z)\}^k$, we have $\xi(F^n(z)) = \{\xi(z)\}^{k^n}$ for any n. Therefore $m' \leq |\xi(z)|^{k^n} \leq M'$ for any $n \geq 1$. Thus $\xi(z)$ is a constant of absolute value 1 with $\xi^k = \xi$. □

Remark In the proof above we could have chosen $\beta(z) + 2m\pi i$ in place of $\beta(z)$. Then we would have got $\exp(\frac{2m\pi i}{k-1})\varphi(p)$ in place of $\varphi(p)$. In the case $c_0 = 1$, we have $g(z)/y^k = 1 + c_1/y + \cdots$. So we can choose a $\beta(z)$ of the form $c_1/y + \cdots$. Then

$$\varphi(z) = y\exp(c_1/y + \cdots) = y + c_1 + O(1/y).$$

Since the function φ is defined using the forward iteration of F, it should have been denoted as φ^+. We drop the suffix $^+$ for the sake of simplicity.

Proposition 7.3.2 *There exists a closed 1-form ω on I^+ such that $F^*\omega = k\omega$ and that $\omega = \frac{d\varphi}{\varphi}$ on V^+.*

Proof We define $\omega = \frac{d\varphi}{\varphi}$ on V^+. By differentiating the relation $F^*\varphi(z) = \varphi(z)^k$ we have $F^* d\varphi = k\varphi(z)^{k-1} d\varphi$. By dividing by $F^*\varphi(z) = \varphi(z)^k$ we obtain $F^* \frac{d\varphi}{\varphi} = k\frac{d\varphi}{\varphi}$. Thus $F^*\omega = k\omega$ on V^+. In general we define $\omega = \frac{1}{k^n}(F^n)^*\omega$ on $F^{-n}(V^+)$ $(n = 1, 2, \ldots)$. □

We can study the fundamental group $\pi_1(I^+)$ of the set I^+ of escaping points by using this holomorphic closed 1-form ω. Let k be an integer ≥ 2. We denote the set of all rational numbers whose denominator is a power of k by

$$\mathbb{Z}\left[\frac{1}{k}\right] = \left\{ \frac{m}{k^n} \,\middle|\, m, n \in \mathbb{Z} \right\}.$$

This set is an infinitely generated commutative group under addition. In what follows we will show that the fundamental group $\pi_1(I^+)$ is isomorphic to $\mathbb{Z}[\frac{1}{k}]$, where k is the degree of F.

For a closed curve C in I^+, we set $\alpha(C) = \frac{1}{2\pi i}\int_C \omega$. Since ω is a closed 1-form, the number $\alpha(C)$ is determined by the homotopy class of C.

Proposition 7.3.3 *For any closed curve C in the set I^+ of escaping points the following assertions hold:*

(1) *$\alpha(F(C)) = k\alpha(C)$, where $F(C)$ denotes the image of the curve C under F;*
(2) *$\alpha(C) \in \mathbb{Z}[\frac{1}{k}]$;*
(3) *the curve C is zero-homotopic in I^+ if and only if $\alpha(C) = 0$;*
(4) *for any $r \in \mathbb{Z}[\frac{1}{k}]$, there exists a closed curve C in I^+ such that $\alpha(C) = r$.*

Proof Assertion (1) holds since

$$\alpha(F(C)) = \frac{1}{2\pi i}\int_{F(C)} \omega = \frac{1}{2\pi i}\int_C F^*\omega = \frac{1}{2\pi i}\int_C k\omega = k\alpha(C).$$

To show the other assertions, consider first the case where the closed curve C is in V^+. In V^+, we can express $\varphi(z)$ in the form $ye^{\gamma(z)}$. Hence

$$\omega = \frac{d\varphi}{\varphi} = \frac{dy}{y} + d\gamma.$$

This shows that for a closed curve C in V^+, the integral

$$\frac{1}{2\pi i}\int_C \omega = \frac{1}{2\pi i}\int_C \frac{dy}{y}$$

is an integer (i.e., the winding number of C around the x-axis $y = 0$). This vanishes if and only if C is zero-homotopic in V^+. Let C_0 denote the closed curve in V^+ defined by

$$C_0 : [0, 1] \ni t \mapsto (x, y) = (0, 2Re^{2\pi it}) \in V^+.$$

Then we have $\alpha(C_0) = \frac{1}{2\pi i}\int_{C_0} \omega = 1$. We note that $\alpha(mC_0) = m$ for a closed curve mC_0 $(m \in \mathbb{Z})$.

Let C be any closed curve in I^+. Choose a positive integer n so that $F^n(C)$ is contained in V^+, in other words so that C is in $F^{-n}(V^+)$. Since $\alpha(F^n(C))$ is an integer, we have $\alpha(C) = \frac{1}{k^n}\alpha(F^n(C)) \in \mathbb{Z}[\frac{1}{k}]$. This shows assertion (2). Further, if $\alpha(C) = 0$, then $\alpha(F^n(C)) = 0$ and hence $F^n(C)$ is zero-homotopic in V^+. Since F^n maps $F^{-n}(V^+)$ onto V^+ biholomorphically, the curve C is zero-homotopic in $F^{-n}(V^+)$. Assertion (3) is thus shown.

Finally, for any $\frac{m}{k^n} \in \mathbb{Z}[\frac{1}{k}]$, we choose a closed curve $F^{-n}(mC_0)$. Then $\alpha(F^{-n}(mC_0)) = \frac{1}{k^n}\alpha(mC_0) = \frac{m}{k^n}$. This proves (4). □

Let a be a point in the set I^+ of escaping points and let $\pi_1(I^+, a)$ denote the fundamental group of I^+ with base point a. The above proposition shows that α is an isomorphism of $\pi_1(I^+, a)$ onto $\mathbb{Z}[\frac{1}{k}]$. Since the fundamental group $\pi_1(I^+, a)$ is commutative, the isomorphism of $\pi_1(I^+, a)$ and the fundamental group $\pi_1(I^+, b)$ with another base point $b \in I^+$ is uniquely determined. So we will not explicitly write the base point of the fundamental group. The above results are summarized in the following theorem.

Theorem 7.3.4 (Hubbard and Oberste-Vorth) *The map α is a group isomorphism of $\pi_1(I^+)$ onto $\mathbb{Z}[\frac{1}{k}]$. Denoting by F_* the automorphism of $\pi_1(I^+)$ induced by the map $F|I^+ : I^+ \to I^+$ and by k the automorphism of $\mathbb{Z}[\frac{1}{k}]$ defined by $r \mapsto kr$, we have $\alpha \circ F_* = k\alpha$.*

Remark We could have proved that $\pi_1(I^+)$ is isomorphic to $\mathbb{Z}[\frac{1}{k}]$ without using the holomorphic closed 1-form ω. Let us explain this point, in order to make clear the geometric meaning. The set I^+ is the union of the increasing sequence

$$V^+ \subset F^{-1}(V^+) \subset F^{-2}(V^+) \subset \cdots \subset F^{-n}(V^+) \subset \cdots.$$

This sequence of inclusion maps induces a sequence of homomorphisms of fundamental groups

$$\pi_1(V^+) \to \pi_1(F^{-1}(V^+)) \to \cdots \to \pi_1(F^{-n}(V^+)) \to \cdots; \tag{7.1}$$

$\pi_1(V^+)$ is a cyclic group $\cong \mathbb{Z}$ generated by the homotopy class of C_0. Since each $F^{-n}(V^+)$ is biholomorphic to V^+ by F^n, the group $\pi_1(F^{-n}(V^+))$ is also a cyclic group $\cong \mathbb{Z}$. The inclusion map $\iota_n : F^{-(n-1)}(V^+) \subset F^{-n}(V^+)$ induces a homomorphism

$$\iota_{n*} : \pi_1(F^{-(n-1)}(V^+)) \cong \mathbb{Z} \to \pi_1(F^{-n}(V^+)) \cong \mathbb{Z}.$$

This is injective and given by $\mathbb{Z} \ni m \mapsto km \in \mathbb{Z}$. The fundamental group $\pi_1(I^+)$ is given as the inductive limit of the sequence (7.1). This gives another proof of Theorem 7.3.4.

Now we will consider the analytic continuation of the holomorphic function $\varphi(z)$ defined on the domain V^+ to I^+. Using the function equation $\varphi(F(z)) = \{\varphi(z)\}^k$, we want to define $\varphi(z) = \{\varphi(F(z))\}^{1/k}$ on $F^{-1}(V^+)$. This kth root is not single valued. (In fact, $\{\varphi(z)\}^{1/k} = y^{1/k}e^{\alpha(z)/k}$ is not single valued on V^+.) Hence the analytic continuation of $\varphi(z)$ to $F^{-1}(V^+)$ gives rise to a function on a k-fold covering manifold over $F^{-1}(V^+)$. Similarly the analytic continuation to $F^{-2}(V^+)$ gives rise to a function defined on a k^2-fold covering manifold. By repetition of this procedure we will obtain an infinitely many valued function over I^+.

In order to precisely describe this situation, let us begin with the construction of a covering manifold $\hat{I}^+$ over I^+. Let $\mathcal{H} = \alpha^{-1}(\mathbb{Z})$ be the subgroup of $\pi_1(I^+)$ that corresponds to $\mathbb{Z} \subset \mathbb{Z}[\frac{1}{k}]$ under the isomorphism $\alpha : \pi_1(I^+) \to \mathbb{Z}[\frac{1}{k}]$. The covering manifold $\hat{\pi} : \hat{I}^+ \to I^+$ corresponding to this subgroup $\mathcal{H}$ is constructed as follows.

We choose and fix a base point a in V^+. In the set of all pairs (z, C) of points z and paths in I^+ that have a as initial point and z as end point, we define an equivalence relation by

$$(z, C) \sim (z', C') \Leftrightarrow z = z' \text{ and } [CC'^{-1}] \in \mathcal{H}.$$

The set of all equivalence classes $[z, C]$ is denoted by $\hat{I}^+$. The correspondence $(z, C) \mapsto z$ defines a map $\hat{\pi} : \hat{I}^+ \to I^+$. The topology on $\hat{I}^+$ is introduced so that $\hat{\pi}$ is a local homeomorphism. The fundamental group of $\hat{I}^+$ is isomorphic to $\mathcal{H} \cong \mathbb{Z}$. Further the structure of a complex manifold is defined so that $\hat{\pi}$ is a holomorphic map.

Now we define a subdomain $\hat{V}^+$ of $\hat{I}^+$ by

$$\hat{V}^+ = \{[z, C] \in \hat{I}^+ \mid C \text{ is a path in } V^+\}.$$

Then the map $\hat{\pi}$ maps $\hat{V}^+$ onto V^+ injectively. In fact, if $p = [z, C]$, $p' = [z, C'] \in \hat{V}^+$ and $\hat{\pi}(p) = \hat{\pi}(p') = z$, then CC'^{-1} is a closed curve in V^+ such that $[CC'^{-1}] \in \mathcal{H}$; hence $p = p'$.

Let us define a holomorphic function $\hat{\varphi}(p)$ on the covering manifold $\hat{I}^+$. For a $p = [z, C] \in \hat{I}^+$, we let $\hat{\varphi}(p) = \varphi(a) \exp \int_C \omega$. This does not depend on the choice of path C that determines p. In fact, if $p = [z, C] = [z, C']$, then $[CC'^{-1}] \in \mathcal{H}$; hence $\int_C \omega - \int_{C'} \omega$ is a multiple of $2\pi i$.

The following proposition shows that $\hat{\varphi}$ is a holomorphic function on the covering manifold $\hat{I}^+$, obtained by analytic continuation of φ along all curves in I^+.

Proposition 7.3.5 (1) *For $p \in \hat{V}^+$, we have $\hat{\varphi}(p) = \varphi(\hat{\pi}(p))$.*

(2) *If p, p' are points in $\hat{I}^+$ such that $\hat{\pi}(p) = \hat{\pi}(p')$ and $\hat{\varphi}(p) = \hat{\varphi}(p')$, then $p = p'$.*

Proof (1) Suppose that $p = [z, C] \in \hat{V}^+$, then

$$\hat{\varphi}(p) = \varphi(a) \exp \int_C \omega = \varphi(a) \exp \int_C \frac{d\varphi}{\varphi} = \varphi(z) = \varphi(\hat{\pi}(p)).$$

(2) Suppose that $z = \hat{\pi}(p) = \hat{\pi}(p')$, and write $p = [z, C]$, $p' = [z, C']$. Then CC'^{-1} is a closed curve.

Since $\hat{\varphi}(p) = \hat{\varphi}(p')$, the integrals $\int_C \omega$ and $\int_{C'} \omega$ differ by a multiple of $2\pi i$. Therefore $\alpha(CC'^{-1}) \in \mathbb{Z}$. This implies that $[CC'^{-1}] \in \mathcal{H}$ and hence $p = p'$. □

We will define a holomorphic map $\hat{F} : \hat{I}^+ \to \hat{I}^+$ that satisfies the conditions $\hat{\pi}(\hat{F}(p)) = F(\hat{\pi}(p))$ and $\hat{F}(\hat{V}^+) \subset \hat{V}^+$. We choose and fix a path l in V^+ with initial point a and end point $F(a)$. If $p \in I^+$ is expressed by $[z, C]$, then the path C connects the points a and z. Hence $F(C)$ connects $F(a)$ and $F(z)$. The composite $lF(C)$ connects a and $F(z)$. We define the map $\hat{F}$ by $\hat{F}(p) = [F(z), lF(C)]$. We can easily verify that this definition is independent of the choice of the path C. The map $\hat{F}$ is a covering map (which has neither branch point nor relative boundary) from $\hat{I}^+$ onto $\hat{I}^+$ with sheet number k.

Proposition 7.3.6 *The functional equation $\hat{\varphi}(\hat{F}(p)) = \{\hat{\varphi}(p)\}^k$ holds.*

Proof

$$\begin{aligned} \hat{\varphi}(\hat{F}(p)) &= \varphi(a) \exp \int_{lF(C)} \omega \\ &= \varphi(a) \left(\exp \int_l \omega \right) \left(\exp \int_{F(C)} \omega \right) \\ &= \varphi(F(a)) \exp k \int_C \omega \\ &= \left\{ \varphi(a) \exp \int_C \omega \right\}^k = \{\hat{\varphi}(p)\}^k. \end{aligned}$$

□

Now we will show that the covering manifold $\hat{I}^+$ is biholomorphic to $\mathbb{C} \times (\mathbb{C} - \bar{\Delta})$ where $\bar{\Delta}$ denotes the closed unit disk in $\mathbb{C}$. For this purpose we will construct a holomorphic function $\hat{\psi}$ on $\hat{I}^+$ so that the pair $(\hat{\psi}, \hat{\varphi})$ defines a biholomorphic map $\hat{I}^+ \to \mathbb{C} \times (\mathbb{C} - \bar{\Delta})$. First we define a subdomain U^+ of V^+ by

$$U^+ = \left\{ z \in V^+ \,\middle|\, |\varphi(z)| > M \max\{R, |x|\} \right\}.$$

Here M is a sufficiently large constant. We have $F(U^+) \subset U^+$.

Lemma 7.3.7 *There exists a holomorphic function $\psi(z)$ on U^+ satisfying the following conditions:*

(i) *$\psi(z)$ satisfies the function equation*

$$\psi(F(z)) = \frac{\delta}{k}\psi(z) + Q(\varphi(z)),$$

where $Q(t)$ is a polynomial of degree $k+1$ of a single variable;

(ii) *the map $\Phi = (\psi, \varphi) : U^+ \to \mathbb{C}^2$ is an injective holomorphic map such that the image $\Phi(U^+)$ contains the domain*

$$\left\{ (s, t) \in \mathbb{C}^2 \,\middle|\, |t| > M, |s| < \rho |t|^k \right\}$$

where ρ is a sufficiently large constant.

Proof The holomorphic map $z = (x, y) \mapsto (x, t) = (x, \varphi(z))$ maps U^+ onto

$$\left\{ (x, t) \in \mathbb{C}^2 \,\middle|\, |t| > M \max\{R, |x|\} \right\}$$

biholomorphically. Hence we can regard $(x, t) = (x, \varphi(z))$ as a coordinate system on U^+. Let $(x, t) \mapsto (x, y) = (x, \lambda(x, t))$ be the inverse map, whose Jacobian determinant is

$$\frac{\partial(x, y)}{\partial(x, t)} = \frac{\partial \lambda}{\partial t}.$$

Now we define a function s on U^+ by

$$s(x, t) = t \int_0^x \frac{\partial \lambda}{\partial t}(\xi, t) d\xi,$$

and regard (s, t) as a new coordinate system. Then we have

$$\frac{\partial(s, t)}{\partial(x, t)} = t \frac{\partial \lambda}{\partial t}.$$

Therefore the Jacobian determinant of the change of coordinates $(s, t) \mapsto (x, y)$ is given by

$$\frac{\partial(x, y)}{\partial(s, t)} = \frac{\partial(x, y)}{\partial(x, t)} \bigg/ \frac{\partial(s, t)}{\partial(x, t)} = \frac{1}{t}.$$

Now we express the map F with respect to the coordinate system (s, t). This amounts to considering the composed map

$$(s, t) \mapsto (x, y) \stackrel{F}{\mapsto} (x_1, y_1) \mapsto (s_1, t_1). \tag{7.2}$$

Here we have $t_1 = t^k$ since $t = \varphi(x, y), t_1 = \varphi(x_1, y_1)$. The Jacobian determinant of the map $(s, t) \mapsto (s_1, t_1)$ is

$$\frac{\partial(s_1, t_1)}{\partial(s, t)} = kt^{k-1}\frac{\partial s_1}{\partial s}.$$

On the other hand, this is the product of the Jacobian determinants of the maps in (7.2):

$$\frac{\partial(s_1, t_1)}{\partial(x_1, y_1)}\frac{\partial(x_1, y_1)}{\partial(x, y)}\frac{\partial(x, y)}{\partial(s, t)} = t_1\delta\frac{1}{t} = \delta t^{k-1}.$$

Hence $\partial s_1/\partial s = \delta/k$. Thus we conclude that the map (7.2) is of the form

$$(s, t) \mapsto (s_1, t_1) = \left(\frac{\delta}{k}s + Q(t), t^k\right) \tag{7.3}$$

where $Q(t)$ is a holomorphic function for $|t| > MR$. To study the properties of the function $Q(t)$, we will look at the image of $(0, t)$ by the map (7.2). From

$$(0, t) \mapsto (0, \lambda(0, t)) \stackrel{F}{\mapsto} (\lambda(0, t), P(\lambda(0, t)) \mapsto (Q(t), t^k)$$

we have

$$Q(t) = t^k \int_0^{\lambda(0,t)} \frac{\partial\lambda}{\partial t}(\xi, t^k)d\xi.$$

Since $\lambda(0, t) - t$ and $\frac{\partial\lambda}{\partial t}$ are bounded functions, we have

$$Q(t) = t^{k+1} + q_k t^k + \cdots + q_0 + \frac{q_{-1}}{t} + \frac{q_{-2}}{t^2} + \cdots.$$

We decompose $Q(t)$ as

$$Q(t) = Q_+(t) + Q_-(t),$$

with

$$Q_+(t) = q_n t^{k+1} + \cdots + q_0, \quad Q_-(t) = \frac{q_{-1}}{t} + \frac{q_{-2}}{t^2} + \cdots.$$

Let

$$R(t) = \frac{k}{\delta} Q_-(t) + \left(\frac{k}{\delta}\right)^2 Q_-(t^k) + \cdots + \left(\frac{k}{\delta}\right)^{n+1} Q_-(t^{k^n}) + \cdots.$$

This sequence is convergent and $R(t)$ satisfies the function equation

$$\frac{\delta}{k} R(t) - R(t^k) = Q_-(t).$$

We set $\tilde{s} = s + R(t)$ and express F with respect to the coordinate system $(\tilde{s}, t)$. Then we have

$$(\tilde{s}, t) \mapsto \left(\frac{\delta}{k}(\tilde{s} - R(t)) + Q(t) + R(t^k), t^k\right) = \left(\frac{\delta}{k}\tilde{s} + Q_+(t), t^k\right).$$

Let us write s, $Q(t)$ in place of $\tilde{s}$, $Q_+(t)$. Then we can assume that $Q(t)$ in (7.3) is a polynomial of degree $k+1$. Therefore the function $s = \psi(z)$ on U^+ satisfies the condition of the lemma. □

Now we denote by $\hat{U}^+$ the subdomain of $\hat{V}^+$ that corresponds to U^+ by π:

$$\hat{U}^+ = \{p \in \hat{V}^+ \mid \pi(p) \in U^+\}.$$

Lemma 7.3.8 *The sequences $\{\hat{F}^{-n}(\hat{V}^+)\}_{n=0}^{\infty}$ and $\{\hat{F}^{-n}(\hat{U}^+)\}_{n=0}^{\infty}$ are increasing and*

$$\hat{I}^+ = \bigcup_{n=0}^{\infty} \hat{F}^{-n}(\hat{V}^+) = \bigcup_{n=0}^{\infty} \hat{F}^{-n}(\hat{U}^+).$$

Proof First we will prove the assertion concerning $\hat{V}^+$. By considering the inverse image of the relation $\hat{F}(\hat{V}^+) \subset \hat{V}^+$ by the map $\hat{F}^n$, we find that $\hat{F}^{-(n-1)}(\hat{V}^+) \subset \hat{F}^{-n}(\hat{V}^+)$. Now, for any point $p = [z, C] \in \hat{I}^+$ (where C is a path in I^+ connecting a and z), there exists an n such that $F^n(C) \subset V^+$ by Proposition 7.2.6. Let $\hat{F}^n(p) = [F^n(z), C']$. Since $C' = lF(l) \cdots F^{n-1}(l)F(C)$ is a path in V^+, we have $\hat{F}^n(p) \in \hat{V}^+$. Hence $p \in \hat{F}^{-n}(\hat{V}^+)$. The assertion for $\hat{U}^+$ can be proved from the fact that $F(U^+) \subset U^+$ and that $F^m(V^+) \subset U^+$ for sufficiently large m. □

Lemma 7.3.9 *There exists a holomorphic function $\hat{\psi}$ on $\hat{I}^+$ such that $\hat{\psi}(p) = \psi(\hat{\pi}(p))$ for $p \in \hat{U}^+$ and that*

$$\hat{\psi}(\hat{F}(p)) = \frac{\delta}{k}\hat{\psi}(p) + Q(\hat{\varphi}(p)) \tag{7.4}$$

for $p \in \hat{I}^+$.

Proof We define $\hat{\psi}$ for $p \in \hat{U}^+$ by $\hat{\psi}(p) = \psi(\hat{\pi}(p))$. Then $\hat{\psi}$ satisfies the functional equation (7.4). By using this equation, we can extend $\hat{\psi}$ to the whole of $\hat{I}^+$. Specifically, we define, on each $\hat{F}^{-n}(U^+)$,

$$\hat{\psi}(p) = \left(\frac{k}{\delta}\right)^n \hat{\psi}(\hat{F}^n(p)) - \left(\frac{k}{\delta}\right)^n Q(t^n) - \cdots - \frac{k}{\delta}Q(t), \quad t = \hat{\varphi}(p).$$

□

We define a holomorphic map $\hat{\Phi}$ from the covering manifold $\hat{I}^+$ to $\mathbb{C}\times(\mathbb{C}-\bar{\Delta})$ by $\hat{\Phi} = (\hat{\psi}, \hat{\varphi})$. Further we define a holomorphic map G of $\mathbb{C} \times (\mathbb{C} - \bar{\Delta})$ to itself by

$$G(s, t) = \left(\frac{\delta}{k}s + Q(t), t^k\right).$$

Then $\hat{F}$ is conjugate to G by $\hat{\Phi}$, i.e., $\hat{\Phi} \circ \hat{F} = G \circ \hat{\Phi}$.

Theorem 7.3.10 *The map $\hat{\Phi}$ is a biholomorphic map from $\hat{I}^+$ onto $\mathbb{C} \times (\mathbb{C} - \bar{\Delta})$.*

Proof First we note that G is a k-sheeted covering map from $\mathbb{C} \times (\mathbb{C} - \bar{\Delta})$ to itself and that $\mathbb{C} \times (\mathbb{C} - \bar{\Delta}) = \bigcup_{n=0}^{\infty} G^{-n}(W)$ where $W = \hat{\Phi}(\hat{U}^+)$. To show that $\hat{\Phi}$ maps $\hat{I}^+$ bijectively onto $\mathbb{C} \times (\mathbb{C} - \bar{\Delta})$, it suffices to show that, for all positive integers n, the map $\hat{\Phi} : F^{-n}(\hat{U}^+) \to G^{-n}(W)$ is bijective. If $p \in F^{-n}(\hat{U}^+)$, then $\hat{\Phi}(p) \in G^{-n}(W)$ since $G^n \circ \hat{\Phi}(p) = \hat{\Phi} \circ \hat{F}^n(p) \in W$. Next, to show the injectivity of $\hat{\Phi}$, suppose that $\hat{\Phi}(p) = \hat{\Phi}(p')$ for two points $p, p' \in \hat{I}^+$. Then we have $\hat{\Phi}\circ\hat{F}^n(p) = G^n\circ\hat{\Phi}(p) = G^n\circ\hat{\Phi}(p') = \hat{\Phi}\circ\hat{F}^n(p')$. Since $\hat{\Phi}$ is injective on $\hat{U}^+$ we have $\hat{F}^n(p) = \hat{F}^n(p')$, hence $F^n \circ \hat{\pi}(p) = \hat{\pi} \circ \hat{F}^n(p) = \hat{\pi} \circ \hat{F}^n(p') = F^n \circ \hat{\pi}(p')$. Since F^n is a biholomorphic map, we have $\hat{\pi}(p) = \hat{\pi}(p')$. By Proposition 7.3.5 (2), we conclude that $p = p'$. Now, for each $p \in \hat{U}^+$, the set $\hat{F}^{-n}(p)$ is mapped into $G^{-n}(\hat{\Phi}(p))$. Since both $\hat{F}^{-n}(p)$ and $G^{-n}(q)$ consist of k^n points, this map is surjective. Thus $F^{-n}(\hat{U}^+)$ is mapped onto $G^{-n}(W)$. □

Since the universal covering manifold of $\mathbb{C} \times (\mathbb{C} - \bar{\Delta})$ is biholomorphic to $\mathbb{C} \times \Delta$, we conclude as follows.

Corollary *The universal covering manifold of the set I^+ of escaping points is biholomorphic to $\mathbb{C} \times \Delta$.*

Let us conclude this section by mentioning the Green function, which is an analogue of the Green function given in §1.4.

Proposition 7.3.11 *The pluriharmonic function* $\log|\varphi(z)|$ *on the domain* V^+ *can be extended to a positive pluriharmonic function* $h(z)$ *on* I^+. *Further we define* $h(z) = 0$ *on* $K^+ = \mathbb{C}^2 - I^+$. *Then* $h(z)$ *is a continuous plurisubharmonic function* ≥ 0 *on* $\mathbb{C}^2$ *satisfying the functional equation* $h(F(z)) = kh(z)$.

Corollary *Let* c_0 *be the coefficient of the leading term (of degree* k*) of the polynomial* $P(y)$. *Let* $L_a = \{(a, y) \mid y \in \mathbb{C}\}$. *Then, for every* $a \in \mathbb{C}$, *the set* $L_a \cap K^+$ *is a compact subset of* L_a *with logarithmic capacity* $|c_0|^{1/(k-1)}$.

See §8.3 for the proof.

Generalized Hénon maps will be studied from the point of view of potential theory of several complex variables in Chapters 8 and 9.

7.4 The set of non-escaping points and symbolic dynamics

The properties of the set K^+ of non-escaping points for a generalized Hénon map F depend on the parameters that determine F. (See Figure 7.2.) The set K^+ may or may not have interior points. For example, if F has an attracting fixed point as in the example in §6.3, then the basin of attraction is contained in K^+ and hence has non-empty interior. If the absolute value of the Jacobian determinant of F is greater than 1, then K^+ is of Lebesgue measure 0, hence has no interior points. In this section we will define the 'horseshoe condition'. We will see that under this condition the structure of K^+ is described by symbolic dynamics.

In general, to study local properties of K^+, it suffices to look at the set $D \cap K^+$. In fact, for any point $z \in K^+$, we can choose an $n \geq 1$ so that $F^n(z) \in D \cap K^+$ and the neighborhood of the point z is mapped biholomorphically to that of the point $F^n(z)$. Now we will show that the set $D \cap K^+$ can be approximated by the sequence $\{D \cap F^{-n}(D)\}_{n=1}^{\infty}$.

Lemma 7.4.1 *Suppose that two polynomial automorphisms* F *and* $\hat{F}$ *satisfy condition (A) for a positive number* R. *Let* $\tilde{F} = \hat{F} \circ F$ *be their composition map. Then we have* $D \cap \tilde{F}^{-1}(D) \subset D \cap F^{-1}(D)$.

Proof By condition (A) we have $F(D) \subset W^+$, $\hat{F}^{-1}(D) \subset W^-$. Hence $F(D) \cap \hat{F}^{-1}(D) \subset W^+ \cap W^- = D$. By mapping by F^{-1}, we obtain the required assertion. □

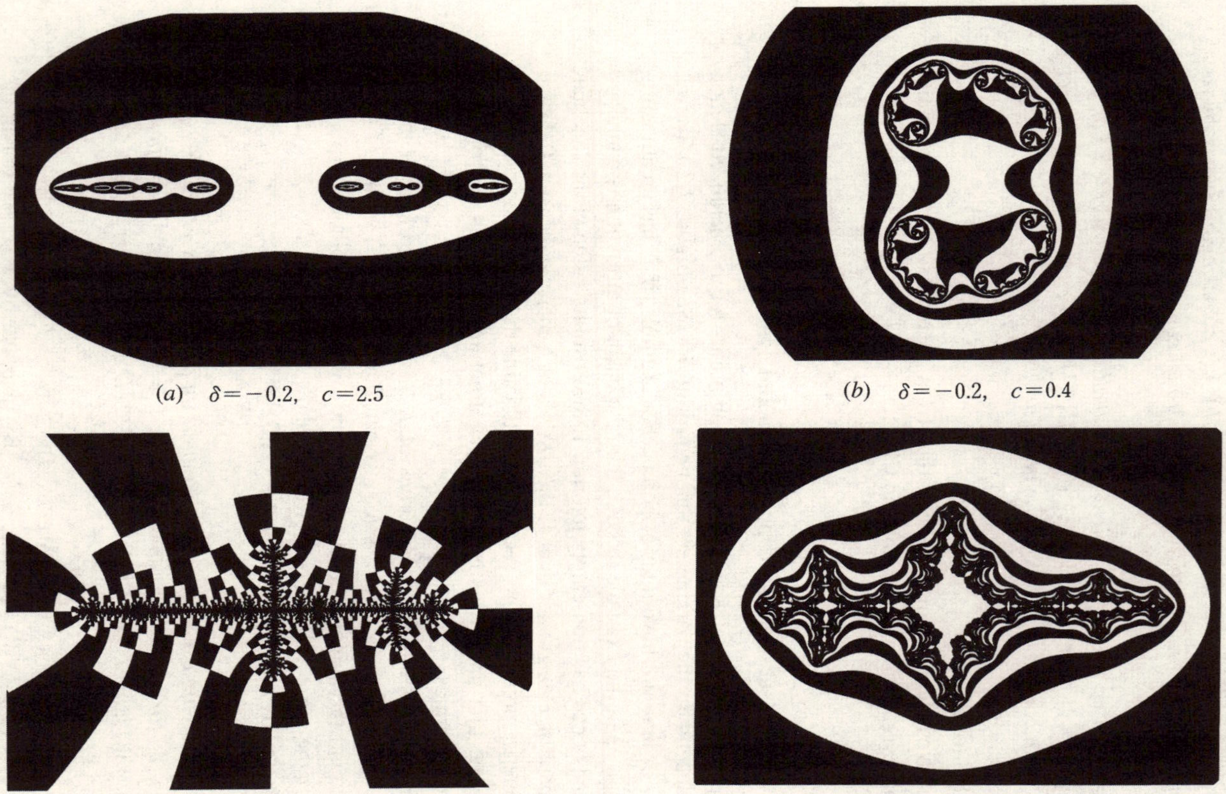

(*a*) $\delta = -0.2, \quad c = 2.5$

(*b*) $\delta = -0.2, \quad c = 0.4$

(*c*) $\delta = -0.3, \quad c = -1.4$

(*d*) $\delta = -0.3, \quad c = -1.0$

Fig. 7.2. Slices of the escaping and non-escaping sets of $F_{\delta,c}$ by $\{x = 0\}$

Proposition 7.4.2 *Suppose that a polynomial automorphism F satisfies condition (A) for a positive number R. Then the sequence $\{D \cap F^{-n}(D)\}_{n=1}^{\infty}$ is decreasing and $D \cap K^+ = \bigcap_{n=1}^{\infty}(D \cap F^{-n}(D))$.*

Proof The first assertion can be proved by induction using Lemma 7.4.1. If $z \in \bigcap_{n=1}^{\infty}(D \cap F^{-n}(D))$, then its forward orbit is bounded and hence $z \in K^+$. This shows that $\bigcap_{n=1}^{\infty}(D \cap F^{-n}(D)) \subset D \cap K^+$. In general, the forward orbit of a point $z \in D$ is contained in W^+. Further, if $z \in D \cap K^+$, then $F^n(z) \notin \overline{V}^+ \subset I^+$ for any n. This shows that $D \cap K^+ \subset \bigcap_{n=1}^{\infty}(D \cap F^{-n}(D))$. □

Next we will look at each $D \cap F^{-n}(D)$.

Proposition 7.4.3 *Suppose that a polynomial automorphism F satisfies condition (A) for some R.*

(1) *For every integer $n \geq 1$, we have*

$$D \cap F^{-n}(D) = \left\{z = (x, y) \in \mathbb{C}^2 \,\middle|\, |x| < R, |g_n(z)| < R\right\}.$$

(2) *There exists a positive number $R' < R$ such that for every integer $n \geq 1$, we have*

$$D \cap F^{-n}(D) \subset \left\{z \in D \cap F^{-(n-1)}(D) \,\middle|\, |g_{n-1}(z)| < R'\right\}.$$

Proof (1) First we will show, in the case $n = 1$, that

$$D \cap F^{-1}(D) = \left\{z = (x, y) \in \mathbb{C}^2 \,\middle|\, |x| < R, |g(z)| < R\right\}.$$

Let E denote the right hand side of this equation. It is clear that $D \cap F^{-1}(D) \subset E$. To show the converse inclusion, suppose that $z = (x, y) \in E$. Then $z \in W^+$ since $|x| < R$. Therefore $F(z) \in F(W^+) \subset W^+$. Since $|g(z)| < R$, we have $F(z) \in W^-$ and hence $z \in F^{-1}(W^-) \subset W^-$. Therefore both z and $F(z)$ are contained in $W^+ \cap W^- = D$. This shows that $z \in D \cap F^{-1}(D)$. For general n, the assertion holds since F^n satisfies condition (A) for R.

(2) Let R' denote the maximum of $|y|$ on the set $D \cap F^{-1}(D)$. We have $R' < R$. In fact it is clear that $R' \leq R$. The points $z = (x, y)$ with $|x| \leq R, |y| = R$ are contained in the set $F^{-1}(V^+)$ and hence not contained in $F^{-1}(\bar{D})$. Thus, $z = (x, y) \in D \cap F^{-1}(D)$ implies $|y| < R'$. For general n, let $z \in D \cap F^{-n}(D)$. Then $F^{n-1}(z) = (f_{n-1}(z), g_{n-1}(z)) \in D \cap F^{-1}(D)$, and hence $|g_{n-1}(z)| < R'$. This shows (2). □

In general, the open set $D \cap F^{-1}(D)$ consists of connected components. Concerning this point, let us make the following definition.

Definition A polynomial automorphism F of $\mathbb{C}^2$ is said to satisfy the *horseshoe condition for a positive number R* if it satisfies condition (A) and the following condition, (B).

Each connected component of $D \cap F^{-1}(D)$ is mapped biholomorphically by $G : (x, y) \mapsto (\xi, \eta) = (x, g(x, y))$ onto the bidisk $\Delta^2 = \{(\xi, \eta) \mid |\xi|, |\eta| < R\}$.

If $D \cap F^{-1}(D)$ has k connected components, F is said to satisfy the *k-fold horseshoe condition.*

Remark Since the open set $D \cap F^{-1}(D)$ is mapped biholomorphically onto $F(D) \cap D$ by F, condition (B) is equivalent to the following condition: each connected component of $F(D) \cap D$ is mapped biholomorphically onto the bidisk Δ^2 by the map $G \circ F^{-1}(x, y) = (f_{-1}(x, y), y)$. This shows that the horseshoe condition is symmetric with respect to the interchange of F and F^{-1} and the coordinates x, y. In what follows we will first study sufficient conditions that guarantee the horseshoe condition. And then, we will study the structure of $D \cap K^+$ and K^+ under the horseshoe condition.

Proposition 7.4.4 *A polynomial automorphism F satisfies the horseshoe condition for some R if and only if condition (A) and the following condition, (B′), hold:*

$$\frac{\partial g}{\partial y} \text{ does not vanish on } D \cap F^{-1}(D).$$

Proof First we remark that the Jacobian determinant of the map G is equal to $\partial g/\partial y$.

If condition (B) holds, then the Jacobian determinant of G does not vanish. Hence condition (B′) holds.

When condition (A) holds, we have $D \cap F^{-1}(D) = \{|x| < R, |g(x, y)| < R\}$ by Proposition 7.4.3, therefore $G : D \cap F^{-1}(D) \to \Delta^2$ is a proper map. If in addition condition (B′) holds, then G is a covering map (without critical points). Since Δ^2 is simply connected, G maps each connected component of $D \cap F^{-1}(D)$ biholomorphically onto Δ^2 by the monodromy theorem. □

Any generalized Hénon map $F(x, y) = (f(x, y), g(x, y)) = (y, P(y) - \delta x)$ satisfies condition (A) for a sufficiently large R. However, it does not always satisfy condition (B) simultaneously. We will now study this condition. Let $\{\alpha_\nu\}$ be the set of critical points of the polynomial $P(y)$.

Proposition 7.4.5 *A generalized Hénon map F satisfies the horseshoe condition for R if and only if F satisfies condition (A) and at the same time*

$$R < \frac{1}{|\delta| + 1} \min_{\nu} |P(\alpha_\nu)|$$

holds. In this case F satisfies the k-fold horseshoe condition, where $k = \deg F = \deg P(y)$.

Proof Condition (B′) is equivalent to saying that the inequality $|g(z)| \geq R$ holds on the set

$$\left\{ z = (x, y) \,\middle|\, |x| < R, \frac{\partial g}{\partial y}(z) = 0 \right\} = \bigcup_{\nu} \{(x, \alpha_\nu) \mid |x| < R\}.$$

Now, for every ν,

$$\inf_{|x|<R} |g(x, \alpha_\nu)| = \inf_{|x|<R} |P(\alpha_\nu) - \delta x| = \max\{|P(\alpha_\nu)| - |\delta| R, 0\} \geq R$$

if and only if $|P(\alpha_\nu)| > (1 + |\delta|)R$. This proves our assertion. In this case, the inverse image of the map G consists of k points. Hence F satisfies the k-fold horseshoe condition. □

Proposition 7.4.6 *An Hénon map $F_{\delta,c}(x, y) = (y, y^2 + c - \delta x)$ satisfies the (2-fold) horseshoe condition for some R if*

$$|c| > 2(1 + |\delta|)^2.$$

Proof By the remark after the proof of Proposition 7.2.2, condition (A) holds if R is greater than the positive root of the equation $Q(t) = t^2 - (1 + |\delta|)t - |c| = 0$. Since the critical value of $y^2 + c$ is c, the condition in the previous proposition reduces to $R < |c|/(|\delta| + 1)$. Thus, there exists an R for which the horseshoe condition is satisfied if $|c|/(|\delta| + 1)$ is greater than the positive root of $Q(t) = 0$. This is equivalent to the condition $Q\,(|c|/(|\delta| + 1)) > 0$. This is equivalent to the inequality of the proposition to be proved. □

Proposition 7.4.7 *Let $P(y)$ be a polynomial of degree k and let δ be a non-zero constant. If $|a|$ is sufficiently large, then the generalized Hénon map*

$$F_a : (x, y) \to (y, P(y) + a - \delta x)$$

satisfies the (k-fold) horseshoe condition.

Proof Choose a positive number R_0 so that

$$|P(y)| > 3(|\delta| + 1)|y| \quad \text{for } |y| > R_0.$$

We have

$$|P(y)+a| \geq |P(y)| - |a| \geq (|\delta|+1)|y| + 2(|\delta|+1)|y| - |a|.$$

Hence condition (A) is satisfied if

$$R > \max\left\{R_0, \frac{|a|}{2(|\delta|+1)}\right\}.$$

Let $\{\alpha_\nu\}$ denote the critical points of the polynomial $P(y)$. Then

$$\min|P(\alpha_\nu)+a| \geq \min(|a| - |P(\alpha_\nu)|) = |a| - \max|P(\alpha_\nu)|.$$

Therefore, if we take R so that

$$R < \frac{|a| - \max|P(\alpha_\nu)|}{|\delta|+1},$$

then the condition in Proposition 7.4.5 holds. Hence, if $|a|$ is sufficiently large there exists an R satisfying the horseshoe condition. □

Now suppose that two polynomial automorphisms F and $\hat{F}$ satisfy the k-fold and $\hat{k}$-fold horseshoe conditions, respectively, for the same $R > 0$. We denote by $D(\kappa)$ ($\kappa = 0, \ldots, k-1$) the k connected components of the open set $D \cap F^{-1}(D)$ and by $\hat{D}(\lambda)$ ($\lambda = 0, \ldots, \hat{k}-1$) the $\hat{k}$ connected components of $D \cap \hat{F}^{-1}(D)$. (See Figure 7.3.)

Theorem 7.4.8 *Under the condition as above, the composed map $\tilde{F} = \hat{F} \circ F$ satisfies the $k\hat{k}$-fold horseshoe condition for the same R. The set $D \cap \tilde{F}^{-1}(D)$ has $k\hat{k}$ connected components*

$$D(\kappa) \cap F^{-1}(\hat{D}(\lambda)) \quad (\kappa = 0, \ldots, k-1; \lambda = 0, \ldots, \hat{k}-1).$$

Proof We have already shown in §7.2 that the composed map $\tilde{F}$ satisfies condition (A) for R. The open set $D \cap \tilde{F}^{-1}(D)$ is mapped biholomorphically onto $F(D) \cap \hat{F}^{-1}(D)$ by F. Since, by Lemma 7.4.1, the set $F(D) \cap \hat{F}^{-1}(D)$ is contained in D, we have

$$F(D) \cap \hat{F}^{-1}(D) = (F(D) \cap D) \cap (D \cap \hat{F}^{-1}(D)).$$

Hence $F(D) \cap \hat{F}^{-1}(D)$ is the union of the sets $F(D(\kappa)) \cap \hat{D}(\lambda)$. Now we will show that each $F(D(\kappa)) \cap \hat{D}(\lambda)$ is non-empty and mapped biholomorphically onto Δ^2 by the map $(x, y) \mapsto (f_{-1}(x, y), \hat{g}(x, y))$. To show this, it suffices to prove that, for each $(\xi, \eta) \in \Delta^2$, there exists a unique point $z \in F(D(\kappa)) \cap \hat{D}(\lambda)$ such that $(f_{-1}(z), \hat{g}(z)) = (\xi, \eta)$. Since each $\hat{D}(\lambda)$ is mapped biholomorphically onto Δ^2 by $(x, y) \mapsto (x, \hat{g}(x, y))$, the set $\hat{C}_\eta =$

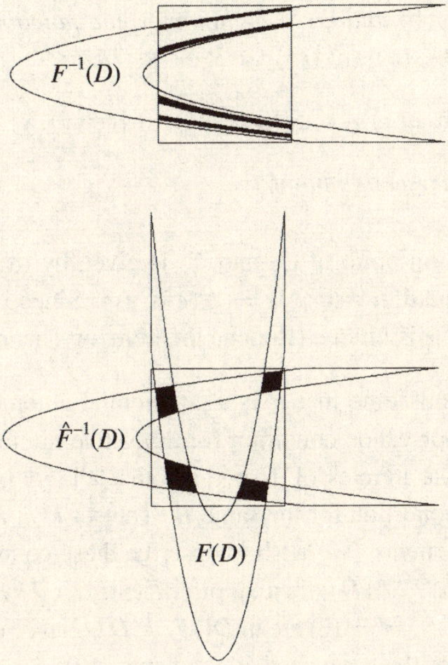

Fig. 7.3. $D \cap \tilde{F}^{-1}(D)$ and $F(D) \cap \hat{F}^{-1}(D)$

$\{z \in \hat{D}(\lambda) \mid \hat{g}(z) = \eta\}$ is expressed as the graph of a holomorphic function $\psi(x)$, i.e.,

$$\hat{C}_\eta = \big\{(x, \psi(x)) \,\big|\, |x| < R\big\}.$$

Here, $\psi(x)$ is a holomorphic function on $|x| < R$ with $|\psi(x)| < R'$. Similarly, since each $F(D(\kappa))$ is mapped biholomorphically onto Δ^2 by $(x, y) \mapsto (f_{-1}(x, y), y)$, the set $C_\xi = \{z \in F(D(\kappa)) \mid f_{-1}(z) = \xi\}$ is expressed as the graph of a holomorphic function $\varphi(y)$, i.e.,

$$C_\xi = \big\{(\varphi(y), y) \,\big|\, |y| < R\big\}$$

Here, $\varphi(y)$ is a holomorphic function on $|y| < R$ with $|\varphi(y)| < R'$. By Lemma 7.4.9 below, we see that C_ξ and $\hat{C}_\eta$ have exactly one intersection point. If we map this relation by F^{-1}, we conclude that each $D(\kappa) \cap F^{-1}(\hat{D}(\lambda))$ is non-empty and mapped biholomorphically onto Δ^2 by $(x, y) \mapsto (x, \hat{g}(F(x, y)))$. $\square$

Lemma 7.4.9 *Let $\varphi(\zeta)$ and $\psi(\zeta)$ be holomorphic functions defined on $\Delta = \{|\zeta| < R\}$ such that $|\varphi(\zeta)|, |\psi(\zeta))| < R' < R$. Then*

$$C_1 = \{(x, \varphi(x)) \mid x \in \Delta\} \text{ and } C_2 = \{(\psi(y), y) \mid y \in \Delta\}$$

have exactly one intersection point.

Proof An intersection point of C_1 and C_2 is given by $(\alpha, \varphi(\alpha))$, where α is a solution of the equation $\psi(\varphi(\zeta)) = \zeta$ $(\zeta \in \Delta)$. Since $|\psi(\varphi(\zeta))| < R'$ for $\zeta \in \Delta$, the solution α is unique (Rouché theorem, or Theorem 6.3.5). □

From now on we assume that F is a polynomial automorphism of $\mathbb{C}^2$ satisfying the k-fold horseshoe condition for a positive number R. By applying Theorem 7.4.8 to the iterates of F, we see that all F^n $(n \geq 1)$ satisfy the k^n-fold horseshoe condition for the same R. The set $D \cap F^{-n}(D)$ consists of k^n connected components. We wish to describe these connected components. If a point z is in $D \cap F^{-n}(D)$, then all of $z, F(z), \ldots, F^n(z)$ are in D. Hence the points $z, F(z), \ldots, F^{n-1}(z)$ are in $D \cap F^{-1}(D)$. Hence each of these points is contained in one of the connected components $D(0), \ldots, D(k-1)$. We can classify the points of $D \cap F^{-n}(D)$ using this fact. A sequence $(s_0, \ldots, s_{n-1})$ with terms $s_k = 0, 1, \ldots, k-1$ is said to be a *symbol sequence of length n*. The set of all symbol sequences of length n is denoted by $\{0, \ldots, k-1\}^n$. For a symbol sequence $(s_0, \ldots, s_{n-1}) \in \{0, \ldots, k-1\}^n$ we define

$$\begin{aligned} D(s_0, \ldots, s_{n-1}) &= \{z \mid F^j(z) \in D(s_j), j = 0, \ldots, n-1\} \\ &= D(s_0) \cap F^{-1}(D(s_1)) \cap \cdots \cap F^{-(n-1)}(D(s_{n-1})). \end{aligned}$$

By applying 7.4.8 repeatedly we obtain the following proposition.

Proposition 7.4.10 *The iterates F^n $(n \geq 1)$ satisfy the k^n-fold horseshoe condition for R. Each $D \cap F^{-n}(D)$ consists of k^n connected components $D(s_0, \ldots, s_{n-1})$, $(s_0, \ldots, s_{n-1}) \in \{0, \ldots, k-1\}^n$.*

Let us note that each $D(s_0, \ldots, s_{n-1})$ is mapped biholomorphically onto the 2-fold disk Δ^2 by the map $G_n(z) = (x, g_n(z))$. With the above consideration we can describe the structure of the set $D \cap K^+$. As in §1.1, we define the symbol sequence space Σ_k to be the set of all infinite symbol sequences $s = (s_0, s_1, \ldots, s_j, \ldots)$ such that $s_j = 0, 1, \ldots, k-1$. We introduce the topology on Σ_k by defining the open sets to be the sets that can be expressed as unions of

$$U(a_0, \ldots, a_n) = \{s = (s_0, s_1, \ldots) \in \Sigma_k \mid s_j = a_j \ (0 \leq j \leq n)\}.$$

With this topology, the symbol sequence space Σ_k is a Cantor set (a compact and totally disconnected Hausdorff space without isolated points; see §1.4).

A symbol sequence $s = (s_0, \ldots, s_n, \ldots) \in \Sigma_k$ defines a decreasing sequence

$$D(s_0) \supset D(s_0, s_1) \supset D(s_0, s_1, s_2) \supset \cdots.$$

We consider its limit:

$$\Gamma^+(s) = \bigcap_{n=0}^{\infty} D(s_0, \ldots, s_n) = \{z \mid F^n(z) \in D(s_n),\ n = 0, 1, 2, \ldots\}.$$

It is clear that

$$D \cap K^+ = \bigcup_{s \in \Sigma_k} \Gamma^+(s).$$

Proposition 7.4.11 *For every symbol sequence $s \in \Sigma_k$, the set $\Gamma^+(s)$ is expressed as the graph of a holomorphic function $\gamma(x, s)$ on $\Delta = \{x \mid |x| < R\}$ with $|\gamma(x, s)| < R'$, i.e.,*

$$\Gamma^+(s) = \{(x, \gamma(x, s)) \mid x \in \Delta\}.$$

Proof We choose and fix a symbol sequence $s = (s_0, \ldots, s_n, \ldots) \in \Sigma_k$. Let L_a denote the complex line $\{x = a\}$ for $a \in \Delta$. We will show that the set $L_a \cap \Gamma^+(s)$ consists of a single point. The set $L_a \cap \Gamma^+(s)$ is the intersection of the decreasing sequence

$$L_a \cap D(s_0) \supset L_a \cap D(s_0, s_1) \supset \cdots \supset L_a \cap D(s_0, \ldots, s_{n-1}) \supset \cdots.$$

For each n, $g_n(a, y)$ maps $L_a \cap D(s_0, \ldots, s_{n-1})$ onto the disk $\{|\eta| < R\}$ conformally. Further $g_n(a, y)$ maps $L_a \cap D(s_0, \ldots, s_n)$ into the disk $\{|\eta| < R'\}$ by Proposition 7.4.3 (2). Hence, from Lemma 6.3.7, we conclude that $L_a \cap \Gamma^+(s)$ consists of a single point. Hence the set $\Gamma^+(s)$ is expressed as a graph of a function $\gamma(x, s)$. We wish to show that $\gamma(x, s)$ thus defined is a holomorphic function of x. To do this, we consider the set

$$\Gamma(s_0, \ldots, s_{n-1}) = \{z \in D(s_0, \ldots, s_{n-1}) \mid g_n(z) = 0\}$$

for each n. This set can be expressed as the graph of a holomorphic function $\gamma(x, s_0, \ldots, s_{n-1})$ with $|\gamma(x, s_0, \ldots, s_{n-1})| < R'$:

$$\Gamma(s_0, \ldots, s_{n-1}) = \{(x, \gamma(x, s_0, \ldots, s_{n-1})) \mid x \in \Delta\}.$$

When $n \to \infty$ the sequence $\gamma(x, s_0, \ldots, s_{n-1})$ converges to $\gamma(x, s)$ for every $x \in \Delta$. Since the sequence $\gamma(x, s_0, \ldots, s_{n-1})$, $n = 1, 2, \ldots$, is uniformly bounded and is a normal family, its convergence is uniform on compact sets in Δ. Hence $\gamma(x, s)$ is holomorphic in x. □

Let $\varphi : D \cap K^+ \to \Sigma_k$ denote the map that maps $z \in D \cap K^+$ to the symbol sequence $s = (s_0, s_1, \ldots)$ such that $z \in \Gamma^+(s)$. This map φ is continuous. In fact for any $U(a_0, \ldots, a_n)$, its inverse image under φ is $\varphi^{-1}(U(a_0, \ldots, a_n)) = D(a_0, \ldots, a_n) \cap K^+$, which is an open subset of $D \cap K^+$. This shows that the inverse image of any open subset of Σ_k is an open subset of $D \cap K^+$. We define the map $\Phi : D \cap K^+ \to \Delta \times \Sigma_k$ by $\Phi(x, y) = (x, \varphi(x, y))$; and the map $\Psi : \Delta \times \Sigma_k \to D \cap K^+$ by $\Psi(x, s) = (x, \gamma(x, s))$.

Theorem 7.4.12 *The maps Φ and Ψ are the inverses of each other that give a homeomorphism between $D \cap K^+$ and $\Delta \times \Sigma_k$.*

Proof It is clear that Φ and Ψ are bijective and inverse of each other. The map Φ is continuous since φ is continuous. To show that Ψ is continuous, it suffices to show that, if $s^{(j)} \to s$ $(j \to \infty)$ in the space Σ_k of symbol sequences, then the sequence of functions $\gamma(x, s^{(j)})$ converges to $\gamma(x, s)$ uniformly on compact sets in Δ. But this can be easily verified from the proof of Proposition 7.4.11. □

Now F maps $D \cap K^+$ into itself. We study this map. We define the shift $\sigma : \Sigma_k \to \Sigma_k$ by $s = (s_0, s_1, s_2, \ldots) \mapsto \sigma(s) = (s_1, s_2, \ldots)$. This is a k-to-one map. In fact the inverse image of $s = (s_0, s_1, s_2, \ldots)$ by σ consists of k points $(1, s_0, s_1, s_2, \ldots), \ldots, (k, s_0, s_1, s_2, \ldots)$.

Theorem 7.4.13 *The equation $\sigma(\varphi(z)) = \varphi(F(z))$ holds. Therefore the set $\Gamma^+(s)$ is mapped into $\Gamma^+(\sigma(s))$ by F.*

Proof If $\varphi(z) = s = (s_0, s_1, s_2, \ldots)$, then $z \in D(s_0)$, $F(z) \in D(s_1)$, $F^2(z) \in D(s_2), \ldots$, by definition. This implies that $\varphi(F(z)) = (s_1, s_2, \ldots) = \sigma(s)$. □

Now we look at the structure of the set K^+ of non-escaping points. For $s \in \Sigma_k$ the sequence

$$\Gamma^+(s) \subset F^{-1}(\Gamma^+(\sigma(s))) \subset \cdots \subset F^{-n}(\Gamma^+(\sigma^n(s))) \subset \cdots$$

is an increasing sequence of sets homeomorphic to a disk. Therefore

$$C^+(s) = \bigcup_{n=0}^{\infty} F^{-n}(\Gamma^+(\sigma^n(s)))$$

is arcwise connected and simply connected. Moreover we have

$$K^+ = \bigcup_{s \in \Sigma_k} C^+(s).$$

Definition Two symbol sequences $s, s' \in \Sigma_k$ are said to be *equivalent*, denoted by $s \sim s'$, if there exists an integer $m \geq 0$ such that $\sigma^m(s) = \sigma^m(s')$. In other words, the sequences $s = (s_0, \ldots, s_n, \ldots)$ and $s' = (s'_0, \ldots, s'_n, \ldots)$ coincide except for a finite number of terms. The following proposition asserts that $C^+(s)$ and $C^+(s')$ either coincide or have no intersection and hence that the arcwise connected components of K^+ are the sets $C^+(s)$.

Proposition 7.4.14 *The following three conditions are equivalent to one another:*

(i) $C^+(s) = C^+(s')$*;* (ii) $C^+(s) \cap C^+(s') \neq \varnothing$*;* (iii) $s \sim s'$*.*

Proof It is clear that (i) implies (ii). We will show that (ii) implies (iii) and that (iii) implies (i). Suppose that $C^+(s) \cap C^+(s') \neq \varnothing$. Then $F^{-m}(\Gamma^+(\sigma^m(s))) \cap F^{-m}(\Gamma^+(\sigma^m(s'))) \neq \varnothing$ for some $m \geq 0$. This implies that $\Gamma^+(\sigma^m(s)) \cap \Gamma^+(\sigma^m(s')) \neq \varnothing$. Hence $\Gamma^+(\sigma^m(s)) = \Gamma^+(\sigma^m(s'))$ and $\sigma^m(s) = \sigma^m(s')$. Now suppose that $\sigma^m(s) = \sigma^m(s')$. Then for $n \geq m$ we have $F^{-n}(\Gamma^+(\sigma^n(s))) = F^{-n}(\Gamma^+(\sigma^n(s')))$. Hence $C^+(s) = C^+(s')$. □

Remark For each $s \in \Sigma_k$, the set $C^+(s)$ has the structure of a simply connected open Riemann surface. In fact this is conformally equivalent to $\mathbb{C}$.

Theorem 7.4.15 *Each $C^+(s)$ is dense in K^+. Hence K^+ is connected (although not arcwise connected).*

Proof The set $C^+(s) \cap D$ is the union of those connected components $\Gamma^+(s')$ of $K^+ \cap D$ such that $s' \sim s$. Since the symbol sequences s' with $s' \sim s$ are dense in Σ_k, the set $C^+(s) \cap D$ is dense in $K^+ \cap D$. If we map this relation by F^{-n} $(n = 1, 2, \ldots)$, we know that $C^+(s)$ is dense in K^+. □

Now let K^- denote the set of non-escaping points with respect to the backward iteration of F. Since the horseshoe condition holds for F^{-1} with the coordinates x, y exchanged, our discussions for K^+ apply also for K^-. To be specific, if $z \in D \cap K^-$, then $F^{-n}(z) \in D \cap F^{-1}(D)$ for $n = 1, 2, \ldots$. Hence we can determine $s_{-n} \in \{0, \ldots, k-1\}$ so that $F^{-n}(z) \in D(s_{-n})$. We define a continuous map $\varphi^- : D \cap K^- \to \Sigma_k$ by $\varphi^-(z) = (s_{-1}, s_{-2}, \ldots)$. Further we define a homeomorphism $\Phi^- : D \cap K^- \to \Sigma_k \times \Delta$ by $z = (x, y) \mapsto (\varphi^-(z), y)$. In view of the above results we can describe the structure of the set $K = K^+ \cap K^-$. A map from the set $\mathbb{Z}$ of all integers into the set $\{0, 1, \ldots, k-1\}$ can be expressed by an infinite sequence $\hat{s} = (\ldots, s_{-2}, s_{-1}, s_0, s_1, \ldots)$. We will call $\hat{s}$ a *bi-infinite symbol sequence* and denote by $\hat{\Sigma}_k$ the set of all bi-infinite symbol sequences. The shift map $\sigma : \hat{\Sigma}_k \to \hat{\Sigma}_k$ is defined by $\sigma(s)_n = s_{n+1}$. This is an automorphism of $\hat{\Sigma}_k$. We can also express $\hat{s}$ as a pair $\hat{s} = (s^-, s^+)$ of (one-sided) infinite symbol sequences $s^- = (s_{-1}, s_{-2}, \ldots)$ and $s^+ = (s_0, s_1, s_2, \ldots)$. Thus $\hat{\Sigma}_k = \Sigma_k \times \Sigma_k$ is a Cantor set with respect to the product topology.

Definition We define the map $\hat{\varphi} : K = K^+ \cap K^- \to \hat{\Sigma}_k$ by

$$z \mapsto \hat{s} = \hat{\varphi}(z) = (\varphi^-(z), \varphi^+(z)) = (\ldots, s_{-1}, s_0, s_1, \ldots).$$

Theorem 7.4.16 *The map $\hat{\varphi}$ is a homeomorphism of K onto $\hat{\Sigma}_k$, so that K is a Cantor set. Further $F_{|K} : K \to K$ is conjugate to the shift map $\sigma : \hat{\Sigma}_k \to \hat{\Sigma}_k$ by $\hat{\varphi}$, that is $\sigma \circ \varphi(z) = \varphi \circ F(z)$ for $z \in K$.*

Proof To prove that $\hat{\varphi}$ is bijective, it suffices to show that, for every $\hat{s} = (\ldots, s_{-1}, s_0, \ldots) \in \hat{\Sigma}_k$, the inverse image $\varphi^{-1}(\hat{s})$ consists of a single point. Let $s^+ = (s_0, \ldots)$, $s^- = (s_{-1}, \ldots)$. Then

$$\varphi^{-1}(\hat{s}) = (\varphi^-)^{-1}(s^-) \cap (\varphi^+)^{-1}(s^+) = \Gamma^-(s^-) \cap \Gamma^+(s^+)$$

consists of a single point by Lemma 7.4.9. Since $\hat{\varphi}$ is a continuous bijective map from the compact set K, its inverse is continuous. □

Theorem 7.4.17 *The set of all periodic points of F is a dense subset of K.*

Proof It is clear that every periodic point is contained in K. The periodic points of F and those of the shift map σ on $\hat{\Sigma}_k$ are in one-to-one correspondence. Since the set of periodic points of σ is dense in $\hat{\Sigma}_k$, the proposition holds. □

Remark All the periodic points of F are of saddle type. Further, if z is a periodic point of F, i.e., $F^m(z) = z$ for some m, then, $C^+(s^+)$, $C^-(s^-)$ with $s^+ = \varphi^+(z)$, $s^- = \varphi^-(z)$ are stable and unstable manifolds at z of F^m.

8

Pluripotential theory of polynomial automorphisms

This chapter is devoted to the preliminaries which we need to develop the pluripotential theoretic method of the study of dynamics on $\mathbb{C}^2$ of a polynomial automorphism f. Through the studies of several authors such as Bedford and Smillie (1991a), and Fornaess and Sibony (1992), this method turned out to be very decisive in exploiting the dynamics of f.

First, in §8.1, we introduce the notion of positive closed currents in a domain Ω of $\mathbb{C}^2$. Though the theory of positive currents was established in the general n-dimensional case, we confine ourselves to stating only the two-dimensional case, because this is the case we will discuss later. We quote the book of Klimek (1991) for the standard reference to these materials.

In §8.2, Lelong's theorem will be proved. This theorem concerns the solution of the $\partial\bar{\partial}$-problem with growth conditions in $\mathbb{C}^n$. We deal with the case of two dimensions and the minimal growth condition. The reference to the general cases is the book of Lelong and Gauman (1986).

In §8.3, we present some fundamental properties of polynomial automorphisms f. Associated with f, we define the Green functions $G^{\pm}$ and the stable and the unstable currents $\mu^{\pm}$. Some fundamental properties of these materials from the viewpoint of pluripotential theory are discussed in §8.4.

In §8.5 for a positive $(1, 1)$-current T on $\mathbb{C}^2$ satisfying a condition, the sequence $\{T_n^{\pm}\}$ of currents defined by the pullback of T by the n times iteration $f^{\pm n}$ are studied and the fundamental convergence theorem (Theorem 8.5.14) is proved. The convergence theorem was proved by Bedford and Smillie (1991a, 1991c, 1992).

In §8.6, as special cases of this theorem, we state two theorems, one dealing with the case where T is a one-dimensional algebraic set of $\mathbb{C}^2$ (Theorem 8.6.8), and another with the case where T is a one-dimensional local complex submanifold (Theorem 8.6.9). These theorems will be applied in the next chapter to derive some dynamical properties of f.

8.1 Positive currents

In this section, we gather some standard facts on the positive currents and some related materials for our later use. We do not give any proof for them except for a few technical statements.

For a point $z = (x, y) = (z_1, z_2)$ of $\mathbb{C}^2$, we use the Euclidean norm $\|z\| = \sqrt{|x|^2 + |y|^2}$. We define $d^c = i(\bar{\partial} - \partial)$ so that $dd^c = 2i\partial\bar{\partial}$. Let

$$\beta = \frac{1}{4} dd^c \|z\|^2 = \frac{i}{2} dz_1 \wedge d\bar{z}_1 + \frac{i}{2} dz_2 \wedge d\bar{z}_2;$$

then the volume form dV of $\mathbb{C}^2$ is $dV = \beta^2/2$.

Let $\mathbb{N}_0$ be the set of non-negative integers. Let Ω be a non-empty open subset in $\mathbb{C}^2$. For an $r \in \mathbb{N}_0$ or $r = \infty$, $C^r(\Omega)$ denotes the space of all complex valued C^r functions in Ω, and $C_0^r(\Omega)$ denotes the space of all C^r functions in Ω of compact support.

The family of all differential (p, q)-forms whose coefficients belong to $C_0^0(\Omega)$ (respectively $C_0^\infty(\Omega)$) will be denoted by $\mathcal{D}_0^{p,q}(\Omega)$ (respectively $\mathcal{D}^{p,q}(\Omega)$). Note that the symbol 0 of $\mathcal{D}_0^{p,q}(\Omega)$ corresponds to the condition that the coefficients of $\varphi \in \mathcal{D}_0^{p,q}(\Omega)$ are continuous rather than that they are of compact support. Let $\{\Omega_r\}$ be a sequence of relatively compact open subsets of Ω such that $\Omega_r \subset \Omega_{r+1}$ for each r and $\bigcup_r \Omega_r = \Omega$. The set of all forms $\varphi \in \mathcal{D}_0^{p,q}(\Omega)$ (respectively $\mathcal{D}^{p,q}(\Omega)$) whose supports are contained in $\bar{\Omega}_r$ is denoted by $\mathcal{D}_0^{p,q}(\bar{\Omega}_r)$ (respectively $\mathcal{D}^{p,q}(\bar{\Omega}_r)$). We equip each of the spaces $\mathcal{D}_0^{p,q}(\bar{\Omega}_r)$ with the topology of uniform convergence of the coefficients, and then we endow $\mathcal{D}_0^{p,q}(\Omega)$ with the topology of the strict inductive limit of the spaces $\mathcal{D}_0^{p,q}(\bar{\Omega}_r)$. Similarly, we equip each of the spaces $\mathcal{D}^{p,q}(\bar{\Omega}_r)$ with the topology of uniform convergence of the coefficients and all of their derivatives. With this topology, the spaces $\mathcal{D}^{p,q}(\bar{\Omega}_r)$ become Fréchet spaces. Then we equip $\mathcal{D}^{p,q}(\Omega)$ with the topology of the strict inductive limit of the spaces $\mathcal{D}^{p,q}(\bar{\Omega}_r)$.

We interpret the above topology of, say, $\mathcal{D}^{1,1}(\Omega)$ as follows. For a sequence $\{\varphi_r\}_{r\geq 0} \subset \mathcal{D}^{1,1}(\Omega)$ with $\varphi_r = \frac{i}{2}\sum_{j,l=1}^2 \varphi_{jl}^r dz_j \wedge d\bar{z}_l$, we have $\varphi_r \to \varphi_0$ as $r \to \infty$ if and only if the following conditions are satisfied:

(1) there is a compact set $K \subset \Omega$ such that $\operatorname{supp}(\varphi_{jl}^r) \subset K$ for all r, j, l;

(2) $D^\alpha(\varphi_{jl}^r) \to D^\alpha(\varphi_{jl}^0)$ uniformly as $r \to \infty$ for all j, l and α, where $\alpha = (\alpha_1, \alpha_2, \alpha_3, \alpha_4) \in (\mathbb{N}_0)^4$ and

$$D^\alpha = \frac{\partial^{|\alpha|}}{\partial z_1^{\alpha_1} \partial \bar{z}_1^{\alpha_2} \partial z_2^{\alpha_3} \partial \bar{z}_2^{\alpha_4}} \quad \text{with } |\alpha| = \sum_{j=1}^4 \alpha_j.$$

The elements of the dual spaces $(\mathcal{D}^{2-p,2-q}(\Omega))'$ are called *(p,q)-currents*, and those of the dual space $(\mathcal{D}_0^{2-p,2-q}(\Omega))'$ are called *(p,q)-currents of order 0*. Note that, for $p = q = 2$,

$$\mathcal{D}^{2-p,2-q}(\Omega) = C_0^\infty(\Omega) \text{ and } \mathcal{D}_0^{2-p,2-q}(\Omega) = C_0^0(\Omega).$$

And the (2, 2)-currents can be identified with the distributions in $\Omega \subset \mathbb{C}^2$, and the (2, 2)-currents of order 0 with the measures on Ω.

For $T \in (\mathcal{D}^{0,0}(\Omega))'$ and $\varphi \in \mathcal{D}^{0,0}(\Omega)$ (respectively $T \in (\mathcal{D}_0^{0,0}(\Omega))'$ and $\varphi \in \mathcal{D}_0^{0,0}(\Omega)$), the duality is denoted by $T(\varphi)$ or $\int_\Omega T \wedge \varphi$. The duality for other bidegrees (p,q) can be similarly described so that we can regard a current (respectively a current of order 0) T as a differential form with the distribution (respectively measure) coefficients. For example, for a (1, 1)-current $T \in (\mathcal{D}^{1,1}(\Omega))'$, define the (2, 2)-current $T_{jl}dV$ by

$$T_{jl}dV(\psi) = (-1)^{j+l}T\left(\psi \frac{i}{2} dz_{(3-j)} \wedge d\bar{z}_{(3-l)}\right)$$

for any j, l and all functions $\psi \in \mathcal{D}^{0,0}(\Omega)$. Then, we have

$$T(\varphi) = \int_\Omega T \wedge \varphi = \sum_{j,l=1}^{2} (-1)^{j+l} T_{jl}dV(\varphi_{(3-j)(3-l)})$$

for $\varphi = \frac{i}{2}\sum_{j,l=1}^2 \varphi_{jl} dz_j \wedge d\bar{z}_l \in \mathcal{D}^{1,1}(\Omega)$, so we can write

$$T = \frac{i}{2} \sum_{j,l=1}^{2} T_{jl} dz_j \wedge d\bar{z}_l.$$

Observe that, if T is a (p,q)-current in Ω, ψ is a (s,t)-form in Ω with coefficients in $C^\infty(\Omega)$ and $\max\{p+s,\ q+t\} \le 2$, then the formula

$$(T \wedge \psi)(\varphi) = T(\psi \wedge \varphi) \quad \text{for} \quad \varphi \in \mathcal{D}^{2-p-s,2-q-t}(\Omega)$$

defines a $(p+s, q+t)$-current $T \wedge \psi$.

We define the differential operators $\partial, \bar{\partial}$, $d = \partial + \bar{\partial}$ and $d^c = i(\bar{\partial} - \partial)$ acting on the currents. For a (p,q)-current T, ∂T is defined by

$$\int_\Omega \partial T \wedge \varphi = (-1)^{p+q+1} \int_\Omega T \wedge \partial\varphi \quad \text{for all } \varphi \in \mathcal{D}^{1-p,2-q}(\Omega); \tag{8.1}$$

$\bar{\partial} T$ is defined similarly by substituting $\bar{\partial}$ in place of ∂ and $\varphi \in \mathcal{D}^{2-p,1-q}(\Omega)$ in place of $\varphi \in \mathcal{D}^{1-p,2-q}(\Omega)$ in formula (8.1). A current T is called *closed* if $dT = 0$.

Next, we introduce the notion of positive currents.

Definition First, we deal with forms of constant complex coefficients. A $(2,2)$-form $\varphi = \tau dV$ is real if $\tau \in \mathbb{R}$, and is called positive if $\tau \geq 0$, where dV is the volume form. A $(1,1)$-form $\varphi = \frac{i}{2}\sum_{j,l=1}^{2} \varphi_{jl} dz_j \wedge d\bar{z}_l$ is real if and only if (φ_{jl}) is a Hermitian $(2,2)$ matrix. A real $(1,1)$-form φ is called positive, and denoted by $\varphi \geq 0$, if, for any $(1,0)$-form $t = \sum_{j=1}^{2} t_j dz_j$, $\varphi \wedge (\frac{i}{2} t \wedge \bar{t})$ is a positive $(2,2)$-form: letting $s_1 = t_2$ and $s_2 = -t_1$, $\sum_{j,l=1}^{2} \varphi_{jl} s_j \bar{s}_l \geq 0$, that is, the Hermitian matrix (φ_{jl}) is positive semi-definite.

Next, a (p,p)-form $\varphi \in \mathcal{D}^{p,p}(\Omega)$ or $\varphi \in \mathcal{D}_0^{p,p}(\Omega)$ is said to be *positive*, and also denoted by $\varphi \geq 0$, if $\varphi(z) \geq 0$ at each $z \in \Omega$. A (p,p)-current $T \in (\mathcal{D}^{2-p,2-p}(\Omega)')$ (respectively a (p,p)-current of order 0 $T \in (\mathcal{D}_0^{2-p,2-p}(\Omega))')$ is called *positive* if $T(\varphi) \geq 0$ for all $\varphi \in \mathcal{D}^{2-p,2-p}(\Omega)$ with $\varphi \geq 0$ (respectively for all $\varphi \in \mathcal{D}_0^{2-p,2-p}(\Omega)$ with $\varphi \geq 0$).

Note that the positive forms φ are *real*, $\varphi = \bar{\varphi}$, and the positive currents are *real*, $\overline{T(\bar{\varphi})} = T(\varphi)$, for all forms φ. We furnish the space of all (p,q)-currents, respectively the space of currents of order 0, with the *weak$*$ topology*. A sequence of (p,q)-currents is convergent in this topology if it converges pointwise on $\mathcal{D}^{2-p,2-q}(\Omega)$, respectively $\mathcal{D}_0^{2-p,2-q}(\Omega)$. A set $\mathcal{T}$ of $(1,1)$-currents of order 0 on Ω is said to be sequentially compact in the weak$*$ topology if every sequence in $\mathcal{T}$ contains a convergent subsequence. A sufficient condition for $\mathcal{T}$ to be sequentially compact in the weak$*$ topology is that, for each relatively compact open set $G \subset \Omega$, there is a constant $C > 0$ such that $|T(\varphi)| \leq C\|\varphi\|_G$ for all $T \in \mathcal{T}$ and all $\varphi = \frac{i}{2}\sum_{j,l=1}^{2} \varphi_{jl} dz_j \wedge d\bar{z}_l \in \mathcal{D}^{1,1}(\bar{G})$, where $\|\varphi\|_G = \max_{j,l} \max_{z\in G} |\varphi_{jl}(z)|$. This is a consequence of the Banach–Alaoglu theorem. We use it in §8.5.

If Ω is an open set and E is a subset of Ω, then we will write $E \subset\subset \Omega$ to mean that E is relatively compact in Ω.

Lemma 8.1.1 *If a distribution T on $\Omega \subset \mathbb{C}^2$ is positive, then T is of order* 0, *hence is a positive measure.*

Proof It suffices to show that T can be extended to the space $C_0^0(\Omega)$, which contains the domain $C_0^\infty(\Omega)$ of T as a dense subspace. Take a relatively compact open set $G \subset\subset \Omega$, and $\{\varphi_r\}, \{\tilde{\varphi}_r\} \subset \mathcal{D}^{0,0}(\bar{G})$ which are convergent to $\varphi \in C_0^0(\Omega)$ in $C_0^0(\Omega)$. We shall show that $\lim_r T(\varphi_r)$ exists and $\lim_r T(\varphi_r) = \lim_r T(\tilde{\varphi}_r)$.

Take a $\chi \in C_0^\infty(\Omega)$ satisfying $\chi \equiv 1$ on $\bar{G}$, and we have

$$|\varphi_r - \tilde{\varphi}_s| \leq \|\varphi_r - \tilde{\varphi}_s\|_G \chi \text{ on } \Omega,$$

where $\|\psi\|_G = \sup_{z\in G} |\psi(z)|$. Then, by positivity of T, we have

$$0 \le T(\|\varphi_r - \tilde{\varphi}_s\|_G \chi \pm (\varphi_r - \tilde{\varphi}_s)),$$

so $|T(\varphi_r) - T(\tilde{\varphi}_s)| \le \|\varphi_r - \tilde{\varphi}_s\|_G T(\chi)$. Hence $\lim_r T(\varphi_r)$ exists (which can be seen by setting $\tilde{\varphi}_r = \varphi_r$) and $\lim_r T(\varphi_r) = \lim_r T(\tilde{\varphi}_r)$. □

Definition For a measure T on a domain Ω,

$$|T|_\Omega = \sup_\varphi |T(\varphi)|$$

where $\varphi \in C_0^0(\Omega)$ with $\sup_{z\in\Omega} |\varphi(z)| \le 1$.

Lemma 8.1.2 *Suppose that a positive measure λ and a distribution $T = T_1 + iT_2$ on Ω (where $T_1 = \operatorname{Re} T$ and $T_2 = \operatorname{Im} T$) satisfy*

$$|T_j(\varphi)| \le \lambda(\varphi) \quad (j = 1, 2;\ \varphi \ge 0, \varphi \in C_0^\infty(\Omega)).$$

Then, (*a*) *T is a measure,* (*b*) $|T|_\Omega \le 2\sqrt{2}|\lambda|_\Omega$.

Proof (a) Take a relatively compact open set $G \subset\subset \Omega$, and two sequences $\{\psi_r\}, \{\tilde{\psi}_r\} \subset \mathcal{D}(\bar{G})$ which are convergent to $\psi \in C_0^0(\Omega)$ in $C_0^0(\Omega)$. We shall show that $\lim_r T_1(\psi_r)$ exists and $\lim_r T_1(\psi_r) = \lim_r T_1(\tilde{\psi}_r)$, which then proves that $T_1 = \operatorname{Re} T$ is a real measure.

Take a $\chi \in C_0^\infty(\Omega)$ satisfying $\chi \ge 0$ and $\chi \equiv 1$ on $\bar{G}$, so that we have $|\psi_r - \tilde{\psi}_s| \le \|\psi_r - \tilde{\psi}_s\|_G \chi$ on Ω. Since $T_1(\varphi) \le \lambda(\varphi)$ for $\varphi \ge 0$, we have

$$T_1(\|\psi_r - \tilde{\psi}_s\|_G \chi - \psi_r + \tilde{\psi}_s) \le \lambda(\|\psi_r - \tilde{\psi}_s\|_G \chi - \psi_r + \tilde{\psi}_s),$$

and since $-\lambda(\chi) \le T_1(\chi)$,

$$-2\|\psi_r - \tilde{\psi}_s\|_G \lambda(\chi) + \lambda(\psi_r) - \lambda(\tilde{\psi}_s) \le T_1(\psi_r) - T_1(\tilde{\psi}_s).$$

Similarly, $-\lambda(\varphi) \le T_1(\varphi)$ for $\varphi \ge 0$ implies

$$T_1(\psi_r) - T_1(\tilde{\psi}_s) \le 2\|\psi_r - \tilde{\psi}_s\|_G \lambda(\chi) - (\lambda(\psi_r) - \lambda(\tilde{\psi}_s)).$$

Therefore $\lim_r T_1(\psi_r)$ exists and $\lim_r T_1(\psi_r) = \lim_r T_1(\tilde{\psi}_r)$.

(b) First we take a real $\psi \in C_0^0(\Omega)$ and set $\psi_+ = \max\{\psi, 0\}$ and $\psi_- = \max\{-\psi, 0\}$, then $\psi = \psi_+ - \psi_-$. We have

$$T_1(\psi) = T_1(\psi_+) - T_1(\psi_-) \le \lambda(\psi_+) + \lambda(\psi_-) = \lambda(|\psi|).$$

Combining with the other similar estimate, we have

$$-\lambda(|\psi|) \le T_1(\psi) \le \lambda(|\psi|).$$

Next, for a general ψ, letting $\psi = \operatorname{Re}\psi + i \operatorname{Im}\psi = \psi_1 + i\psi_2$,

$$|T_j(\psi)|^2 \le |\lambda(|\psi_1|)|^2 + |\lambda(|\psi_2|)|^2 \le 2|\lambda(|\psi|)|^2 \quad (j = 1, 2).$$

Thus we have $|T_j(\psi)| \le \sqrt{2}|\lambda(|\psi|)|$, hence $|T(\psi)| \le 2\sqrt{2}|\lambda(|\psi|)|$. □

Lemma 8.1.3 *Let* $T = \sum_{j,l=1}^{2} T_{jl}\frac{i}{2}dz_j \wedge d\bar{z}_l$ *be a positive* (1, 1)*-current on an open set* Ω *in* $\mathbb{C}^2$. *Then* $T_{jl}dV$ *is a measure, and*

$$\operatorname{supp}(T) = \operatorname{supp}(T \wedge \beta) \quad \textit{and} \quad |T_{jl}dV|_\Omega \le \sqrt{2}|T \wedge \beta|_\Omega,$$

where β *is as above.*

Proof Take an arbitrary $\varphi_{11} \in \mathcal{D}^{0,0}(\Omega)$ with $\varphi_{11} \ge 0$. Then, by the positivity of T, we have $T_{22}dV(\varphi_{11}) = T(\varphi_{11}\frac{i}{2}dz_1 \wedge d\bar{z}_1) \ge 0$, which proves that $T_{22}dV$ is a positive distribution, hence, by Lemma 8.1.1, a positive measure. Similarly, $T_{11}dV$ is a positive measure.

Let $\varphi \in \mathcal{D}^{0,0}(\Omega)$ with $\varphi \ge 0$. Then, by the positivity of T, we have

$$T_{11}dV(\varphi) \mp T_{12}dV(\varphi) \mp T_{21}dV(\varphi) + T_{22}dV(\varphi)$$

$$= T\left(\varphi\frac{i}{2}(dz_1 \pm dz_2) \wedge (d\bar{z}_1 \pm d\bar{z}_2)\right) \ge 0, \tag{8.2}$$

$$|T_{12}dV(\varphi) + T_{21}dV(\varphi)| \le (T_{11} + T_{22})dV(\varphi), \tag{8.3}$$

which proves, by Lemma 8.1.2, that the real part of $T_{12}dV$ is a measure.

By using $\varphi\frac{i}{2}(dz_1 \pm i dz_2) \wedge (d\bar{z}_1 \mp i d\bar{z}_2)$ in (8.2), we have

$$|i(T_{12}dV(\varphi) - T_{21}dV(\varphi))| \le (T_{11} + T_{22})dV(\varphi), \tag{8.4}$$

which proves that the imaginary part of $T_{12}dV$ is a measure.

Since $T \wedge \beta = (T_{11} + T_{22})dV$, (8.3) and (8.4) indicate that $\operatorname{supp}(T) \subset \operatorname{supp}(T \wedge \beta)$. The opposite inclusion being clear, we have $\operatorname{supp}(T) = \operatorname{supp}(T \wedge \beta)$. Finally, the norm estimate follows from (8.3) and (8.4) by using Lemma 8.1.2 (b). □

Example 8.1.1 *Let* Ω *be an open set in* $\mathbb{C}^2$ *and let* u *be a plurisubharmonic function in* Ω. *Then* u *is locally integrable –* $u \in (\mathcal{D}(\Omega)^{2,2})'$ *– and* $dd^c u$ *is a positive closed* (1, 1)*-current. Conversely for a positive closed* (1, 1)*-current* T *in, say* $\mathbb{C}^2$, *there is a plurisubharmonic function* u *in* $\mathbb{C}^2$ *such that* $dd^c u = T$.

The subject of the next section is to establish the same type fact as in this, Example 8.1.1, for $\Omega = \mathbb{C}^2$ but taking account of the minimal growth condition.

Example 8.1.2 *Let Ω be an open set in $\mathbb{C}^2$ and let Y be an analytic subset of Ω of pure dimension* 1. *Then, Y defines the* current of integration $[Y]$ *as follows. Let $\varphi \in \mathcal{D}^{1,1}(\Omega)$ and let $\tilde{Y}$ be the set of regular points of Y. Then,* $[Y]$ *is defined by $\int_\Omega [Y] \wedge \varphi = \int_{\tilde{Y}} \varphi$.*

This $[Y]$ is a positive closed $(1,1)$-current in Ω. Furthermore, if there is a holomorphic function h in Ω such that $Y = \{z \mid h(z) = 0\}$ and that the zero multiplicity of h of each irreducible component of Y is equal to 1, *then we have the following, the* Poincaré–Lelong formula*:*

$$[Y] = \frac{1}{2\pi} dd^c \log |h|.$$

Lelong's theorem (Theorem 8.2.9) belongs to the pluripotential theory on $\mathbb{C}^2$; however, its proof in the next section needs the real potential theory on $\mathbb{R}^4 = \mathbb{C}^2$. So we shall summarize here some classes of functions which appear in the potential theory and the pluripotential theory.

Definition Let $\mathcal{PSH}(\Omega)$ be the set of all plurisubharmonic functions in Ω and $\mathcal{PH}(\Omega)$ be the set of all pluriharmonic functions in Ω (cf. §6.1).

Let Δ be the real Laplacian. A real valued $u \in C^2(\Omega)$ is called *harmonic* if $\Delta u = 0$ in Ω. The set of all harmonic functions in Ω is denoted by $\mathcal{H}(\Omega)$. An upper semi-continuous function $u : \Omega \to \mathbb{R} \cup \{-\infty\}$ which is not identically $-\infty$ on any connected component of Ω is called *subharmonic* in Ω, if for every relatively compact open subset G of Ω and every function $h \in \mathcal{H}(\Omega) \cap C(\bar{G})$, the following maximum principle is true:

$$u \leq h \text{ on } \partial G \implies u \leq h \text{ on } G.$$

The set of all subharmonic functions is denoted by $\mathcal{SH}(\Omega)$.

Note that $\mathcal{PSH}(\Omega) \subset \mathcal{SH}(\Omega)$. For example, for a real $u \in C^2(\Omega)$, $u \in \mathcal{SH}(\Omega)$ iff $\Delta u \geq 0$ in Ω, whereas $u \in \mathcal{PSH}(\Omega)$ iff $dd^c u \geq 0$ in Ω.

Further, we need the usual smoothing procedure for a subharmonic function and a positive closed $(1,1)$-current on an open set Ω in $\mathbb{C}^2$. For an $\varepsilon > 0$, we set $\Omega_\varepsilon = \{z \in \Omega \mid \mathrm{d}(z, \partial\Omega) > \varepsilon\}$, where d denotes the Euclidean distance. As a special case, we have $\Omega_\varepsilon = \mathbb{C}^2$ if $\Omega = \mathbb{C}^2$.

Let $\rho(z)$ be a C^∞ function with $\rho \geq 0$, depending only on $\|z\|$, and with compact support contained in $\{\|z\| < 1\}$, such that $\int_{\mathbb{C}^2} \rho(z) dV(z) = 1$. For $\varepsilon > 0$, set $\rho_\varepsilon(z) = \varepsilon^{-4} \rho(z/\varepsilon)$.

Theorem 8.1.4 (a) *Let $\varphi \in \mathcal{SH}(\Omega)$. Then,*

$$\varphi_\varepsilon(z) = \varphi * \rho_\varepsilon(z) = \int_{\mathbb{C}^2} \varphi(t) \rho_\varepsilon(t - z) dV(t)$$

is in $\mathcal{SH}(\Omega_\varepsilon) \cap C^\infty(\Omega_\varepsilon)$. In particular, when $\varphi \in \mathcal{PSH}(\Omega)$, $\varphi_\varepsilon \in \mathcal{PSH}(\Omega_\varepsilon) \cap C^\infty(\Omega_\varepsilon)$. The family $\{\varphi_\varepsilon(z)\}$ converges decreasingly to $\varphi(z)$ as ε decreases to 0.

(b) *Let T be a positive closed $(1, 1)$-current in Ω. Then,*

$$T_\varepsilon(z) = T * \rho_\varepsilon(z) = \sum_{j,l=1}^{2} \left(\int_{\mathbb{C}^2} T_{jl}(t)\rho_\varepsilon(t - z)dV(t) \right) \frac{i}{2} dz_j \wedge d\bar{z}_l$$

are C^∞ positive closed $(1, 1)$-forms in Ω_ε and $\{T_\varepsilon\}$ converges in the sense of weak$*$ convergence to T as $\varepsilon \to 0$.

Theorem 8.1.5 (a) *For a real function u in Ω, $u \in \mathcal{H}(\Omega)$ if and only if $u \in \mathcal{SH}(\Omega)$ and $-u \in \mathcal{SH}(\Omega)$. If a locally integrable real function v satisfies $\Delta v = 0$ in the sense of distributions, then there is a $u \in \mathcal{H}(\Omega)$ such that $u = v$ in the sense of distributions.*

(b) *For a real function u in Ω, $u \in \mathcal{PH}(\Omega)$ if and only if $u \in \mathcal{PSH}(\Omega)$ and $-u \in \mathcal{PSH}(\Omega)$. If a locally integrable real function v satisfies $dd^c v = 0$ in the sense of currents, then there is a $u \in \mathcal{PH}(\Omega)$ such that $u = v$ in the sense of distributions.*

8.2 Lelong's theorem

In this section, after Lelong and Gruman (1986), we shall prove Lelong's theorem on the $\partial\bar{\partial}$-problem with the minimal growth condition for a positive closed $(1, 1)$-current on $\mathbb{C}^2$. Lelong's theorem was established in the general n-dimensional case with diverse growth conditions.

The following theorem implies that the Newton kernel

$$N_z(a) = \frac{1}{4\pi^2} \frac{-1}{\|z - a\|^2} \qquad (z, a \in \mathbb{C}^2,\ z \neq a)$$

is the fundamental solution for the Laplacian Δ.

Theorem 8.2.1 *Let $\psi \in C_0^2(\mathbb{C}^2)$. Then,*

$$\psi(z) = \int_{\mathbb{C}^2} N_z(a) dd^c \psi(a) \wedge \beta(a) = \int_{\mathbb{C}^2} N_z(a) \Delta_a \psi(a) dV(a).$$

Proof Set $B(z, \varepsilon) = \{a \in \mathbb{C}^2 \mid \|a - z\| \leq \varepsilon\}$. Then, Stokes' theorem gives

$$\begin{aligned}&\int_{\partial B(z,\varepsilon)} N_z(a) d^c \psi(a) \wedge \beta(a) \\ &= -\int_{\mathbb{C}^2 - B(z,\varepsilon)} N_z(a) dd^c \psi(a) \wedge \beta(a) + \int_{\partial B(z,\varepsilon)} \psi(a) d^c N_z(a) \wedge \beta(a).\end{aligned}$$

Here the left hand side tends to zero, whereas the second term of the right hand side equals

$$\frac{1}{4\pi^2}\int_{\partial B(z,\varepsilon)} \psi(a)\frac{d^c\|z-a\|^2}{\|z-a\|^4}\wedge\beta(a)$$
$$= \frac{1}{4\pi^2\varepsilon^4}\int_{\partial B(z,\varepsilon)} \psi(a)d^c\|z-a\|^2\wedge\beta(a)$$

and converges to $\psi(z)$ as $\varepsilon\to 0$, so we have the conclusion. □

Let $B(r_0, r) = \{z \mid r_0 < \|z\| \leq r\}$, $B(r) = \{z \mid \|z\| \leq r\}$ and $S(r) = \{z \mid \|z\| = r\}$. We define

$$\alpha = \frac{1}{4\pi}dd^c\log\|z\|^2 = \frac{1}{\pi\|z\|^4}\left(\|z\|^2\beta - \frac{i}{2}\partial\|z\|^2\wedge\bar\partial\|z\|^2\right)$$

on $\mathbb{C}^2 - \{0\}$. If we restrict α and β to $S(r)$, then $\partial\|z\|^2\wedge\bar\partial\|z\|^2 = 0$ on $S(r)$, so $\alpha = \dfrac{1}{\pi r^2}\beta$ on $S(r)$. Let $\lambda = \log(1+\|z\|^2)$ and

$$\omega = \frac{1}{4}dd^c\lambda = \frac{i}{2}\frac{(1+\|z\|^2)\partial\bar\partial\|z\|^2 - \partial\|z\|^2\wedge\bar\partial\|z\|^2}{(1+\|z\|^2)^2}.$$

Then, we have $\omega = \dfrac{1}{1+r^2}\beta$ on $S(r)$.

Lemma 8.2.2 *Let T be a positive closed $(1,1)$-current on $\mathbb{C}^2$.*

(a) $\displaystyle\int_{B(r_0,r)} T\wedge\alpha = \frac{1}{\pi r^2}\int_{B(r)} T\wedge\beta - \frac{1}{\pi r_0^2}\int_{B(r_0)} T\wedge\beta.$

(b) $\displaystyle\int_{B(r)} T\wedge\omega = \frac{1}{1+r^2}\int_{B(r)} T\wedge\beta.$

Proof (a) First we assume that T is of class C^∞, then by the Poincaré lemma, there is a 1-form φ of class C^∞ on $\mathbb{C}^2$ such that $T = d\varphi$. So by Stokes' theorem,

$$\begin{aligned}\int_{B(r_0,r)} T\wedge\alpha &= \int_{B(r_0,r)} d(\varphi\wedge\alpha) = \int_{S(r)}\varphi\wedge\alpha - \int_{S(r_0)}\varphi\wedge\alpha\\ &= \frac{1}{\pi r^2}\int_{S(r)}\varphi\wedge\beta - \frac{1}{\pi r_0^2}\int_{S(r_0)}\varphi\wedge\beta\\ &= \frac{1}{\pi r^2}\int_{B(r)} T\wedge\beta - \frac{1}{\pi r_0^2}\int_{B(r_0)} T\wedge\beta.\end{aligned}$$

For the general case, letting $\varepsilon > 0$, we have the formula (a) for T_ε. So letting $\varepsilon\to 0$, we obtain (a) for T.

The formula (b) can be proved similarly. □

The function $e(a, z) = \dfrac{1}{\|a\|^2} - \dfrac{1}{\|a - z\|^2}$ is called the *canonical potential of genus* 0.

Lemma 8.2.3 (a) *For* $a \neq 0$, $e(a, z) \leq \dfrac{1}{\|a\|^2}$;

(b) $|e(a, z)| \leq 9\dfrac{\|z\|}{\|a\|^3}$ *if* $a \neq 0$ *and* $\|z\| \leq \dfrac{4}{9}\|a\|$;

(c) *for* $a \neq 0$, $e(a, z) \leq \dfrac{13\|z\|}{\|a\|^2(\|a\| + \|z\|)}$.

Proof Part (a) holds trivially. For (b), set $\tau = 4/9$ and $u = \|z\|/\|a\|$.

If $0 < u \leq \tau$, then

$$\|a - z\|^{-2} \leq (\|a\| - \|z\|)^{-2} = \|a\|^{-2}\left(1 + \sum_{s=1}^{\infty}(s+1)u^s\right).$$

Using $s + 1 \leq 3(3/2)^{s-1}$ $(s \geq 1)$, we have

$$|e(a, z)| \leq 3\|a\|^{-2}u\sum_{s=1}^{\infty}(3/2)^{s-1}\tau^{s-1} = 9\|a\|^{-2}u.$$

(c) If $\|z\| \leq 4\|a\|/9$, using (b), $e(a, z) \leq 9\|z\|/\|a\|^3$. Since $\|a\| + \|z\| \leq 13\|a\|/9$, we have

$$e(a, z) \leq \frac{13\|z\|}{\|a\|^2(\|a\| + \|z\|)}.$$

If $\|z\| \geq 4\|a\|/9$, using (a) and $13\|z\|/9 \geq 4(\|a\| + \|z\|)/9$, we have

$$e(a, z) \leq \frac{1}{\|a\|^2} \leq \frac{13}{4}\frac{1}{\|a\|^2}\frac{\|z\|}{\|a\| + \|z\|}.$$

□

In the rest of this section, for a positive closed (1, 1)-current T, we assume

$$C = \int_{\mathbb{C}^2} T \wedge \omega < +\infty, \tag{8.5}$$

$$\operatorname{supp}(T) \cap B(r_0) = \varnothing \quad \text{for some } r_0 > 0. \tag{8.6}$$

Then by Lemma 8.2.2 (a) and (b), we have $\displaystyle\int_{B(r_0,\infty)} T \wedge \alpha = \frac{C}{\pi}$.

Theorem 8.2.4 *Let T be a positive closed $(1, 1)$-current satisfying (8.5) and (8.6). And, for $r \geq r_0$, let*

$$I_r(z) = \frac{1}{2\pi} \int_{B(r_0,r)} e(a,z) T(a) \wedge \beta(a).$$

Then, $I(z) = \lim_{r\to\infty} I_r(z)$ converges uniformly on every compact set in $\mathbb{C}^2$.

Proof Let $\|z\| \leq R$ and $R' > 9R/4$. Then by Lemma 8.2.3 (b),

$$\frac{1}{2\pi} \int_{\|a\|>R'} |e(a,z)| T(a) \wedge \beta(a) \leq \frac{9R}{2\pi} \int_{\|a\|>R'} \frac{1}{\|a\|^3} T(a) \wedge \beta(a).$$

Since

$$\int_{B(r,2r)} \frac{1}{\|a\|^3} T(a) \wedge \beta(a) \leq \frac{1}{r^3} \int_{B(2r)} T \wedge \beta \leq \frac{8C}{r}$$

for $(2r)^2 \geq 1$, where we used Lemma 8.2.2 (b), letting $R' = 2^m > 1$ we have

$$\int_{\|a\|>R'} \frac{1}{\|a\|^3} T(a) \wedge \beta(a) = \sum_{s=m}^{\infty} \int_{B(2^s, 2^{s+1})} \frac{1}{\|a\|^3} T(a) \wedge \beta(a) \leq \frac{16C}{R'}.$$

□

Note that $I(z) = \frac{1}{2\pi} \int_{\mathbb{C}^2} e(a,z) T(a) \wedge \beta(a)$ by (8.6).

Proposition 8.2.5 $\Delta I(z) dV(z) = 2\pi T(z) \wedge \beta(z)$.

Proof Let $\varphi \in C_0^\infty(\mathbb{C}^2)$. Then, by Theorem 8.2.1 and (8.6),

$$\begin{aligned}
\int_{\mathbb{C}^2} \varphi(z) \Delta I(z) dV(z) &= \int_{\mathbb{C}^2} \Delta\varphi(z) \frac{1}{2\pi} \int_{\mathbb{C}^2} e(a,z) T(a) \wedge \beta(a) dV(z) \\
&= \frac{4\pi^2}{2\pi} \int_{\mathbb{C}^2} T(a) \wedge \beta(a) \frac{1}{4\pi^2} \int_{\mathbb{C}^2} \Delta\varphi(z) e(a,z) dV(z) \\
&= 2\pi \int_{\mathbb{C}^2} \varphi(a) T(a) \wedge \beta(a).
\end{aligned}$$

□

Lemma 8.2.6 $\sup_{\|z\|=r} I(z) \leq 13C \left\{ 2\log\left(\frac{r_0 + r}{r_0}\right) + \frac{(1+r_0^2)}{r_0^2} \right\}.$

Proof By Lemma 8.2.3 (c),

$$\sup_{\|z\|=r} I(z) \leq 13 \int_{B(r_0,\infty)} \frac{r}{\|a\|^2(\|a\| + r)} T(a) \wedge \beta(a).$$

Without loss of generality, assume T is of class C^∞; then, we can take a C^∞ 1-form φ with $T = d\varphi$. Then the integral in the right hand side but over $B(r_0, t)$ in place of $B(r_0, \infty)$ equals

$$
\begin{aligned}
&\int_{B(r_0,t)} d\left(\frac{r}{\|a\|^2(\|a\|+r)}\varphi(a)\wedge\beta(a)\right)\\
&\quad + \int_{B(r_0,t)}\left(\frac{2r}{\|a\|^3(\|a\|+r)}+\frac{r}{\|a\|^2(\|a\|+r)^2}\right)d\|a\|\wedge\varphi(a)\wedge\beta(a)\\
&= \int_{S(t)}\frac{r}{\|a\|^2(\|a\|+r)}\varphi(a)\wedge\beta(a) - \int_{S(r_0)}\frac{r}{\|a\|^2(\|a\|+r)}\varphi(a)\wedge\beta(a)\\
&\quad + \int_{r_0}^t\left(\frac{2r}{s^3(s+r)}+\frac{r}{s^2(s+r)^2}\right)ds\int_{S(s)}\varphi(a)\wedge\beta(a)\\
&\le \frac{r}{t^2(t+r)}\int_{B(t)}T(a)\wedge\beta(a)\\
&\quad + \int_{r_0}^t\left(\frac{2r}{s^3(s+r)}+\frac{r}{s^2(s+r)^2}\right)ds\int_{B(s)}T(a)\wedge\beta(a).
\end{aligned}
$$

Since (8.6) and Lemma 8.2.2 (b) imply $\displaystyle\int_{B(t)} T\wedge\beta \le C(1+t^2)$, a direct calculation shows that the right hand side of the above inequality has the desired bound when $|t|\to\infty$. □

Since $T = \frac{i}{2}\sum_{j,l=1,2} T_{jl}dz_j\wedge d\bar z_l$ satisfies $2i\partial\bar\partial I\wedge\beta = 2\pi T(z)\wedge\beta(z)$ by Proposition 8.2.5, we have, letting $T' = (i/\pi)\partial\bar\partial I - T$,

$$T'_{11} + T'_{22} = 0, \tag{8.7}$$

and

$$\partial T' = \bar\partial T' = 0. \tag{8.8}$$

Lemma 8.2.7 *If a* $(1,1)$*-current* T' *satisfies (8.7) and (8.8), then* T' *is equal to a* $(1,1)$*-form with harmonic coefficients.*

Proof First, assume that T' is of class C^2. Then (8.8) implies

$$\frac{\partial T'_{pq}}{\partial z_j} = \frac{\partial T'_{jq}}{\partial z_p} \quad\text{and}\quad \frac{\partial T'_{jq}}{\partial \bar z_j} = \frac{\partial T'_{jj}}{\partial \bar z_q},\quad\text{so}\quad \frac{\partial^2 T'_{pq}}{\partial z_j\partial\bar z_j} = \frac{\partial^2 T'_{jj}}{\partial z_p\partial\bar z_q},$$

hence $\displaystyle\frac{\partial^2 T'_{pq}}{\partial z_1\partial\bar z_1} + \frac{\partial^2 T'_{pq}}{\partial z_2\partial\bar z_2} = \frac{\partial^2(T'_{11}+T'_{22})}{\partial z_p\partial\bar z_q} = 0$ by (8.7).

Next, for the general T, T'_ε satisfies (8.7) and (8.8), so T'_ε is a (1, 1)-form with harmonic coefficients. Theorem 8.1.4 (a) yields $(T'_{\varepsilon'})_\varepsilon = (T'_\varepsilon)_{\varepsilon'} = T'_\varepsilon$, so letting $\varepsilon \to 0$, we have $T'_{\varepsilon'} = T'$. □

Lemma 8.2.8 *The* (1, 1)*-current* $T' = (i/\pi)\partial\bar{\partial} I - T$ *satisfies* $T' = 0$.

Proof Lemma 8.2.7 implies that $T'_{jl}(z)$ is a harmonic function, so by Theorem 8.1.4 (a) applied for $\pm T'_{jl}$,

$$T'_{jl} = T'_{jl} * \rho_\varepsilon = \frac{2}{\pi}\frac{\partial^2 I}{\partial z_j \partial \bar{z}_l} * \rho_\varepsilon - T_{jl} * \rho_\varepsilon. \tag{8.9}$$

First we will investigate the first term of (8.9). Then,

$$\begin{aligned}\frac{\partial^2 I}{\partial z_j \partial \bar{z}_l} * \rho_\varepsilon(z) &= \int \frac{\partial^2 I}{\partial z'_j \partial \bar{z}'_l}(z + z')\rho_\varepsilon(z')dV(z') \\ &= \int I(z + z')\frac{\rho_\varepsilon(z')}{\partial z'_j \partial \bar{z}'_l}dV(z').\end{aligned}$$

Let $M = \max\left|\dfrac{\partial^2 \rho}{\partial z_j \partial \bar{z}_l}\right|$. Then we have $\left|\dfrac{\partial^2 \rho_\varepsilon(z')}{\partial z'_j \partial \bar{z}'_l}\right| \le M\varepsilon^{-6}$. Specifying $\varepsilon = r$, fixing $|z| \le r$ and setting $M_I(r) = \max_{\|z\| \le r} |I(z)|$,

$$\left|\frac{\partial^2 I}{\partial z_j \partial \bar{z}_l} * \rho_r(z)\right| \le \frac{M}{r^6}\int_{\|z'\|<r} |I(z + z')|dV(z') \le MM_I(2r)\frac{\pi^2 r^4}{2r^6},$$

which tends to 0 as $r \to \infty$, where we used Lemma 8.2.6.

Next, we shall study the second term of (8.9). Set $M_1 = \max \rho$. Then, by Lemma 8.1.3,

$$\begin{aligned}|T_{jl} * \rho_\varepsilon(z)| &\le M_1\varepsilon^{-4}\int_{t \in B(z,\varepsilon)} |T_{jk}(t)|dV(t) \\ &= M_1\varepsilon^{-4}|T_{jl}(t)dV|_{B(z,\varepsilon)} \le M_1\varepsilon^{-4}\sqrt{2}|T \wedge \beta|_{B(z,\varepsilon)}.\end{aligned}$$

For $\|z\| \le r$, letting $\varepsilon = r$, $|T_{jl} * \rho_r(z)| \le M_1 r^{-4}\sqrt{2}|T \wedge \beta|_{B(2r)} \to 0$ as $r \to \infty$, where we used (8.5) and Lemma 8.2.2 (b). Therefore, combining the above two investigations, we see that the harmonic function $T'_{jl}(z)$ satisfies $|T'_{jl}(z)| \to 0$ as $\|z\| \to 0$. Hence, by the maximum principle we have $T'_{jl}(z) = 0$. □

Theorem 8.2.9 *Let T be a positive closed* $(1,1)$*-current satisfying (8.5) and (8.6). Then, $I(z)$ satisfies $\frac{i}{\pi}\partial\bar{\partial} I = T$ and*

$$\sup_{\|z\|=r} I(z) \le 13C\left\{2\log\left(\frac{r_0+r}{r_0}\right) + \frac{(1+r_0^2)}{r_0^2}\right\}.$$

8.3 Fundamental properties of polynomial automorphisms

From the viewpoint of dynamics, the study of polynomial automorphisms of $\mathbb{C}^2$ reduces to that of finite compositions of generalized Hénon maps due to Theorem 7.1.4. First, we will fix a polynomial automorphism f which we investigate through Chapters 8 and 9.

Let us denote by π_1 and π_2 the projections of $\mathbb{C}^2$ to $\mathbb{C}$ defined by

$$\pi_1(x, y) = x, \quad \pi_2(x, y) = y.$$

Using this notation, for a generalized Hénon map $g(x, y) = (y, p(y) - \delta x)$, we have $\pi_1 \circ g(x, y) = y$ and $\pi_2 \circ g(x, y) = p(y) - \delta x$.

Definition Let $g_1, \ldots, g_m$ be m generalized Hénon maps, where each g_j is defined by a complex number $\delta_j \neq 0$ and a polynomial $p_j(y)$ of degree k_j ($k_j \geq 2$). Define

$$f = g_m \circ g_{m-1} \circ \cdots \circ g_1, \quad \delta = \delta_1\delta_2\cdots\delta_m \quad \text{and} \quad k = k_1k_2\cdots k_m.$$

We can see easily the following.

Proposition 8.3.1 *The Jacobian determinant J_f of f is the constant δ. The degree of $\pi_1 \circ f$ is less than k and the degree of $\pi_2 \circ f$ is equal to k. The number of the fixed points of f^n is between 1 and k^n.*

We shall study the dynamics on $\mathbb{C}^2$ of f. Besides the sets $K^{\pm}$ and K introduced in §7.2, we define the sets $J^{\pm}$ and J. We use the filtrations D_R and $V_R^{\pm}$ defined in §7.2.

Definition

$$K^+ = \{z \in \mathbb{C}^2 \mid \{f^n(z) \mid n \in \mathbb{N}\} \text{ is bounded in } \mathbb{C}^2\},$$

$$K^- = \{z \in \mathbb{C}^2 \mid \{f^{-n}(z) \mid n \in \mathbb{N}\} \text{ is bounded in } \mathbb{C}^2\},$$

$$J^+ = \partial K^+, \quad J^- = \partial K^-, \quad K = K^+ \cap K^-, \quad J = J^+ \cap J^-.$$

Proposition 8.3.2 *The sets K, $K^{\pm}$, J, and $J^{\pm}$ are closed sets and invariant under f. The sets K and J are compact.*

Proof The $K^{\pm}$ are invariant under f and, by Proposition 7.2.3, closed. So all the remaining sets are closed and invariant by f. As was shown in §7.2, K is compact, so J is also compact. □

Proposition 8.3.3 *The family $\{f^n\}_{n=1}^{\infty}$ constitutes a normal family in* Int K^+, *and so does the family $\{f^{-n}\}_{n=1}^{\infty}$ in* Int K^-.

Here we say that the family of maps $\{f^n\}$ constitutes a normal family if both $\{\pi_1 \circ f^n\}_{n=1}^{\infty}$ and $\{\pi_2 \circ f^n\}_{n=1}^{\infty}$ are normal families.

Proof By Proposition 7.2.6, Montel's theorem assures us that the family $\{f^n\}_{n=1}^{\infty}$ constitutes a normal family in Int K^+. □

We set $z = (x_0, x_1)$ and define (x_i, x_{i+1}) for positive integers i by

$$(x_0, x_1) \stackrel{g_1}{\to} (x_1, x_2) \stackrel{g_2}{\to} \cdots \stackrel{g_m}{\to} (x_m, x_{m+1}) \stackrel{g_1}{\to} (x_{m+1}, x_{m+2}) \stackrel{g_2}{\to} \cdots,$$

where we apply $g_1, \ldots, g_m$ periodically. (So in particular, we have $f^n(z) = (x_{mn}, x_{mn+1})$.)

Lemma 8.3.4 *Let R be sufficiently large. Set $N = D_R \cup g_1(D_R) \cup \cdots \cup g_m(D_R)$, and take $R_1 > R$ with $N \subset \left\{(x, y) \in \mathbb{C}^2 \,\middle|\, |x| \le R_1\right\}$. If $(x_i, x_{i+1}) \in \overline{D_R \cup V_R^-}$ $(0 \le i \le j-1)$, then $|x_i| \le \max\{R_1, |x_0|\}$ $(0 \le i \le j)$.*

Proof We proceed by induction on j. The implication is clearly true for $j = 1$. We assume that the implication is true for $j - 1$ $(j \ge 2)$ and that $(x_i, x_{i+1}) \in \overline{D_R \cup V_R^-}$ $(0 \le i \le j-1)$, and we shall show $|x_j| \le \max\{R_1, |x_0|\}$. If $(x_{j-1}, x_j) \in \overline{V_R^-}$, then $|x_j| \le |x_{j-1}| \le \max\{R_1, |x_0|\}$. To the contrary, if $(x_{j-1}, x_j) \in D_R$, then we have $(x_j, x_{j+1}) \in N$, hence $|x_j| \le R_1 \le \max\{R_1, |x_0|\}$. □

We will introduce the minimal growth conditions for the plurisubharmonic functions and the positive closed (1, 1)-current on $\mathbb{C}^2$.

Definition Set $\omega = \frac{1}{4} dd^c \log(1 + \|z\|^2)$ as in §8.2, and

$\mathcal{L} = \{v \in \mathcal{PSH}(\mathbb{C}^2) \mid v(z) \le \log^+ \|z\| + M \text{ with some } M > 0\}$,

$$\mathcal{K} = \left\{ T \,\middle|\, \text{positive closed } (1,1)\text{-current on } \mathbb{C}^2 \text{ with } \int_{\mathbb{C}^2} T \wedge \omega < +\infty \right\}.$$

Theorem 8.3.5 (a) *If $v \in \mathcal{L}$, then $dd^c v \in \mathcal{K}$.*

(b) *If $T \in \mathcal{K}$, then there are $c \geq 0$ and $v \in \mathcal{L}$ satisfying $T = cdd^c v$.*

Proof (a) Take $v \in \mathcal{L}$. We use the notation before Lemma 8.2.2. Note that $d^c\|z\|^2 \wedge \beta$ on $S(r)$ is the area element of $S(r)$, so in particular $d^c\|z\|^2 \wedge \beta > 0$ on $S(r)$. Then, for $r \geq 1$, we have

$$\int_{S(r)} v \wedge d^c \log \|z\|^2 \wedge \alpha = \frac{1}{\pi r^4} \int_{S(r)} v \wedge d^c \|z\|^2 \wedge \beta$$

$$\leq (\log r + M) \frac{1}{\pi r^4} \int_{S(r)} d^c \|z\|^2 \wedge \beta = 4\pi (\log r + M).$$

Suppose v is of class C^∞. Fix some $r_0 > 0$. Since $\alpha^2 = 0$ in $B(r_0, r)$ and $dv \wedge d^c \log \|z\|^2 \wedge \alpha = d \log \|z\|^2 \wedge d^c v \wedge \alpha$, using Stokes' theorem we have

$$\int_{S(r)} v \wedge d^c \log \|z\|^2 \wedge \alpha - \int_{S(r_0)} v \wedge d^c \log \|z\|^2 \wedge \alpha$$

$$= \int_{B(r_0,r)} dv \wedge d^c \log \|z\|^2 \wedge \alpha = \int_{B(r_0,r)} \frac{d\|z\|^2}{\|z\|^2} \wedge d^c v \wedge \alpha.$$

Fubini's theorem and Stokes' theorem convert the last integral to

$$\int_{r_0}^{r} \frac{2dr}{\pi r^3} \int_{S(r)} d^c v \wedge \beta = \int_{r_0}^{r} \frac{2dr}{\pi r^3} \int_{B(r)} dd^c v \wedge \beta.$$

Putting $K(r) = \dfrac{2}{\pi r^2} \displaystyle\int_{B(r)} dd^c v \wedge \beta$, $K(r)$ is an increasing function due to Lemma 8.2.2 (a) and satisfies

$$\int_{r_0}^{r} \frac{K(r)}{r} dr \leq 4\pi (\log r + M) + \int_{S(r_0)} v \wedge d^c \log \|z\|^2 \wedge \alpha,$$

so $K(r)$ must be bounded by 4π.

The general case can be proved by applying the above result in the C^∞ case to $T_\varepsilon = dd^c v_\varepsilon$ and by letting ε to 0.

(b) Since the assertion $T \in \mathcal{K}$ is equal to (8.5), in order to apply Lelong's theorem 8.2.9, we need to produce the situation of (8.6). First we assume that $0 \neq T \in \mathcal{K}$ is of class C^∞. Then, there is a $u \in C^\infty \cap \mathcal{PSH}(\mathbb{C}^2)$ such that $T = dd^c u$. Consider $u_1 = u + \log \|z\| \in \mathcal{PSH}(\mathbb{C}^2)$ and, for a constant $M < 0$,

$$u_2 = \begin{cases} \max\{u_1, M\} & \text{in } B(1), \\ u_1 & \text{in } \mathbb{C}^2 - B(1). \end{cases}$$

Since u is continuous and $\log \|0\| = -\infty$, we can choose $M < 0$ and $0 < 2r_0 < 1$ such that $u_2 \in \mathcal{PSH}(\mathbb{C}^2)$, $dd^c u_2 \neq 0$, and $u_2 = M$ on $B(2r_0)$. Since $\alpha^2 = 0$ in $\mathbb{C}^2 - \{0\}$, we have

$$\int_{B(1,r)} dd^c u_2 \wedge \alpha = \int_{B(1,r)} dd^c u_1 \wedge \alpha = \int_{B(1,r)} dd^c u \wedge \alpha,$$

hence $T_2 = dd^c u_2 \in \mathcal{K}$ by Lemma 8.2.2. This T_2 satisfies $\mathrm{supp}(T_2) \cap B(r_0) = \varnothing$. Applying Lelong's theorem, 8.2.9, there are $c > 0$ and $v_2 \in \mathcal{L}$ satisfying $cdd^c v_2 = T_2$. Noting $cv_2 - u_2$ is pluriharmonic, define v by $cv = (cv_2 - u_2) + u \in \mathcal{PSH}(\mathbb{C}^2)$. Since $cv = cv_2 - \log \|z\|$ in $\mathbb{C}^2 - B(1)$, $v_2 \in \mathcal{L}$ yields $v \in \mathcal{L}$, and we have $cdd^c v = dd^c u = T$.

Next, for the general case $0 \neq T \in \mathcal{K}$, there is a $u \in \mathcal{PSH}(\mathbb{C}^2)$ such that $dd^c u = T$. Fix $\varepsilon > 0$ so that $0 \neq T_\varepsilon = dd^c u_\varepsilon$. Then, we can see $T_\varepsilon \in \mathcal{K}$. So there are $c > 0$ and $v_1 \in \mathcal{L}$ such that $cdd^c v_1 = T_\varepsilon$. Since $u \leq u_\varepsilon$, we have $u - (u_\varepsilon - cv_1) \leq cv_1$. Define v by $cv = u - (u_\varepsilon - cv_1)$. Then $cdd^c v = dd^c u = T$ and $v \leq v_1$ implies $v \in \mathcal{L}$. □

In the remainder of this chapter, we chiefly use the maximum norm $|z| = \max\{|x|, |y|\}$ for a point $z = (x, y) \in \mathbb{C}^2$. By restricting the classes $\mathcal{L}$ and $\mathcal{K}$, we define two classes $\mathcal{L}_x, \mathcal{L}_y$ of plurisubharmonic functions and two classes $\mathcal{K}_x, \mathcal{K}_y$ of positive closed currents.

Definition

$\mathcal{L}_x = \{v \in \mathcal{L} \mid \text{on some } \overline{V_R^-}, v(x, y) - \log |x| \in \mathcal{PH} \text{ and is bounded}\}$,

$\mathcal{L}_y = \{v \in \mathcal{L} \mid \text{on some } \overline{V_R^+}, v(x, y) - \log |y| \in \mathcal{PH} \text{ and is bounded}\}$,

$\mathcal{K}_x = \{T \in \mathcal{K} \mid \mathrm{supp}(T) \cap \overline{V_R^-} = \varnothing \text{ for some } R\}$,

$\mathcal{K}_y = \{T \in \mathcal{K} \mid \mathrm{supp}(T) \cap \overline{V_R^+} = \varnothing \text{ for some } R\}$.

Proposition 8.3.6 (a) *If $v \in \mathcal{L}_y$, then $dd^c v \in \mathcal{K}_y$.*

(b) *For a non-zero $T \in \mathcal{K}_y$, there are $c > 0$ and $v \in \mathcal{L}_y$ such that $T = cdd^c v$. Here, c is determined uniquely and v is determined uniquely up to additive real constants.*

Proof Part (a) is clear. Let us prove (b). By Theorem 8.3.5 (b), there are $c_1 > 0$ and $\varphi \in \mathcal{L}$ with $T = c_1 dd^c \varphi$, and $\mathrm{supp}(T) \cap \overline{V_R^+} = \varnothing$ implies that φ is pluriharmonic on $\overline{V_R^+}$.

In the two-dimensional complex projective space $\mathbb{P}^2$ with homogeneous coordinates $\zeta_0 : \zeta_1 : \zeta_2$, we identify our $\mathbb{C}^2(x, y)$ by $\mathbb{C}^2 = \mathbb{P}^2 - \{\zeta_0 = 0\}$

and $x = \zeta_1/\zeta_0$, $y = \zeta_2/\zeta_0$. Let

$$\xi = \frac{\zeta_0}{\zeta_2}, \eta = \frac{\zeta_1}{\zeta_2} \quad \text{and} \quad \sigma = \frac{\zeta_0}{\zeta_1}, \tau = \frac{\zeta_2}{\zeta_1};$$

then

$$\begin{aligned} \overline{V_R^+} &= \{(\xi, \eta) \,|\, 0 < R|\xi| \le 1, |\eta| \le 1\} \\ &= \{(\sigma, \tau) \,|\, |\tau| \ge R|\sigma| > 0, |\tau| \ge 1\}. \end{aligned}$$

Set

$$\begin{aligned} \psi(\xi, \eta) &= \varphi(\eta/\xi, 1/\xi) \quad \text{in} \quad U = \{(\xi, \eta) \in \mathbb{C}^* \times \mathbb{C}\}, \text{ and} \\ \chi(\sigma, \tau) &= \varphi(1/\sigma, \tau/\sigma) \quad \text{in} \quad V = \{(\sigma, \tau) \in \mathbb{C}^* \times \mathbb{C}\}. \end{aligned}$$

Then the function $\psi(\xi, \eta)$ satisfies $\psi \in \mathcal{PSH}(U)$ and its restriction to $\overline{V_R^+}$ belongs to $\mathcal{PH}(\overline{V_R^+})$ with $\psi(\xi, \eta) \le \log|1/\xi| + M$ for some $M \in \mathbb{R}$. So for η being fixed in $\{|\eta| \le 1\}$, there are a constant $\alpha(\eta)$ with $0 \le \alpha(\eta) \le 1$ and a bounded harmonic function $h(\xi, \eta)$ in $\{0 < |\xi| \le 1/R\}$ such that

$$\psi(\xi, \eta) = \alpha(\eta) \log\left|\frac{1}{\xi}\right| + h(\xi, \eta) \quad \text{in} \quad \left\{0 < |\xi| \le \frac{1}{R}\right\}. \tag{8.10}$$

We shall show that the function $\alpha(\eta)$ in $\{|\eta| \le 1\}$ is equal to a constant c_2 with $0 \le c_2 \le 1$. Take $r > R$ and define

$$A(\eta) = \int_{|\xi|=1/r} \frac{\partial \psi(\xi, \eta)}{\partial \xi} \frac{i d\xi}{\pi} \quad \text{for} \quad \eta \in \mathbb{C};$$

then $A(\eta)$ is subharmonic in $\eta \in \mathbb{C}$ and, since $h(\xi, \eta)$ is continued up to $\xi = 0$ as a harmonic function, (8.10) implies $A(\eta) = \alpha(\eta)$ in $\{|\eta| \le 1\}$. Next, $A(\eta)$ can be continued up to $\eta = \infty$ so that $A(\eta)$ is subharmonic at $\eta = \infty$, which implies $A(\eta) = c_2$ (a constant) with $0 \le c_2 \le 1$.

In fact, τ being fixed in $\{|\tau| \ge 1\}$,

$$\chi(\sigma, \tau) = \alpha\left(\frac{1}{\tau}\right) \log\left|\frac{\tau}{\sigma}\right| + h\left(\frac{\sigma}{\tau}, \frac{1}{\tau}\right),$$

where $h(\sigma/\tau, 1/\tau)$ is harmonic in $\{|\sigma| \le |\tau|/R\}$, so

$$\alpha\left(\frac{1}{\tau}\right) = \int_{|\sigma|=1/R} \frac{\partial \chi(\sigma, \tau)}{\partial \sigma} \frac{i d\sigma}{\pi} \quad \text{in} \quad \{|\tau| \ge 1\}.$$

But, since $\chi(\sigma, \tau) = \varphi(1/\sigma, \tau/\sigma) \in \mathcal{PSH}(V)$,

$$B(\tau) = \int_{|\sigma|=1/R} \frac{\partial \chi(\sigma, \tau)}{\partial \sigma} \frac{d\sigma}{i\pi}$$

is subharmonic in $\{|\tau| < \infty\}$ and $B(\tau) = \alpha(1/\tau)$ in $\{|\tau| \geq 1\}$, which shows that $\alpha(\eta)$ is subharmonic up to $\eta = \infty$.

As a consequence, the function $h(\xi, \eta) = \psi(\xi, \eta) - c_2 \log |1/\xi|$, being pluriharmonic in $\overline{V_R^+}$ and bounded up to $\xi = 0$, is a bounded pluriharmonic function in $\overline{V_R^+}$. If $c_2 = 0$, then φ must be a bounded plurisubharmonic function in the whole of $\mathbb{C}^2$, therefore identically equal to a constant, and so $T = c_1 dd^c \varphi \equiv 0$, which contradicts the hypothesis. Therefore, $c_2 > 0$. Let $c = c_1 c_2$, $u(x, y) = (1/c_2) h(1/y, x/y)$ in $\overline{V_R^+}$ and $v = \varphi/c_2$; then, $v = \log |y| + u$ in $\overline{V_R^+}$, and $T = cdd^c v$ in $\mathbb{C}^2$, which proves the existence part of the proposition.

Finally, we prove the uniqueness. Let $T = c_1 dd^c v_1 = c_2 dd^c v_2$ with $c_1 \geq c_2$. Then since $c_1 v_1 - c_2 v_2$ is pluriharmonic in $\mathbb{C}^2$,

$$\Psi(x, y) = (c_1 - c_2) \log |y| - (c_1 v_1 - c_2 v_2) \tag{8.11}$$

is plurisubharmonic in $\mathbb{C}^2$ and is bounded pluriharmonic in some $\overline{V_R^+}$. By the maximum principle, Ψ is a bounded plurisubharmonic function in $\mathbb{C}^2$, hence must be a constant. On the other hand, the assumption $c_1 > c_2$ yields $\Psi(x, 0) = -\infty$; a contradiction. This proves the uniqueness of the constant c. Substituting $c_1 = c_2$ into (8.11), $v_1 - v_2$ is seen to be a constant. □

Definition Let k denote the degree of f, and $\log^+ t = \max\{\log t, 0\}$. Set

$$G_n^{\pm}(z) = \frac{1}{k^n} \log^+ |f^{\pm n}(z)| \quad (z \in \mathbb{C}^2).$$

We denote by $c_{0j} y^{k_j}$ the terms of the highest degree of the generalized Hénon maps g_j in the definition of f. We set

$$c_0 = \prod_{j=1}^{m} |c_{0j}|^{k_{j+1} \cdots k_m} \quad (\text{where } k_{j+1} \cdots k_m = 1 \text{ when } j = m).$$

Lemma 8.3.7 (a) *Let* $0 < \varepsilon < |c_{0j}|$. *Then for a sufficiently large* $R > 0$,

$$(|c_{0j}| - \varepsilon)|y|^{k_j} < |\pi_2 \circ g_j(x, y)| < (|c_{0j}| + \varepsilon)|y|^{k_j} \qquad (x, y) \in \overline{V_R^+}.$$

(b) *Let* $0 < \varepsilon < \min_j |c_{0j}|$. *Then for a sufficiently large* $R > 0$,

$$(c_0 - \varepsilon)|y|^k < |\pi_2 \circ f(x, y)| < (c_0 + \varepsilon)|y|^k \qquad (x, y) \in \overline{V_R^+}.$$

Proof Part (a) is clear. We prove (b). Let

$$N_1 = \prod_{j=1}^{m} (|c_{0j}| - \varepsilon_1)^{k_{j+1} \cdots k_m} \text{ and } N_2 = \prod_{j=1}^{m} (|c_{0j}| + \varepsilon_1)^{k_{j+1} \cdots k_m}.$$

By choosing $0 < \varepsilon_1$ so that $0 < \varepsilon_1 < \min_j |c_{0j}|$, we have $c_0 - \varepsilon < N_1$ and $N_2 < c_0 + \varepsilon$. In view of (a) and Proposition 7.2.2, take $R > 0$ so that

$$(|c_{0j}| - \varepsilon_1)|y|^{k_j} < |\pi_2 \circ g_j(x, y)| < (|c_{0j}| + \varepsilon_1)|y|^{k_j},$$

and $g_s \circ g_{s-1} \circ \cdots \circ g_1(x, y) \in V_R^+$ for every $(x, y) \in \overline{V_R^+}$. By estimating $|\pi_2 \circ g_s \circ g_{s-1} \circ \cdots \circ g_1|$ in sequence from $s = 1$ to $s = m$, we obtain $N_1|y|^k \le |\pi_2 \circ g_m \cdots \circ g_1(z)| \le N_2|y|^k$. □

Lemma 8.3.8 *Set $0 < \varepsilon < c_0$. Then we can choose a sufficiently large $R > 0$ so that, for every $n \in \mathbb{N}$ and $(x, y) \in \overline{V_R^+}$, we have*

$$0 < \log|y| + L_1 < G_n^+(x, y) = \frac{1}{k^n} \log|\pi_2 \circ f^n(x, y)| < \log|y| + L_2,$$

$$L_1 = \frac{k^n - 1}{k^n(k-1)} \log(c_0 - \varepsilon), \quad L_2 = \frac{k^n - 1}{k^n(k-1)} \log(c_0 + \varepsilon).$$

Proof Let $(x_n, y_n) = f^n(x, y)$, $M_1 = c_0 - \varepsilon$ and $M_2 = c_0 + \varepsilon$. By Proposition 7.2.2, $(x, y) \in \overline{V_R^+}$ yields $(x_n, y_n) \in \overline{V_R^+}$, so $|y_n| \ge R > 1$, hence $G_n^+(x, y) = k^{-n} \log|y_n|$. And inductively, we obtain

$$M_1^{1+k+\cdots+k^{n-1}}|y|^{k^n} < |y_n| < M_2^{1+k+\cdots+k^{n-1}}|y|^{k^n},$$

which proves the assertion. □

Definition For $z \in \mathbb{C}^2$, set $G^\pm(z) = \lim_{n\to\infty} G_n^\pm(z)$.

In the following lemmas, 8.3.9–8.3.11, we prove that the sequence $\{G_n^\pm\}$ converges uniformly on every compact set in $\mathbb{C}^2$. We call $G^\pm$ the *Green function for $K^\pm$*.

Lemma 8.3.9 *For a sufficiently large $R > 0$, $\{G_n^+\}$ converges uniformly on $\overline{V_R^+}$ as $n \to \infty$. The limit function G^+ is a pluriharmonic function on $\overline{V_R^+}$ admitting the expression*

$$G^+(x, y) = \log|y| + \gamma_0 + u(x, y) \qquad \text{on } \overline{V_R^+},$$

where $\gamma_0 = (\log c_0)/(k-1)$, and u is a bounded pluriharmonic function with $\lim_{|y|\to\infty} u(x, y) = 0$.

We call γ_0 the *Robin constant* for G^+.

Proof Set $(x_n, y_n) = f^n(z)$. For $0 < \varepsilon < c_0$, we fix $R > 0$ which gives the inequalities of Lemma 8.3.8. Take $N > 0$ satisfying $|\log(c_0 \pm \varepsilon)| < N$. Then, by Lemma 8.3.7 (b),

$$|G^+_{n+1}(z) - G^+_n(z)| = \left| \frac{1}{k^{n+1}} \log \left| \frac{y_{n+1}}{y_n^k} \right| \right| \leq \frac{1}{k^{n+1}} N$$

which proves that $\{G^+_n\}$ converges uniformly on $\overline{V^+_R}$. As a limit function of uniform convergence of pluriharmonic functions $G^+_n(z)$, G^+ is also pluriharmonic. Since $G^+ - \log|y|$ is bounded pluriharmonic in $\overline{V^+_R}$ by Lemma 8.3.8, this function can be continued up to $\{x \in \mathbb{C}, y = \infty\} \in \mathbb{C} \times \mathbb{P}$ as a pluriharmonic function. Then, $\lim_{|y| \to \infty} (G^+(x, y) - \log|y|)$, being a bounded harmonic function in $x \in \mathbb{C}$, becomes a constant γ. By Lemma 8.3.8, γ satisfies

$$\frac{\log(c_0 - \varepsilon)}{k - 1} \leq \gamma \leq \frac{\log(c_0 + \varepsilon)}{k - 1}.$$

Since we can take ε arbitrarily small we have $\gamma = \gamma_0$. □

Lemma 8.3.10 *As $n \to \infty$, $\{G^+_n\}$ converges uniformly on every compact set in $\mathbb{C}^2 - K^+$. The limit function G^+ is > 0 and pluriharmonic and satisfies $G^+ \circ f = kG^+$ in $\mathbb{C}^2 - K^+$.*

Proof First, fix n. Since $G^+_r = k^{-n} G^+_{r-n} \circ f^n$, Lemma 8.3.9 assures us that $\{G^+_r\}$ converges, as $r \to \infty$, uniformly on $f^{-n}(V^+_R)$ to the pluriharmonic function $k^{-n} G^+ \circ f^n$. We see that G^+ satisfies $G^+(z) = k^{-n} G^+ \circ f^n(z)$ for $z \in f^{-n}(V^+_R)$. Since Lemma 8.3.8 yields

$$G^+_r(z) = \frac{1}{k^n} G^+_{r-n} \circ f^n(z) > \frac{1}{k^n} (\log R + L_1) > 0,$$

we have $G^+(z) \geq k^{-n}(\log R + L_1) > 0$ for $z \in f^{-n}(V^+_R)$.

Now, we apply the above result on $f^{-n}(V^+_R)$ to $\bigcup_{n=0}^{\infty} f^{-n}(V^+_R) = \mathbb{C}^2 - K^+$. (See Proposition 7.2.3.) Then as $r \to \infty$, $\{G^+_r\}$ converges uniformly on every compact set in $\mathbb{C}^2 - K^+$ to the pluriharmonic function $G^+ > 0$. □

Lemma 8.3.11 *The sequence $\{G^+_n\}$ converges uniformly on every compact set of $\mathbb{C}^2$ to the function G^+ satisfying $G^+ = 0$ on K^+. Hence, G^+ is continuous on $\mathbb{C}^2$.*

Proof By Lemma 8.3.10, the assertion holds for $\mathbb{C}^2 - K^+$. It is also clear by Lemma 8.3.4 that $\{G^+_n\}$ converges to 0 uniformly on every compact set of K^+.

Take $R > 0$ so that $K^+ \subset D_R \cup V_R^-$ and that Lemma 8.3.8 holds. By the notation of Lemma 8.3.4, let $R_1 \geq R$. Take an arbitrary $A > R_1$, and set

$$C = \{(x, y) \mid |x| \leq A\} \cap \overline{D_R \cup V_R^-}, \quad B = \sup_{z \in C} |\pi_2 \circ f(z)|.$$

Then, it is enough to show that $\{G_n^+\}$ converges to G^+ uniformly on C. For that purpose, since Proposition 7.2.3 gives $f^{-n}(V_R^+) \subset f^{-n-1}(V_R^+)$ and $\bigcup_{n=0}^{\infty} f^{-n}(V_R^+) = \mathbb{C}^2 - K^+$, letting

$$C_n = C \cap (f^{-n}(V_R^+) - f^{-n+1}(V_R^+))$$

it suffices to show that $\lim_{n\to\infty} \sup_{C_n} (\sup_{r \geq n} G_r^+) = 0$.

Take $z = (x_0, x_1) \in C_n$, then we have $f^n(z) \in V_R^+$, $f^{n-1}(z) \in \overline{D_R \cup V_R^-}$. Since Proposition 7.2.2 implies $(x_i, x_{i+1}) \in \overline{D_R \cup V_R^-}$ $(0 \leq i \leq (n-1)m)$, Lemma 8.3.4 gives $|\pi_1 \circ f^{n-1}(z)| \leq \max\{R_1, |x_0|\} \leq A$, hence $f^{n-1}(z) \in C$. Putting $M = (\log(c_0 + \varepsilon))/(k-1)$, and using Lemma 8.3.8, we obtain independently of r

$$G_r^+(z) = \frac{1}{k^n} G_{r-n}^+ \circ f^n(z) < \frac{1}{k^n}(\log|\pi_2 \circ f^n(z)| + M) \leq \frac{1}{k^n}(\log B + M).$$

Therefore, $\lim_{n\to\infty} \sup_{C_n} (\sup_{r\geq n} G_r^+) = 0$. □

Remark If we use another norm of $\mathbb{C}^2$ in the definition of $G_n^{\pm}$, we obtain the same limit functions $G^{\pm}$. There is another proof of Lemma 8.3.11, where one homogenizes f into a birational map of $\mathbb{P}^2$. See Fornaess and Sibony (1995) for such a proof and more information.

Lemma 8.3.12 *G^+ is plurisubharmonic on $\mathbb{C}^2$.*

Proof By Lemma 8.3.11, we know that G^+ is continuous. Furthermore, by Lemma 8.3.10, G^+ is pluriharmonic outside K^+, and G^+ takes the minimum value 0 on K^+. Therefore, G^+ is plurisubharmonic on $\mathbb{C}^2$. □

Proposition 8.3.13 $G^+ \in \mathcal{L}_y$.

Proof In view of Lemmas 8.3.12 and 8.3.9, it remains to show that

$$G^+(z) \leq \log^+|z| + M_1 \quad (z \in \mathbb{C}^2).$$

Let $D = \{z \in \mathbb{C}^2 \mid |z| < 1\}$ and $D_1 = D \cup \bigcup_{s=1}^{m-1} (g_s \circ g_{s-1} \circ \cdots \circ g_1)^{-1}(D)$. Then, we can choose $A > 0$ such that the generalized Hénon maps g_j $(1 \leq$

$j \le m$) satisfy

$$|g_j(z)| \le A|z|^{k_j} \quad \text{for } z \in \mathbb{C}^2 - D.$$

Let $A_1 = A^{1+k_m+k_{m-1}k_m+\cdots+k_2\cdots k_m}$, then we obtain

$$|f(z)| \le A_1|z|^k \quad \text{for } z \in \mathbb{C}^2 - D_1.$$

Now we set $M_0 = \max\{k^{-1}\log A_1, \max_{D_1} k^{-1}\log^+|f(z)|\}$. Then,

$$k^{-1}\log^+|f(z)| \le \log^+|z| + M_0 \quad \text{for } z \in \mathbb{C}^2.$$

Inductively, we obtain

$$k^{-n}\log^+|f^n(z)| \le \log^+|z| + \frac{(k^n-1)M_0}{k^{n-1}(k-1)} \le \log^+|z| + 2M_0, \qquad (8.12)$$

which implies the assertion. □

Since the same result holds for G^-, we can summarize Lemmas 8.3.10, 8.3.11, 8.3.12, and Proposition 8.3.13 into the next theorem, where we set $\mathcal{L}_+ = \mathcal{L}_y$ and $\mathcal{L}_- = \mathcal{L}_x$.

Theorem 8.3.14 *The sequence $\{G_n^\pm\}$ converges, as $n \to \infty$, on every compact set of $\mathbb{C}^2$ to a continuous function $G^\pm \in \mathcal{L}_\pm$ satisfying $G^\pm \circ f = k^{\pm 1}G^\pm$. This $G^\pm$ is positive pluriharmonic in $\mathbb{C}^2 - K^\pm$, and is equal to 0 on $K^\pm$.*

Next, we define positive closed (1, 1)-currents $\mu^\pm$ on $\mathbb{C}^2$ by setting

$$\mu^\pm = dd^c G^\pm.$$

In view of Proposition 8.3.13, we can apply Proposition 8.3.6 (a) for $G^\pm$, and we have $\mu^\pm \in \mathcal{K}_\pm$, where we set $\mathcal{K}_+ = \mathcal{K}_y$ and $\mathcal{K}_- = \mathcal{K}_x$. We call $\mu^\pm$ the stable/unstable current for f.

Definition For a (1, 1)-current T on $\mathbb{C}^2$, the pullback f^*T of T by f is defined by the formula

$$\int_{\mathbb{C}^2} \psi \wedge f^*T = \int_{\mathbb{C}^2} \psi \circ f^{-1} \wedge T,$$

for all $\psi \in \mathcal{D}^{1,1}(\mathbb{C}^2)$.

Remark In the formalism of current theory, for a proper holomorphic map $h : X \to Y$ between complex manifolds and for a current S on X, the pushforward h_*S on Y is well defined. Our pullback f^*T is equal to $(f^{-1})_*T$ for the automorphism $f^{-1} : \mathbb{C}^2 \to \mathbb{C}^2$.

If T is given as $T = dd^c\varphi$ by a locally integrable function φ on $\mathbb{C}^2$, then $f^*T = dd^c(\varphi \circ f)$. In view of Theorem 8.3.14,

$$G^{\pm} = k^{\mp 1} G^{\pm} \circ f, \quad \mu^{\pm} = k^{\mp 1} f^* \mu^{\pm}.$$

Theorem 8.3.15 $\quad \operatorname{supp}(\mu^{\pm}) = J^{\pm}$.

Proof By Theorem 8.3.14, $\operatorname{supp}(\mu^+) \subset J^+$. To prove the converse inclusion, suppose $\mu^+ \equiv 0$ in a neighborhood U of a point $z \in J^+$. Then G^+ is pluriharmonic in U. Again by Theorem 8.3.14, $G^+ \geq 0 = G^+(z)$ in U, which implies that G^+ takes its minimum in U at the interior point z, so, by the minimum principle for pluriharmonic functions, $G^+ \equiv 0$ in U. Since $G^+ > 0$ in $\mathbb{C}^2 - K^+$, this contradicts $z \in J^+ = \partial K^+$. □

For a compact Y of the complex plane $\mathbb{C}$, we denote by G_Y^* on $\mathbb{C}$ the Green function for Y. (See §1.4.) As a last topic of this section, we summarize the facts on G_Y^* which will be used later, and investigate some relations between G_Y^* and the function $G^+(x, \cdot)$ in the y-plane with x being fixed (Theorem 8.3.20).

Lemma 8.3.16 *Let v be a subharmonic function v on $\mathbb{C}$. If v is of the form $v(y) = \log|y| + u(y)$ in $\{|y| > R\}$ by a bounded harmonic function $u(y)$ in $\{|y| > R\}$, then $\displaystyle\int_{\mathbb{C}} dd^c v = 2\pi$.*

Proof Take $\chi \in C_0^\infty(\mathbb{C})$ which is identically equal to 1 in a neighborhood of $\{|y| \leq R\}$. Then by Stokes' theorem,

$$\begin{aligned}\int_{\mathbb{C}} dd^c v &= \int_{\mathbb{C}} \chi dd^c v = -\int_{\mathbb{C}} d\chi \wedge d^c v \\ &= -\int_{|y|>R} d(\chi \wedge d^c v) = \int_{|y|=R} \chi \wedge d^c v = \int_{|y|=R} d^c v.\end{aligned}$$

Since a bounded harmonic function in $\{|y| > R\}$ can be extended to the harmonic function around $y = \infty$, $-\int_{\{|y|=R\}} d^c u = \int_{\{|y|>R\}} dd^c u = 0$, whereas we have $\int_{|y|=R} d^c \log|y| = 2\pi$. Therefore, $\int_{\mathbb{C}} dd^c v = 2\pi$. □

The next theorem describes that, since $dd^c h = \Delta h du dv$ via real coordinates (u, v) with $y = u + iv$, $\frac{1}{2\pi}\log|y|$ is the fundamental solution of the Laplacian Δ of the y-plane. Compare with Theorem 8.2.1.

Theorem 8.3.17 *For a measure λ of compact support in $\mathbb{C}$,*

$$dd^c \int_{\mathbb{C}} \log|y-\xi|\lambda(\xi) = 2\pi\lambda(y).$$

For a compact Y of $\mathbb{C}$, we define a Perron family $\mathcal{V}(Y)$ of subharmonic functions in $\mathbb{C}$ satisfying $v \le 0$ on Y, and

$$v(\zeta) \le \log^+|\zeta| + M$$

in $\mathbb{C}$ with some $M > 0$, and set $G_Y(\zeta) = \sup_{v\in\mathcal{V}(Y)} v(\zeta)$ and $G_Y^*(\zeta) = \limsup_{\eta\to\zeta} G_Y(\eta)$.

Proposition 8.3.18 *If $G_Y^*(\zeta) \not\equiv \infty$, then $G_Y^*(\zeta)$ is subharmonic in $\mathbb{C}$ and satisfies $0 \le G_Y^*(\zeta) \le \log^+|\zeta| + M_1$, with some $M_1 > 0$. Furthermore, G_Y^* is harmonic in $\mathbb{C} - Y$ and $G_Y^* > 0$ in the unbounded connected component of $\mathbb{C} - Y$.*

We call G_Y^* and $dd^c G_Y^*$ the Green function and the *equilibrium measure* of Y. Indeed, the Green function $g(y,\infty)$ for the unbounded connected component D of $\mathbb{C} - Y$ in §1.4 satisfies $g(y,\infty) = (G_Y^*)_{|D}$. The logarithmic capacity $\operatorname{cap}(Y)$ of a compact set Y was defined by $\operatorname{cap}(Y) = \exp(-\gamma(Y))$, where $\gamma(Y) = \lim_{|\zeta|\to\infty}(G_Y^*(\zeta) - \log|\zeta|)$ is the Robin constant. Here we set $\operatorname{cap}(Y) = 0$ when $G_Y^* \equiv \infty$.

Proposition 8.3.19 *For the Green function G_Y^* of a compact Y of $\mathbb{C}$, $G_Y^* \equiv 0$ on Y outside a subset of Y of logarithmic capacity* 0.

Proof By Proposition 8.3.18, $G_Y^* \ge 0$. On the other hand, by the definition of G_Y, $G_Y \le 0$ on Y. Since it is known that the set $\{G_Y^* > G_Y\}$ is of logarithmic capacity 0, $G_Y^* \le 0$ on Y outside a subset of logarithmic capacity 0. □

Definition Let Ω be a domain in $\mathbb{C}^2$, v be a plurisubharmonic function in Ω, $T = dd^c v$ and M be a one-dimensional complex closed submanifold in Ω. The restriction $T_{|M}$ of T to M is defined to be the positive closed $(1,1)$-current $dd^c(v_{|M})$ on M, where $v_{|M}$ is the restriction of v to M.

When M is a y-plane $M_x = \{(x,y) \mid y \in \mathbb{C}\}$ for a fixed x, the restriction of T to $\Omega \cap M_x$ is denoted by T_x.

Theorem 8.3.20 *For a fixed x, let $M_x = \{(x,y) \mid y \in \mathbb{C}\}$ and $Y = K^+ \cap M_x$. Then, $G^+(x,\cdot)$ coincides with the Green function of Y on the y-plane M_x:*

$\int_{\mathbb{C}} \mu_x^+ = 2\pi$, $\mathrm{cap}(Y) = e^{-\gamma_0}$ *and*

$$G^+(x, y) = \frac{1}{2\pi} \int_{\mathbb{C}} \log|y - \xi| \mu_x^+(\xi) + \gamma_0.$$

Proof Let $v \in \mathcal{V}(Y)$. Then, $v - G^+(x, \cdot)$ is subharmonic and bounded above in $\mathbb{C} - Y$, and $v - G^+(x, \cdot) \leq 0$ on $\partial(\mathbb{C} - Y)$. So by the maximum principle, $v \leq G^+(x, \cdot)$ in $\mathbb{C} - Y$. Combining with $v \leq G^+(x, \cdot) = 0$ on Y, we see $v \leq G^+(x, \cdot)$ in $\mathbb{C}$, which proves $G_Y^* \leq G^+(x, \cdot)$. In particular, $G^* \not\equiv \infty$ and $\mathrm{cap}(Y) > 0$. Conversely, $G^+(x, \cdot) \in \mathcal{V}(Y)$ implies $G^+(x, \cdot) \leq G_Y^*$. Therefore, $G^+(x, \cdot) = G_Y^*$. Hence the assertions follow by Lemmas 8.3.16, 8.3.9, and Theorem 8.3.17. □

8.4 Pluricomplex Green functions

In this section, we define a positive closed (2, 2)-current $\mu = \mu^+ \wedge \mu^-$ by taking the product of the stable and unstable currents $\mu^\pm$ of f introduced in §8.3.

Next, we introduce the pluricomplex Green function of a subset $E \subset \mathbb{C}^2$, and in particular, consider those of the sets $K^\pm$ or $J^\pm$, and of K. Finally, we show that this μ is the complex equilibrium measure of K.

Let Ω be a domain in $\mathbb{C}^2$. Besides the notation in §8.1, $L^\infty_{loc}(\Omega)$ denotes the set of all locally bounded measurable functions in Ω.

Lemma 8.4.1 *If a* (1, 1)*-current T in Ω and a function $\chi \in C_0^\infty(\Omega)$ satisfy* $\mathrm{supp}(\chi) \cap \mathrm{supp}(dT) = \varnothing$, *then for $G \in C^\infty(\Omega)$,*

$$\int_\Omega \chi dd^c G \wedge T = \int_\Omega G dd^c \chi \wedge T.$$

Proof By the hypothesis on dT, we have

$$\begin{aligned} \int_\Omega \chi dd^c G \wedge T &= -\int_\Omega d\chi \wedge d^c G \wedge T \\ &= -\int_\Omega dG \wedge d^c \chi \wedge T = \int_\Omega G dd^c \chi \wedge T. \end{aligned}$$

□

Definition For a positive (1, 1)-current T in Ω, a locally bounded plurisubharmonic function G, and a non-negative $\psi \in C^\infty(\Omega)$ satisfying $\mathrm{supp}(\psi) \cap$

$\text{supp}(dT) = \varnothing$, a (2, 2)-current $\psi dd^c G \wedge T$ is defined by the formula

$$\int_\Omega \chi\psi dd^c G \wedge T = \int_\Omega G dd^c(\chi\psi) \wedge T \quad (\chi \in C_0^\infty(\Omega)). \tag{8.13}$$

In the right hand side of (8.13), $dd^c(\chi\psi) \wedge T$ is a (2, 2)-form whose coefficient is a measure of compact support, so the integration is well defined.

We show that $\psi dd^c G \wedge T$ is a positive current. Let $\chi \in C_0^\infty(\Omega)$ with $\chi \geq 0$. By Theorem 8.1.4, take a neighborhood $\Omega_1 \subset\subset \Omega$ of $\text{supp}(\chi)$ and a family $\{G_\varepsilon\}$ which is decreasing as $\varepsilon \to 0$ such that $G_\varepsilon \in \mathcal{PSH}(\Omega_1) \cap C^\infty(\Omega_1)$ and $G_\varepsilon \to G$ in Ω_1. Then by the monotone convergence theorem and Lemma 8.4.1,

$$\begin{aligned} \int_\Omega G dd^c(\chi\psi) \wedge T &= \lim_{\varepsilon\to 0} \int_\Omega G_\varepsilon dd^c(\chi\psi) \wedge T \\ &= \lim_{\varepsilon\to 0} \int_\Omega \chi\psi dd^c G_\varepsilon \wedge T \geq 0, \end{aligned}$$

since $T \geq 0$, $dd^c G_\varepsilon \geq 0$, $\psi \geq 0$, and $\chi \geq 0$. (See Klimek (1991), p. 101.)

Remark If T is closed, then we can define $dd^c G \wedge T$ by taking the constant function $\psi = 1$. For two $u_1, u_2 \in \mathcal{PSH}(\Omega) \cap L^\infty_{loc}(\Omega)$, we can define a positive closed (2, 2)-current $dd^c u_1 \wedge dd^c u_2$ by the above definition. One way of applying this definition by letting $\psi = 1$, $T = dd^c u_1$ and $G = u_2$ and the other way by $\psi = 1$, $T = dd^c u_2$ and $G = u_1$ give the same result.

Next we quote two facts. For the proof, see Klimek (1991).

Theorem 8.4.2 *Let $u_1, u_2, v_{1r}, v_{2r} \in \mathcal{PSH}(\Omega) \cap L^\infty_{loc}(\Omega)$ and assume that $\{v_{jr}\}$ are decreasing sequences converging to u_j as $r \to \infty$ $(j = 1, 2)$. Then, as $r \to \infty$*

$$dd^c v_{jr} \to dd^c u_j, \; dd^c v_{1r} \wedge dd^c v_{2r} \to dd^c u_1 \wedge dd^c u_2.$$

Lemma 8.4.3 *For u_1, u_2 in $\mathcal{PH}(\Omega)$,*

$$(dd^c \max\{u_1, u_2\})^2 = 0.$$

We shall apply the above results to our currents $\mu^\pm$ introduced in §8.3. Since $G^\pm \in \mathcal{PSH}(\mathbb{C}^2) \cap L^\infty_{loc}(\mathbb{C}^2)$ by Theorem 8.3.14,

$$\mu^{\pm 2} = (dd^c G^\pm)^2, \quad \mu = \mu^+ \wedge \mu^- = dd^c G^+ \wedge dd^c G^-$$

in $\mathbb{C}^2$ are well defined.

Proposition 8.4.4 $\mu^{\pm 2} = 0$.

Proof Since G^+ is pluriharmonic in $\{G^+ > 0\}$ by Lemma 8.3.9, taking $\varepsilon > 0$ and letting $G^+_\varepsilon = \max\{G^+, \varepsilon\}$, Lemma 8.4.3 yields $(dd^c G^+_\varepsilon)^2 = 0$. By Theorem 8.4.2, $(dd^c G)^2 = \lim_{\varepsilon \to 0}(dd^c G^+_\varepsilon)^2 = 0$. □

The pullbacks by f of (2, 2)-currents are defined similarly as those of (1, 1)-currents (see §8.3).

Proposition 8.4.5 $f^*\mu = \mu$.

Proof First, we note that $f^*\mu = f^*\mu^+ \wedge f^*\mu^-$, since we have

$$\begin{aligned}
\int_{\mathbb{C}^2} f^*\mu \wedge \psi &= \int_{\mathbb{C}^2} dd^c G^+ \wedge dd^c G^- \wedge \psi \circ f^{-1} \\
&= \int_{\mathbb{C}^2} G^+ \wedge dd^c G^- \wedge dd^c(\psi \circ f^{-1}) \\
&= \int_{\mathbb{C}^2} G^+ \circ f \wedge f^*\mu^- \wedge dd^c \psi \\
&= \int_{\mathbb{C}^2} dd^c(G^+ \circ f) \wedge f^*\mu^- \wedge \psi = \int_{\mathbb{C}^2} f^*\mu^+ \wedge f^*\mu^- \wedge \psi
\end{aligned}$$

for $\psi \in \mathcal{D}^{0,0}(\mathbb{C}^2)$. Therefore, by Theorem 8.3.14, $f^*\mu = f^*\mu^+ \wedge f^*\mu^- = k\mu^+ \wedge k^{-1}\mu^- = \mu^+ \wedge \mu^- = \mu$. □

Now, we introduce the pluricomplex Green function of a subset $E \subset \mathbb{C}^2$.

Definition Let $\tilde{\mathcal{L}}$ be the set of all u in $\mathcal{PSH}(\mathbb{C}^2)$ satisfying

$$\log^+|z| - M < u(z) < \log^+|z| + M \text{ on } \mathbb{C}^2$$

with some $M > 0$.

For the class $\mathcal{L}$ defined in §8.3, $\tilde{\mathcal{L}} \subset \mathcal{L}$ holds clearly. Since $\tilde{\mathcal{L}} \subset \mathcal{PSH}(\mathbb{C}^2) \cap L^\infty_{loc}(\mathbb{C}^2)$ by definition, $dd^c u \wedge dd^c v$ can be defined for $u, v \in \tilde{\mathcal{L}}$.

Definition For a subset $E \subset \mathbb{C}^2$, the function

$$L_E(z) = \sup\{u(z) \mid u \in \mathcal{L}, u \leq 0 \text{ on } E\} \quad (z \in \mathbb{C}^2)$$

is called the *pluricomplex Green function* of E. We set

$$L^*_E(z) = \limsup_{w \to z} L_E(w).$$

A subset $E \subset \mathbb{C}^2$ is called *pluripolar*, if each point $z \in E$ has an open neighborhood U and $v \in \mathcal{PSH}(U)$ satisfying $E \cap U \subset \{z \in U \mid v(z) = -\infty\}$. Then, we know the following facts. For the proofs, see, for instance, Klimek (1991).

Theorem 8.4.6 (a) $L_E^* \equiv +\infty$ *when E is pluripolar, and* $L_E^* \in \mathcal{L}$ *otherwise.*
(b) *If E is not pluripolar and bounded, then* $L_E^* \in \tilde{\mathcal{L}}$.

Proposition 8.4.7 *Let* $\Omega \subset \mathbb{C}^2$ *be a domain, and let* $u \in \mathcal{PSH}(\Omega) \cap L^\infty_{loc}(\Omega)$. *Then, for any pluripolar set* $E \subset \Omega$, $\int_E (dd^c u)^2 = 0$.

Theorem 8.4.8 *If E is not pluripolar,* $\{z \mid L_E^*(z) > L_E(z)\}$ *is a pluripolar set.*

The next theorem is a two-dimensional version of Lemma 8.3.16.

Theorem 8.4.9 *For* $v \in \tilde{\mathcal{L}}$, $\displaystyle\int_{\mathbb{C}^2} (dd^c v)^2 = 4\pi^2$.

Here, we consider the case of the pluricomplex Green functions of the sets $K^\pm$ and $J^\pm$.

Proposition 8.4.10 $\quad L_{K^\pm} = L_{K^\pm}^* = G^\pm, \quad L_{J^\pm} = L_{J^\pm}^* = G^\pm.$

Proof Since $G^+ = 0$ on K^+ by Lemma 8.3.11 and $G^+ \in \mathcal{L}$ by Proposition 8.3.13, therefore $G^+ \le L_{K^+} \le L_{J^+}$. To prove $L_{J^+} \le G^+$, take $u \in \mathcal{L}$ such that $u \le 0$ on J^+. By the maximum principle, $u \le 0 = G^+$ on K^+. Next, we compare u and G^+ in $\mathbb{C}^2 - K^+$. Since $u(x, y) \le \log|y| + M_1$ and Lemma 8.3.9 implies

$$\log|y| - M < G^+(x, y) < \log|y| + M \text{ for } (x, y) \in V_R^+,$$

we have

$$u(x, y) - G^+(x, y) \le M + M_1 \text{ for } (x, y) \in V_R^+.$$

So $x = x_0$ being fixed, $u(x_0, y) - G^+(x_0, y)$ is subharmonic and bounded in $\omega = \mathbb{C} - (K^+ \cap \{x = x_0\})$ of the y-plane, hence continued up to $y = \infty$ as a subharmonic function. Since $\partial\omega \subset J^+ \cap \{x = x_0\}$, $u(x_0, y) - G^+(x_0, y) \le 0$ on $\partial\omega$. By the maximum principle, $u(x_0, y) - G^+(x_0, y) \le 0$ on ω, which implies $u \le G^+$ in $\mathbb{C}^2 - K^+$.

Consequently, $u \le G^+$ in $\mathbb{C}^2$, which implies that $L_{J^+} \le G^+$. Now, since $L_{J^+} = L_{K^+} = G^+$ is a continuous function by Lemma 8.3.11, $L_{K^+}^* = L_{K^+}$, $L_{J^+}^* = L_{J^+}$.

The other cases are similar. □

Next we will study the pluricomplex Green function of the set K after some preliminaries.

Lemma 8.4.11 *Let u_1, u_2, u_3 be three pluriharmonic functions in a domain $\Omega \subset \mathbb{C}^2$. Let $v = \max\{u_1, u_2, u_3\}$, $v_1 = \max\{u_1, u_3\}$ and $v_2 = \max\{u_2, u_3\}$. Then, $(dd^c v)^2 = dd^c v_1 \wedge dd^c v_2$ in Ω.*

Proof It is enough to prove the assertion when $u_3 = 0$. Take $\varphi \in C_0^\infty(\Omega)$. Then, setting

$$I_1 = \int_\Omega (v - v_1) dd^c v \wedge dd^c \varphi \quad \text{and} \quad I_2 = \int_\Omega v_1 dd^c(v - v_2) \wedge dd^c \varphi,$$

we have

$$\begin{aligned} \int_\Omega \varphi (dd^c v)^2 &= \int_\Omega v dd^c v \wedge dd^c \varphi \\ &= I_1 + I_2 + \int_\Omega \varphi dd^c v_1 \wedge dd^c v_2. \end{aligned}$$

We show $I_1 = I_2 = 0$. Note that there is a constant $C > 0$ such that $-C\beta \leq dd^c \varphi \leq C\beta$ in $\mathbb{C}^2$, where $\beta = \frac{1}{4} dd^c \|z\|^2$.

First, take any $\varepsilon > 0$, and let $F_\varepsilon = \{z \in \operatorname{supp}(\varphi) \mid u_2(z) \leq \varepsilon\}$. And divide the integration in I_1 over Ω into those over $\Omega - F_\varepsilon$ and over F_ε. Then, since $0 \leq v - v_1 \leq \varepsilon$ on F_ε,

$$\begin{aligned} \left| \int_{F_\varepsilon} (v - v_1) dd^c v \wedge dd^c \varphi \right| &\leq C\varepsilon \int_{F_\varepsilon} dd^c v \wedge \beta \\ &\leq C\varepsilon \int_{\operatorname{supp}(\varphi)} dd^c v \wedge \beta \to 0 \quad \text{as } \varepsilon \to 0. \end{aligned}$$

As to the integration over $\Omega - F_\varepsilon$, let $G_\varepsilon = \{|u_2 - u_1| \leq \varepsilon\}$. Since $v - v_1 = 0$ on $(\Omega - F_\varepsilon) \cap \{z \in \operatorname{supp}(\varphi) \mid u_2(z) - u_1(z) < -\varepsilon\}$, and $v = u_2$ and hence $dd^c v = 0$ on $(\Omega - F_\varepsilon) \cap \{u_2 - u_1 > \varepsilon\}$, we have

$$\int_{\Omega - F_\varepsilon} (v - v_1) dd^c v \wedge dd^c \varphi = \int_{(\Omega - F_\varepsilon) \cap G_\varepsilon} (v - v_1) dd^c v \wedge dd^c \varphi.$$

Since $0 \leq v - v_1 \leq \varepsilon$ on $(\Omega - F_\varepsilon) \cap G_\varepsilon$, as above we see that

$$\left| \int_{(\Omega - F_\varepsilon) \cap G_\varepsilon} (v - v_1) dd^c v \wedge dd^c \varphi \right| \leq C\varepsilon \int_{\operatorname{supp}(\varphi)} dd^c v \wedge \beta \to 0$$

as $\varepsilon \to \infty$. Thus we have proved $I_1 = 0$.

Next, take any $\varepsilon > 0$, and let $E_\varepsilon = \{z \in \operatorname{supp}(\varphi) \mid u_1(z) \leq \varepsilon\}$. Then,

$$I_2 = \int_{\Omega - E_\varepsilon} (v - v_2) dd^c v_1 \wedge dd^c \varphi + \int_{E_\varepsilon} (v - v_2) dd^c v_1 \wedge dd^c \varphi.$$

Since $v_1 = u_1$ and hence $dd^c v_1 = 0$ on $\Omega - E_\varepsilon$, the first integral vanishes. Since $0 \leq v - v_2 \leq \varepsilon$ on E_ε, similarly as before we conclude that

$$\left| \int_{E_\varepsilon} (v - v_2) dd^c v_1 \wedge dd^c \varphi \right| \leq C\varepsilon \int_{\operatorname{supp}(\varphi)} dd^c v_1 \wedge \beta \to 0$$

as $\varepsilon \to 0$, which proves $I_2 = 0$. Thus we have completed the proof. □

Proposition 8.4.12 *Let* $v = \max\{G^+, G^-\}$. *Then,*

$$v \in \tilde{\mathcal{L}}, \quad \operatorname{supp}((dd^c v)^2) \subset K.$$

In particular, K is not pluripolar.

Proof By Lemma 8.3.9, Proposition 8.3.13 and Lemma 8.3.12 for G^+, and by the same type results for G^-, we have $v \in \tilde{\mathcal{L}}$. For $\varepsilon > 0$, set $G_\varepsilon^\pm = \max\{G^\pm, \varepsilon\}$, and $v_\varepsilon = \max\{G_\varepsilon^+, G_\varepsilon^-\}$. Then, by Lemmas 8.3.10 and 8.4.11, $(dd^c v_\varepsilon)^2 = dd^c G_\varepsilon^+ \wedge dd^c G_\varepsilon^-$. Let $z \in \mathbb{C}^2 - K$, and assume, say, $z \in \mathbb{C}^2 - K^+$. Take a neighborhood $\Omega \subset\subset \mathbb{C}^2 - K^+$. Then, for $\varepsilon > 0$ with $\inf\{G^+(z) \mid z \in \Omega\} > \varepsilon$, $G_\varepsilon^+ = G^+$ on Ω, so $(dd^c v_\varepsilon)^2 = 0$ on Ω. Letting $\varepsilon \to 0$, by Theorem 8.4.2 $(dd^c v)^2 = 0$ on Ω. This proves $\operatorname{supp}((dd^c v)^2) \subset K$. By Theorem 8.4.9, $\int_K (dd^c v)^2 = \int_{\mathbb{C}^2} (dd^c v)^2 = 4\pi^2$. Moreover, Proposition 8.4.7 assures us that K is not a pluripolar. □

We need also the following comparison theorem to show the next lemma.

Theorem 8.4.13 *Let $\Omega \subset \mathbb{C}^2$ be a bounded domain and let $u, v \in \mathcal{PSH}(\Omega) \cap L^\infty(\Omega)$. If $\liminf_{z \to w, z \in \Omega} (u(z) - v(z)) \geq 0$ for every $w \in \partial\Omega$, then*

$$\int_{\{u<v\}} (dd^c v)^2 \leq \int_{\{u<v\}} (dd^c u)^2.$$

Lemma 8.4.14 *Let $u, v \in \tilde{\mathcal{L}}$. If $u \leq v$ holds almost everywhere on* $\operatorname{supp}((dd^c v)^2)$ *with respect to the measure $(dd^c v)^2$, then $u \leq v$ in $\mathbb{C}^2$.*

Proof It suffices to prove the lemma for $u + c$, $v + c$ with a constant c instead of u, v, so we can assume that $v(z) \geq w(z)$ with $w(z) = \frac{1}{2}\log(2 + \|z\|^2)$. Suppose $u(z_0) > v(z_0)$ at some point z_0, and we will show a contradiction. Take $\varepsilon > 0$ and $\eta > 0$ satisfying $2\eta < \varepsilon$ so that $S = \{z \mid u(z) + \eta w(z) > (1 + \varepsilon)v(z)\}$ contains z_0. Then, S is a bounded set, and by the sub-mean-value property of plurisubharmonic functions, S has a positive Lebesgue measure. By Theorem 8.4.13, we have

$$0 < \int_S [dd^c\{u(z) + \eta w(z)\}]^2 \le (1+\varepsilon)^2 \int_S (dd^c v)^2.$$

So the set $S \cap \mathrm{supp}((dd^c v)^2)$ has a positive measure with respect to the measure $(dd^c v)^2$. Therefore, at almost all points z of this set,

$$(1+\varepsilon)v(z) < u(z) + \eta w(z) \le v(z) + \eta w(z),$$

and hence $v(z) < (\eta/\varepsilon)w(z) < \frac{1}{2}w(z)$, which contradicts $v(z) \ge w(z)$. □

Further we recall the following.

Theorem 8.4.15 *Let E be a compact set of $\mathbb{C}^2$. If $L_E^* \equiv 0$ on E, then L_E is continuous in $\mathbb{C}^2$.*

Proposition 8.4.16 $L_K = L_K^* = \max\{G^+, G^-\}$.

Proof By Proposition 8.4.12 and Theorem 8.4.6, $L_K^* \in \tilde{\mathcal{L}}$. Set $v = \max\{G^+, G^-\}$. Since Proposition 8.4.10 yields $G^\pm = L_{K^\pm}^* \le L_K^*$, $v \le L_K^*$.

To prove the opposite inequality, we shall apply Lemma 8.4.14 for v and $u = L_K^*$. By Lemma 8.3.11, $v = 0$ on K. Since the set $\{L_K^* > L_K\}$ is pluripolar by Theorem 8.4.8 and has measure 0 with respect to $(dd^c v)^2$ by Proposition 8.4.7, $L_K^* \le L_K \le 0 = v$ almost everywhere with respect to $(dd^c v)^2$ on K. By Lemma 8.4.14 and Proposition 8.4.12, $L_K^* \le v$ in $\mathbb{C}^2$.

Lastly, since Theorem 8.4.15 implies that L_K is continuous, we have $L_K^* = L_K$. □

Finally, we introduce the complex equilibrium measure of a set $E \subset \mathbb{C}^2$.

Definition Let $E \subset \mathbb{C}^2$ be a non-pluripolar bounded set. Then $\mu_E = (dd^c L_E^*)^2$ is called the *complex equilibrium measure* of E.

Theorem 8.4.17 *The following hold:*

(a) μ *is the complex equilibrium measure of K;*

(b) $\int_{\mathbb{C}^2} \mu = 4\pi^2$;

(c) $\mathrm{supp}(\mu) \subset J$.

Proof Let λ be the complex equilibrium measure of K. Then, by Proposition 8.4.16, $\lambda = \lim_{\varepsilon \to +0}(dd^c \max\{G^+, G^-, \varepsilon\})^2$. On the other hand, for $\varepsilon >$

0, let $G^{\pm}_{\varepsilon} = \max\{G^{\pm}, \varepsilon\}$. Then, Theorem 8.4.2 yields $\mu^{\pm} = \lim_{\varepsilon\to 0} dd^c G^{\pm}_{\varepsilon}$, $\mu = \mu^+ \wedge \mu^- = \lim_{\varepsilon\to 0} dd^c G^+_{\varepsilon} \wedge dd^c G^-_{\varepsilon}$. Since Lemma 8.4.11 yields

$$(dd^c \max\{G^+, G^-, \varepsilon\})^2 = dd^c G^+_{\varepsilon} \wedge dd^c G^-_{\varepsilon},$$

we have $\mu = \lambda$, which proves (a).

According to Theorem 8.4.9, we have (b). Since $\operatorname{supp} \mu^{\pm} = J^{\pm}$ by Theorem 8.3.15, (c) holds. □

Remark Due to Bedford and Smillie (1991a), if f is hyperbolic (see §9.3), then $\operatorname{supp} \mu = J$. It is an open question whether $\operatorname{supp} \mu = J$ for any polynomial automorphism f.

8.5 Stable and unstable convergence theorems

This section is devoted to showing some convergence theorems of the sequence of currents $\{T^+_n\}$ which we define just below. Such convergence theorems were proved by Bedford and Smillie (1991a, 1991c, 1992). See also Fornaess and Sibony (1992).

Definition For a $v \in \mathcal{PSH}(\mathbb{C}^2)$ and a positive (1, 1)-current T on $\mathbb{C}^2$, set

$$v^{\pm}_n = \frac{1}{k^n} v \circ f^{\pm n}, \qquad T^{\pm}_n = \frac{1}{k^n} f^{\pm n *} T.$$

Remark When $T = dd^c v$, then clearly we have $T^{\pm}_n = dd^c v^{\pm}_n$.

Two kinds of positive (1, 1)-currents on $\mathbb{C}^2$ will be treated in this section. The one is a current $T \in \mathcal{K}$, and the other is one in the class $\mathcal{M}$ defined as follows.

Definition Let $\mathcal{M}$ be the set of all positive (1, 1)-currents ψT in $\mathbb{C}^2$ defined by a positive (1, 1)-current T in a domain $\Omega \subset \mathbb{C}^2$ and $\psi \in C^{\infty}_0(\Omega)$ with $\psi \geq 0$ satisfying $\operatorname{supp}(\psi) \cap \operatorname{supp}(dT) = \varnothing$.

For $\psi T \in \mathcal{M}$, $\psi T \wedge \mu^{\pm}$ can be defined on $\mathbb{C}^2$ in the following way. Taking $\psi_1 \in C^{\infty}_0(\Omega)$ with $\psi_1 \geq 0$ and $= 1$ in a neighborhood of $\operatorname{supp}(\psi)$, $\psi_1 T$ can be regarded as the positive (1, 1)-current on $\mathbb{C}^2$ that satisfies $\operatorname{supp}(\psi) \cap \operatorname{supp}(d(\psi_1 T)) = \varnothing$. We consider ψ that $\in C^{\infty}_0(\mathbb{C}^2)$ and applying the formula (8.13) for ψ and $\psi_1 T$ in $\mathbb{C}^2$, $\psi(\psi_1 T) \wedge \mu^{\pm}$ can be defined on $\mathbb{C}^2$, which is the definition of $\psi T \wedge \mu^{\pm} = \psi(\psi_1 T) \wedge \mu^{\pm}$. In the case of $T \in \mathcal{K}$, $T \wedge \mu^{\pm}$ is defined in $\mathbb{C}^2$ by simply applying (8.13) with $\psi = 1$ for $\mathbb{C}^2$.

In the following, we discuss the convergence of T_n^+ and $(\psi T)_n^+$ for $T \in \mathcal{K}_y$ and $\psi T \in \mathcal{M}$.

Lemma 8.5.1 *For $\psi T \in \mathcal{M}$ and for a relatively compact open set $G \subset \mathbb{C}^2$, there is a $C_1 > 0$ such that*

$$\left| \int_{\mathbb{C}^2} (\psi T)_n^+ \wedge \varphi \right| \leq C_1 \|\varphi\|$$

for all $\varphi \in \mathcal{D}^{1,1}(\bar{G})$, where $\|\varphi\|$ denotes the maximum in $\mathbb{C}^2$ of the absolute values of the coefficients of φ. A similar result holds for $T \in \mathcal{K}$.

Proof Let $\lambda(z) = \log(1+\|z\|^2)$ and $\omega = \frac{1}{4} dd^c \lambda$ as before. Since ω is a positive $(1,1)$-form in $\mathbb{C}^2$, there is a $C_2 > 0$ such that $-C_2\|\varphi\|\omega \leq \varphi \leq C_2\|\varphi\|\omega$ in $\mathbb{C}^2$. And we have

$$\begin{aligned} \frac{1}{k^n} \left| \int_{\mathbb{C}^2} f^{n*}(\psi T) \wedge \varphi \right| &\leq \frac{C_2\|\varphi\|}{k^n} \int_{\mathbb{C}^2} \psi T \wedge dd^c(\lambda \circ f^{-n}) \\ &= \frac{C_2\|\varphi\|}{k^n} \int_{\mathbb{C}^2} dd^c(\psi T) \wedge (\lambda \circ f^{-n}). \qquad (8.14) \end{aligned}$$

Since $\lambda(z) \leq 2\log^+|z| + 2\log 2$, applying the same type estimate as in (8.12) for f^{-n}, there is an $M > 0$ independent of n such that

$$0 \leq \frac{1}{k^n} \lambda \circ f^{-n}(z) \leq 2\log^+|z| + M \qquad \text{for } z \in \mathbb{C}^2.$$

Therefore, in view of (8.14), we have the desired C_1. □

Next, let $T_z^*(\mathbb{C}^2)$ be the complexification of the real cotangent space of $\mathbb{C}^2$ at a point $z \in \mathbb{C}^2$, and let $J : T_z^*(\mathbb{C}^2) \to T_z^*(\mathbb{C}^2)$ be the $\mathbb{C}$-linear map defined by $J(dx) = idx$, $J(d\bar{x}) = -id\bar{x}$, $J(dy) = idy$ and $J(d\bar{y}) = -id\bar{y}$. For a C^∞ function ψ, C^∞ 1-forms ξ, η and for a $(1,1)$-current $\psi T \in \mathcal{M}$, we have $J\xi \wedge \eta \wedge \psi T = J\eta \wedge \xi \wedge \psi T$ and $d^c\psi = i(\bar{\partial} - \partial)\psi = -J(d\psi)$.

Lemma 8.5.2 *Let $\psi T \in \mathcal{M}$ and let ξ and η be C^∞ real $(1,1)$-forms in $\mathbb{C}^2$. Then,*

$$\left| \int_{\mathbb{C}^2} J\xi \wedge \eta \wedge \psi T \right|^2 \leq \left(\int_{\mathbb{C}^2} J\xi \wedge \xi \wedge \psi T \right) \left(\int_{\mathbb{C}^2} J\eta \wedge \eta \wedge \psi T \right).$$

Proof Set $\xi = pdx + \bar{p}d\bar{x} - rdy - \bar{r}d\bar{y}$ and $\psi T = i(Adx \wedge d\bar{x} + Bdx \wedge d\bar{y} + \bar{B}dy \wedge d\bar{x} + Cdy \wedge d\bar{y})$. Then,

$$J\xi \wedge \xi \wedge \psi T = 2(r\ p) \begin{pmatrix} AB \\ \bar{B}C \end{pmatrix} \begin{pmatrix} \bar{r} \\ \bar{p} \end{pmatrix} i^2 dx \wedge d\bar{x} \wedge dy \wedge d\bar{y}.$$

Therefore, $\langle \xi, \eta \rangle = \int J\xi \wedge \eta \wedge \psi T$ defines an $\mathbb{R}$-inner-product satisfying $\langle \xi, \xi \rangle \geq 0$, and the Schwarz inequality is nothing but our desired one. □

Let $\psi T \in \mathcal{M}$. As a consequence of Lemma 8.5.1, for an arbitrary subsequence of the sequence $\{(\psi T)_n^+\}$, we can select a further subsequence $(\psi T)_{n_j}^+$ that converges in the sense of currents to a positive $(1, 1)$-current ν in $\mathbb{C}^2$ (which is a consequence of the Banach–Alaoglu theorem). The limit ν satisfies $\int_{\mathbb{C}^2} \nu \wedge \omega < +\infty$.

A similar result holds for $T \in \mathcal{K}$: for a current $T \in \mathcal{K}$, all T_n^+ and $\lim_{j\to\infty} T_{n_j}^+$ are in $\mathcal{K}$, for these currents are clearly positive and closed and Lemma 8.5.1 implies the condition $\int_{\mathbb{C}^2} (\lim_{j\to\infty} T_{n_j}^+) \wedge \omega < +\infty$.

As to the case of $\psi T \in \mathcal{M}$ we have the following.

Lemma 8.5.3 *Let $\psi T \in \mathcal{M}$. Then $\nu = \lim_{j\to\infty} (\psi T)_{n_j}^+ \in \mathcal{K}$.*

Proof As stated above, $\int_{\mathbb{C}^2} \nu \wedge \omega < +\infty$, and ν is clearly positive. It remains to show that ν is closed. Let χ be a C^∞ real 1-form with compact support in $\mathbb{C}^2$, and take $\psi_1 \in C_0^\infty(\Omega)$ with $\psi_1 \geq 0$ such that $\psi_1 = 1$ in a neighborhood of $\mathrm{supp}(\psi)$. Then,

$$\frac{1}{k^{n_j}} \int_{\mathbb{C}^2} d\chi \wedge f^{n_j *}(\psi T) = \frac{1}{k^{n_j}} \int_{\mathbb{C}^2} (\chi \circ f^{-n_j}) \wedge d\psi \wedge \psi_1 T.$$

Use Lemma 8.5.2 for $\psi_1 T \in \mathcal{M}$. Then, by Lemma 8.5.1, there is $C_3 > 0$ that is independent of n_j such that

$$\left| \int_{\mathbb{C}^2} (\chi \circ f^{-n_j}) \wedge d\psi \wedge \psi_1 T \right|^2$$

$$\leq \left(\int_{\mathbb{C}^2} ((J\chi \wedge \chi) \circ f^{-n_j}) \wedge \psi_1 T \right) \left(\int_{\mathbb{C}^2} d\psi \wedge d^c\psi \wedge \psi_1 T \right) \leq C_3 k^{n_j},$$

where we use holomorphy of f to have $J(f^{-n_j *}\chi) = f^{-n_j *}(J\chi)$. Thus,

$$\left| \frac{1}{k^{n_j}} \int_{\mathbb{C}^2} d\chi \wedge f^{n_j *}(\psi T) \right| \leq \sqrt{C_3} k^{-n_j/2} \to 0 \quad (j \to \infty).$$

□

Definition For $\psi T \in \mathcal{M}$, let $S(\psi T)$ be the set of all the limits $\nu = \lim_{j\to\infty} (\psi T)_{n_j}^+$ of a convergent subsequence $\{(\psi T)_{n_j}^+\}$. Similarly, $S(T)$ is defined for $T \in \mathcal{K}$.

By the remark after the proof of Lemma 8.5.2, $S(\psi T)$ and $S(T)$ are not empty. If $\nu \in S(\psi T)$, then $\nu_n^+ = \lim_{j\to\infty} (k^{n_j+n})^{-1} f^{n_j+n*}(\psi T)$ implies that $\nu_n^+ \in S(\psi T)$. Similarly, for $\nu \in S(T)$, $\nu_n^+ \in S(T)$. Actually, we will show that $S(T)$ or $S(\psi T)$ consists of only one element.

Lemma 8.5.4 *Let ν be either in $S(\psi T)$ with $\psi T \in \mathcal{M}$ or in $S(T)$ with $T \in \mathcal{K}_y$. Then* $\mathrm{supp}(\nu) \subset K^+$ *and therefore $\nu \in \mathcal{K}_y$.*

Proof We prove only the case $T \in \mathcal{K}_y$. (The case $\psi T \in \mathcal{M}$ can be proved similarly since the support of $\psi T \in \mathcal{M}$ is compact.) Let $z \notin K^+$. Take $R > 0$ so that $\mathrm{supp}(T) \cap V_R^+ = \varnothing$ and take an open subset $\Omega \subset\subset \mathbb{C}^2$ with $z \in \Omega$ such that $\overline{\Omega} \cap K^+ = \varnothing$. By Proposition 7.2.3, there is $N > 0$ such that $f^n(\Omega) \subset V_R^+$ for all n satisfying $n > N$. Therefore, $\Omega \cap \mathrm{supp}(f^{n*}(T)) = \Omega \cap f^{-n}(\mathrm{supp}(T)) = \varnothing$. Since $\mathrm{supp}(\nu) \subset \overline{\bigcup_{n=N}^\infty \mathrm{supp}(f^{n*}T)}$, we have $z \notin \mathrm{supp}(\nu)$, which proves $\mathrm{supp}(\nu) \subset K^+$. We remarked that $\nu \in \mathcal{K}$ in Lemma 8.5.3, so $\nu \in \mathcal{K}_y$. □

Definition For a non-zero $\nu \in \mathcal{K}_y$, a constant $c_\nu > 0$ and $U_\nu \in \mathcal{L}_y$ are defined by the formulas

$$c_\nu dd^c U_\nu = \nu \quad \text{and} \quad \lim_{|y|\to\infty} (U_\nu - \log|y|) = \gamma_0, \tag{8.15}$$

where γ_0 is the Robin constant for G^+.

By Proposition 8.3.6, in the formula $\nu = c_\nu dd^c U_\nu$, the constant $c_\nu > 0$ is unique and $U_\nu \in \mathcal{L}_y$ is unique up to an additive constant. So the second formula in (8.15) determines U_ν uniquely.

Definition For $T \in \mathcal{K}_y$ and for $\psi T \in \mathcal{M}$, set

$$c(T) = \frac{1}{4\pi^2}\int_{\mathbb{C}^2} T \wedge \mu^-, \quad \text{and} \quad c(\psi T) = \frac{1}{4\pi^2}\int_{\mathbb{C}^2} \psi T \wedge \mu^-.$$

Lemma 8.5.5 *Let $T\ (\neq 0) \in \mathcal{K}_y$. Then all $\nu \in S(T)$ share the common constant c_ν which is equal to the constant $c(T)$.*

Similarly, for $\psi T \in \mathcal{M}$ and any $\nu \in S(\psi T)$, $c_\nu = c(\psi T)$.

Proof We deal with the case $T \in \mathcal{K}_y$ only. Take V_R^+ so that $\mathrm{supp}(T) \cap V_R^+ = \varnothing$. Then, $\mathrm{supp}(T_n^+) \cap f^{-n}(V_R^+) = \varnothing$ for $n \geq 1$. Since $V_R^+ \subset f^{-n}(V_R^+)$ by Lemma 7.2.4, $\mathrm{supp}(T_n^+) \cap V_R^+ = \varnothing$, and hence

$$f^{-n}(\mathrm{supp}(T)) \subset \overline{D_R \cup V_R^-}. \tag{8.16}$$

Take the positive closed (1, 1)-current $\theta = \frac{1}{2}dd^c\log(1 + |x|^2)$ which depends only on x on $\mathbb{C}^2$ with the coordinates x and y. Then we have that, for any $\varphi \in C_0^\infty(\mathbb{C}^2)$,

$$\begin{aligned}\int_{\mathbb{C}^2} \varphi\nu \wedge \theta &= c_\nu \int_{\mathbb{C}^2} \varphi dd^c U_\nu \wedge \theta = c_\nu \int_{\mathbb{C}^2} U_\nu dd^c \varphi \wedge \theta \\ &= c_\nu \int_{\mathbb{C}} \theta \int_{\mathbb{C}} U_\nu dd_y^c \varphi = c_\nu \int_{\mathbb{C}} \theta \int_{\mathbb{C}} \varphi dd_y^c U_\nu.\end{aligned}$$

Since $U_\nu \in \mathcal{K}_y$, applying Lemma 8.3.16 to the function $U_\nu(x, \cdot)$ in y, and moving $\varphi \in C_0^\infty(\mathbb{C}^2)$ with $0 \le \varphi \le 1$, we conclude that

$$\int_{\mathbb{C}^2} \nu \wedge \theta = 2\pi c_\nu \int_{\mathbb{C}} \theta = (2\pi)^2 c_\nu. \tag{8.17}$$

On the other hand, since $\mathrm{supp}(T) \subset \overline{D_R \cup V_R^-}$,

$$\begin{aligned}\int_{\mathbb{C}^2} (T)_{n_j}^+ \wedge \theta &= \int_{\mathbb{C}^2} \frac{1}{k^{n_j}} T \wedge dd^c\left(\frac{1}{2}\log(1 + |\pi_1 \circ f^{-n_j}|^2)\right) \\ &= \int_{\overline{D_R \cup V_R^-}} \frac{1}{k^{n_j}} T \wedge dd^c\left(\frac{1}{2}\log(1 + |\pi_1 \circ f^{-n_j}|^2)\right).\end{aligned}$$

Let $z \in \overline{D_R \cup V_R^-}$. Then, $\log^+|x| \le \log^+|z| \le \log R + \log^+|x|$. So (8.16) and the inequality $\log^+|x| \le \frac{1}{2}\log(1 + |x|^2) \le \log^+|x| + \frac{1}{2}\log 2$ yield

$$\lim_{j\to\infty} \frac{1}{k^{n_j}} \frac{1}{2}\log(1 + |\pi_1 \circ f^{-n_j}|^2) = \lim_{j\to\infty} \frac{1}{k^{n_j}} \log^+|f^{-n_j}| = G^-$$

where the last equality follows from Theorem 8.3.14 and the convergence is uniform in $\overline{D_R \cup V_R^-}$. Therefore,

$$\int_{\mathbb{C}^2} \nu \wedge \theta = \lim_{j\to\infty} \int_{\mathbb{C}^2} (T)_{n_j}^+ \wedge \theta = \int_{\mathbb{C}^2} T \wedge \mu^- = 4\pi^2 c(T).$$

Thus, by combining with (8.17), we have obtained $c_\nu = c(T)$.

A similar proof to the above works for the case $\psi T \in \mathcal{M}$, because by taking a large $R > 0$, we have $\mathrm{supp}(\psi T) \subset D_R \subset D_R \cup V_R^-$. □

Lemma 8.5.6 *For $\nu \in \mathcal{K}$ and $x \in \mathbb{C}$, we denote by ν_x the restriction of ν to the plane $\{(x, y) \mid y \in \mathbb{C}\}$. Then for either $\nu \in S(T)$ with a non-zero $\hat{T} = T \in \mathcal{K}_y$ or $\nu \in S(\psi T)$ with a non-zero $\hat{T} = \psi T \in \mathcal{M}$,*

$$\int_{\mathbb{C}} \nu_x = 2\pi c(\hat{T}), \quad U_\nu(x, y) = \frac{1}{2\pi c(\hat{T})} \int_{\mathbb{C}} \log|y - \xi| \nu_x(\xi) + \gamma_0.$$

Proof The first formula follows from Lemma 8.3.16. So combining this with Theorem 8.3.17, the second follows. □

Lemma 8.5.7 *For $v \in \mathcal{L}_y$, v_n^+ converges to G^+ uniformly on every compact set in $\mathbb{C}^2 - K^+$.*

Proof Take V_R^+ and $M > 0$ so that $-M < v(z) - \log|\pi_2(z)| < M \quad (z \in V_R^+)$. Then,

$$G_n^+(z) - \frac{M}{k^n} < v_n^+(z) < G_n^+(z) + \frac{M}{k^n} \quad (z \in f^{-n}(V_R^+)).$$

Since $\bigcup_{n=0}^{\infty} f^{-n}(V_R^+) = \mathbb{C}^2 - K^+$, Lemma 8.3.10 implies the assertion. □

The next lemma is a variant of Proposition 7.2.3.

Lemma 8.5.8 *There are $C_1 > 0$, $\eta > 1$ such that, letting*

$$B = \{(x, y) \mid |y| \geq C_1(|x|^{1/\eta} + 1)\},$$

we have

$$K^+ \cap B = \varnothing \quad \textit{and} \quad \bigcup_{n=0}^{\infty} f^{-n}(B) = \mathbb{C}^2 - K^+.$$

Proof Use the notation $f = g_m \circ \cdots \circ g_1$, where g_1 is determined by a complex number $\delta_1 \neq 0$ and a polynomial $p_1(y)$ of degree k_1 (≥ 2). Let $c_{01} y^{k_1}$ be the degree k_1 term of $p_1(y)$. We take $R > 0$ such that Proposition 7.2.2 holds for g_1, $|c_{01}| R^{k_1}/2 > R$ and that $|p_1(y)| \geq |c_{01}| |y|^{k_1}/2$ for $|y| \geq R$. Let $R_1 > |c_{01}| R^{k_1}/2$ and

$$A = \{(x, y) \in \mathbb{C}^2 \mid |c_{01}||y|^{k_1} \geq 2(1 + |\delta_1|)|x| + 2R_1\}.$$

Then, we can see $A \subset \{|y| > R\}$. We show $g_1(A) \subset V_R^+$. When $(x, y) \in A \cap V_R^+$, Proposition 7.2.2 yields $g_1(x, y) \in V_R^+$. When $(x, y) \in A \cap (\mathbb{C}^2 - V_R^+)$,

$$|p_1(y) - \delta_1 x| \geq \frac{|c_{01}|}{2}|y|^{k_1} - |\delta_1|\,|x| > \begin{cases} R, \\ |x| \geq |y|, \end{cases}$$

which implies $g_1(x, y) \in V_R^+$.

Therefore, by Proposition 7.2.3, $A \cap K^+ = \varnothing$. Take $C_1 > 0$ satisfying $C_1^{k_1}|c_{01}| \geq 2(1 + |\delta_1|)$ and $C_1^{k_1}|c_{01}| \geq 2R_1$, and let $\eta = k_1$. Then, $|y| \geq C_1(|x|^{1/\eta} + 1)$ yields $|c_{01}|\,|y|^{\eta} \geq C_1^{\eta}|c_{01}|(|x| + 1) \geq 2(1 + |\delta_1|)|x| + 2R_1$, which implies $B \subset A$. Thus we have proved $K^+ \cap B = \varnothing$. Since $V_R^+ \subset B$ for sufficiently large R, Proposition 7.2.3 gives the second formula. □

Lemma 8.5.9 *Take $C_1 > 0$ and $\eta > 1$ as in Lemma 8.5.8 and set $B_1 = \{(x, y) \mid |y| \geq 2C_1(|x|^{1/\eta} + 1)\}$. Let ν be either in $S(T)$ with a non-zero $T \in \mathcal{K}_y$ or in $S(\psi T)$ with a non-zero $\psi T \in \mathcal{M}$. Then, for every (x, y) in B_1,*

$$-2\frac{C_1}{|y|}(|x|^{1/\eta} + 1) \leq U_\nu(x, y) - \gamma_0 - \log|y| \leq \frac{C_1}{|y|}(|x|^{1/\eta} + 1).$$

Proof Set $r = C_1(|x|^{1/\eta} + 1)$. Then, by Lemmas 8.5.8 and 8.5.4,

$$\mathrm{supp}(\nu_x) \subset \{|y| \leq r\}.$$

For $|y| > r$ and $|\xi| \leq r$, we have $|y| - r \leq |y - \xi| \leq |y| + r$, so Lemma 8.5.6 yields

$$\log(|y| - r) \leq U_\nu(x, y) - \gamma_0 \leq \log(|y| + r),$$

and hence

$$-\left(\frac{r}{|y|}\right) \Big/ \left(1 - \frac{r}{|y|}\right) \leq U_\nu(x, y) - \gamma_0 - \log|y| \leq \frac{r}{|y|}.$$

Here, if $|y| > 2r$, then $-2r/|y| \leq U_\nu(x, y) - \gamma_0 - \log|y| \leq r/|y|$. □

Lemma 8.5.10 *Let ν be as in Lemma 8.5.9 and set $\nu(n) = k^n f^{-n*}\nu$ for each $n \in \mathbb{N}$ so that $\nu(n)^+_n = \nu$. Then we have $(U_{\nu(n)})^+_n = U_\nu$.*

Proof For convenience, we write out the case of $\nu \in S(T)$. Then, we also have $\nu(n) \in S(T)$. So using $c(T) \neq 0$, $(U_{\nu(n)})^+_n$ and $U_\nu \in \mathcal{L}_y$ satisfy

$$c(T)dd^c(U_{\nu(n)})^+_n = \nu(n)^+_n = \nu \quad \text{and} \quad c(T)dd^c U_\nu = \nu.$$

Hence, by Proposition 8.3.6, $(U_{\nu(n)})^+_n$ and U_ν coincide up to an additive constant, which we show to be zero.

Indeed, in some V^+_R, there is a bounded $u_n \in \mathcal{PH}(\overline{V^+_R})$ such that

$$U_{\nu(n)}(x, y) - \gamma_0 - \log|y| = u_n(x, y) \quad \text{and} \quad \lim_{|y|\to\infty} u_n(x, y) = 0.$$

So if $(x, y) \in f^{-n}(V^+_R)$,

$$(U_{\nu(n)})^+_n(x, y) - k^{-n}\gamma_0 - k^{-n}\log|y_n| = k^{-n}u_n(x_n, y_n).$$

But, by Lemma 8.3.8, as $|y| \to \infty$,

$$k^{-n}\log|y_n| - \frac{k^n - 1}{k^n}\gamma_0 - \log|y| \to 0,$$

hence,

$$(U_{\nu(n)})^+_n(x, y) - \gamma_0 - \log|y| \to 0.$$

On the other hand, by the definition of U_ν,

$$U_\nu(x, y) - \gamma_0 - \log|y| \to 0$$

as $|y| \to \infty$, which concludes the proof of $(U_{\nu(n)})^+_n = U_\nu$. □

Lemma 8.5.11 *Let ν be as in Lemma 8.5.9. Then,*

$$U_\nu = G^+ \text{ in } \mathbb{C}^2 - K^+.$$

Proof Let $z = (x, y) \in \mathbb{C}^2 - K^+$ and $\varepsilon > 0$. For every $n \in \mathbb{N}$, let $\nu(n)$ be as in Lemma 8.5.10. For a sufficiently large n, $(x_n, y_n) = f^n(z)$ satisfies $|x_n| \le |y_n|$ and $(x_n, y_n) \in B_1$, so by Lemma 8.5.9,

$$\left|k^{-n} U_{\nu(n)}(x_n, y_n) - k^{-n}\gamma_0 - k^{-n}\log|y_n|\right| \le \frac{2C_1}{k^n|y_n|}(|x_n|^{1/\eta} + 1) < \varepsilon.$$

On the other hand, Lemma 8.3.9 and $G^+ \circ f^n = k^n G^+$ give

$$\left|G^+(z) - k^{-n}\gamma_0 - k^{-n}\log|y_n|\right| < \varepsilon$$

for a large enough n. So these two inequalities and Lemma 8.5.10 yield

$$|G^+(z) - U_\nu(z)| < 2\varepsilon.$$

□

It is well known that a compact set of zero capacity has zero Lebesgue measure. We need the following formula.

Proposition 8.5.12 *Let $Y \subset \mathbb{C}$ be compact. Then,*

$$m(Y) \le \pi e \operatorname{cap}(Y)^2,$$

where m is the Lebesgue measure.

Proof Since $\operatorname{cap}(Y) = 0$ implies $m(Y) = 0$, assume $m(Y) > 0$. Define $R > 0$ by $\pi R^2 = m(Y)$ and let $B = \{|z| \le R\}$ so that $m(Y) = m(B)$. Then since

$$\int_B \log|y| dm \le \int_{B\cap Y} \log|y| dm + \log R(m(B) - m(B \cap Y)),$$

$$\int_Y \log|y| dm \ge \int_{B\cap Y} \log|y| dm + \log R(m(Y) - m(B \cap Y)),$$

we obtain $\int_Y \log|y|dm \geq \int_B \log|y|dm$, and since $\displaystyle\int_B \log|y|dm$ equals $m(Y)\log\sqrt{m(Y)/\pi e}$, we have

$$\frac{1}{m(Y)}\int_Y \log|y|dm \geq \log\sqrt{m(Y)/\pi e}. \tag{8.18}$$

Let $v(y) = \dfrac{1}{m(Y)}\displaystyle\int_Y \log|y-\xi|dm(\xi)$. Then, (8.18) gives

$$v(y) = \frac{1}{m(Y)}\int_{\{-y+\xi|\xi\in Y\}} \log|t|dm(t) \geq \log\sqrt{m(Y)/\pi e}. \tag{8.19}$$

Let G_Y^* and $dd^cG_Y^*$ be the Green function and the equilibrium measure of Y, and let $\sigma = \frac{1}{2\pi}dd^cG_Y^*$; then, σ is the probability measure on Y and $G^*(y) = \int_Y \log|y-\xi|\sigma(\xi) + \gamma(Y)$. Hence, Proposition 8.3.19 yields

$$\begin{aligned}\int_Y v(y)\sigma(y) &= \frac{1}{m(Y)}\int_Y\left(\int_Y \log|y-\xi|\sigma(y)\right)dm(\xi)\\ &= \frac{1}{m(Y)}\int_Y (G_Y^*(\xi)-\gamma(Y))dm(\xi) = -\gamma(Y).\end{aligned}$$

Thus, $-\gamma(Y) \geq \inf_{y\in Y} v(y)$ and (8.19) implies $-\gamma(Y) \geq \log\sqrt{m(Y)/\pi e}$, which proves the assertion. □

Lemma 8.5.13 *Let ν be as in Lemma 8.5.9. Then, $U_\nu = 0$ in* Int K^+.

Proof Since $U_\nu \in \mathcal{L}_y$ and $U_\nu = G^+$ in $\mathbb{C}^2 - K^+$ by Lemma 8.5.11, the maximum principle, applied to $U(x,\cdot)$ with x being fixed, gives $U_\nu \leq 0$ on K^+. So it suffices to show that the existence of a constant $\alpha > 0$ and a non-empty open set $\Omega \subset\subset \operatorname{Int} K^+$ with $U_\nu \leq -\alpha$ in Ω leads to a contradiction.

Take D_R so that $\bigcup_{n=1}^\infty f^n(\Omega) \subset D_R$. Let $\nu(n) = k^n f^{-n*}\nu$. Then by Lemma 8.5.10,

$$f^n(\Omega) \subset \{U_{\nu(n)} \leq -k^n\alpha\} \cap D_R$$

for every $n \in \mathbb{N}$. We shall estimate the measure of the last set. With x being fixed in $\{|x| \leq R\}$, let

$$Y_x = \{y \mid U_{\nu(n)}(x,y) \leq -k^n\alpha\} \cap \{|y| < R\}.$$

Then by the definition of the capacity and Proposition 8.5.12,

$$\operatorname{cap}(Y_x) \leq \exp(-\gamma_0 - k^n\alpha), \quad m(Y_x) \leq \pi\exp(1 - 2\gamma_0 - 2k^n\alpha).$$

So by Proposition 8.3.1 and Fubini's theorem,

$$|\delta|^{2n} m(\Omega) = m(f^n(\Omega)) \leq \int_{|x|\leq R} m(Y_x) dv_x \leq \pi^2 R^2 \exp(1 - 2\gamma_0 - 2k^n \alpha),$$

which is a contradiction if $m(\Omega) > 0$. □

Theorem 8.5.14 *For $T \in \mathcal{K}_y$ or for $\psi T \in \mathcal{M}$,*

$$\lim_{n\to\infty} \frac{1}{k^n} f^{n*}(T) = c(T)\mu^+ \quad or \quad \lim_{n\to\infty} \frac{1}{k^n} f^{n*}(\psi T) = c(\psi T)\mu^+,$$

respectively.

Proof We only write out the case $\psi T \in \mathcal{M}$ and we may assume $\psi T \neq 0$. We show $U_\nu = G^+$ for any $\nu \in S(\psi T)$, which then implies $S(\psi T) = \{c(\psi T)\mu^+\}$. By Lemmas 8.5.11 and 8.5.13, we have proved $U_\nu = G^+$ in $\mathbb{C}^2 - J^+$. So by the continuity of G^+ and the upper semi-continuity of U_ν, $U_\nu \geq G^+$ in $\mathbb{C}^2$. Let $\varepsilon > 0$ and let $D = \{y \mid G^+(x, y) < \varepsilon\}$ in the y-plane with x being fixed. Since $U_\nu(x, \cdot) = G^+(x, \cdot) = \varepsilon$ in ∂D, by the maximum principle, $U_\nu(x, \cdot) \leq \varepsilon$ in D. Since x and ε are arbitrary, $U_\nu \leq 0$ in K^+, hence $U_\nu = G^+$ in $\mathbb{C}^2$. □

8.6 Currents for analytic subsets

In this section, we apply the stable and unstable convergence theorems, 8.5.14, to some currents of integration defined by one-dimensional algebraic sets in $\mathbb{C}^2$ or by one-dimensional local complex submanifolds of $\mathbb{C}^2$. As before, we fix a polynomial automorphism f, and related g_j, k_j, δ_j and c_{0j} $(1 \leq j \leq m)$.

Let $\mathbb{C}[x, y]$ be the ring of all polynomials in x, y with complex coefficients. For a polynomial $p = \sum_{a,b} p_{ab} x^a y^b \in \mathbb{C}[x, y]$, the maximum of $\{a + b \mid p_{ab} \neq 0\}$ is called the degree of p. For such a p, define $\lambda_j, \mu_j : \mathbb{C}[x, y] \to \mathbb{N}_0 = \mathbb{N} \cup \{0\}$ $(1 \leq j \leq m)$ by

$$\lambda_j(p) = \max\{a + k_j b \mid p_{ab} \neq 0\}, \quad \mu_j(p) = \max\{k_j a + b \mid p_{ab} \neq 0\}.$$

This $\lambda_j(p)$ is called the *λ_j-weight* of p, and the polynomial

$$\sum_{a+k_j b=\lambda_j(p)} p_{ab} x^a y^b$$

is called the *leading part of p in λ_j-weight*. The *leading part of p in μ_j-weight* is defined similarly.

Lemma 8.6.1 *Let $p, q \in \mathbb{C}[x, y]$.*

(a) $\lambda_j(p+q) \le \max\{\lambda_j(p), \lambda_j(q)\}, \quad \mu_j(p+q) \le \max\{\mu_j(p), \mu_j(q)\}$.

(b) $\lambda_j(pq) = \lambda_j(p) + \lambda_j(q), \quad \mu_j(pq) = \mu_j(p) + \mu_j(q)$.

Proof We shall prove only the statement on λ_j. Let p_1 and q_1 be the leading parts of polynomials p and q in λ_j-weight. Then, in the first formula of (a), the equality holds if either $\lambda_j(p) \neq \lambda_j(q)$, or $\lambda_j(p) = \lambda_j(q)$ and $p_1 + q_1 \neq 0$, whereas the strict inequality holds if $\lambda_j(p) = \lambda_j(q)$ and $p_1 + q_1 = 0$.

Since $pq = p_1q_1 + p_1(q - q_1) + (p - p_1)q_1 + (p - p_1)(q - q_1)$, the leading part of pq in λ_j-weight is equal to p_1q_1, which proves (b). □

Lemma 8.6.2 *Let $l \ge 0$ be an integer and $p = x^{n-k_jl}y^l$ be a monomial. Then, $\lambda_j(p) = n$ and $\mu_j(p \circ g_j) = n$. Furthermore, the leading part of $p \circ g_j$ in μ_j-weight is equal to*

$$\sum_{i=0}^{l} \binom{l}{i} (-\delta_j)^i c_{0j}^{l-i} x^i y^{n-k_j i}.$$

Proof Since $p \circ g_j = y^{n-k_jl}(p_j(y) - \delta_j x)^l$, $\mu_j(y) = 1$ and $\mu_j(p_j(y) - \delta_j x) = k_j$, so Lemma 8.6.1 yields $\mu_j(p \circ g_j) = n$. The leading part of $p \circ g_j$ in μ_j-weight is equal to $y^{n-k_jl}(c_{0j}y^{k_j} - \delta_j x)^l$. □

Lemma 8.6.3 *Let s be a positive integer and let $\{l_\kappa\}$ be s integers with $0 \le l_1 < l_2 < \cdots < l_s$. Then the following matrix A is non-singular:*

$$A = \begin{pmatrix} 1 & l_1 & l_1(l_1-1) & \ldots & l_1(l_1-1)\cdots(l_1-s+2) \\ 1 & l_2 & l_2(l_2-1) & \ldots & l_2(l_2-1)\cdots(l_2-s+2) \\ \ldots & \ldots & \ldots & & \\ 1 & l_s & l_s(l_s-1) & \ldots & l_s(l_s-1)\cdots(l_s-s+2) \end{pmatrix}.$$

Proof By adding the second column to the third, the third column becomes ${}^t(l_1^2 \ l_2^2 \ \ldots \ l_s^2)$. Next, by adding 3 times the third column and (-2) times the second column to the fourth, the fourth column becomes ${}^t(l_1^3 \ l_2^3 \ \ldots \ l_s^3)$. Perform a similar process until the sth column yields the Vandermonde matrix, which is non-singular under the assumption. □

Lemma 8.6.4 *Let j be an integer such that $1 \le j \le m$, p be a non-constant polynomial with $\lambda_j(p) = N$, and s be the number of terms of the leading part of p in λ_j-weight.*

Then $p \circ g_j$ has non-zero $x^i y^{N-k_j i}$ term for some i with $0 \le i \le s-1$. In particular,

$$\mu_j(p \circ g_j) = N.$$

Proof Let us write the leading part of p in λ_j-weight as

$$\sum_{\kappa=1}^{s} b_\kappa x^{N-k_j l_\kappa} y^{l_\kappa} \quad (b_\kappa \ne 0, 1 \le \kappa \le s),$$

where l_κ are integers with $0 \le l_1 < \cdots < l_s \le [N/k_j]$. By Lemma 8.6.2, $\mu_j(p \circ g_j) \le N$ and all the terms of $p \circ g_j$ with μ_j-weight equal to N are

$$\sum_{\kappa=1}^{s} b_\kappa \sum_{i=0}^{l_\kappa} \binom{l_\kappa}{i} (-\delta_j)^i c_{0j}^{l_\kappa - i} x^i y^{N-k_j i}.$$

We deduce a contradiction by assuming that, for all i with $0 \le i \le s-1$, the coefficients of $x^i y^{N-k_j i}$ are equal to 0. For $0 \le i \le s-1$, let

$$\sum_{i \le l_\kappa} b_\kappa c_{0j}^{l_\kappa} \binom{l_\kappa}{i} (-\delta_j c_{0j}^{-1})^i = 0.$$

By using the (s, s) matrix A_1 whose (κ, ι) component is equal to either

$$\binom{l_\kappa}{\iota - 1} (-\delta_j c_{0j}^{-1})^{\iota - 1}$$

or 0 according as $l_\kappa \ge \iota - 1$ or not, and the row vector

$$b = (b_1 c_{0j}^{l_1} \; \ldots \; b_s c_{0j}^{l_s}),$$

the above equalities can be written as $bA_1 = (0 \; \ldots \; 0)$. Multiplying the ιth column of A_1 by $(\iota - 1)!/(-\delta_j c_{0j}^{-1})^{\iota-1}$, we obtain the matrix A in Lemma 8.6.3, so A_1 is a non-singular matrix, therefore $(b_1 \; \ldots \; b_s) = (0 \; \ldots \; 0)$, which is a contradiction. □

Lemma 8.6.5 *Let j be an integer such that $1 \le j \le m$, p be a non-constant polynomial and s be the number of terms of the leading part of p in λ_j-weight. Then the number of terms of the leading part of $p \circ g_j$ in λ_{j-1}-weight is at most $1 + (s-1)/k_{j-1}$. Here, we consider that $j - 1 = m$ when $j = 1$.*

Proof Let $\lambda_j(p) = N$ and $\lambda_{j-1}(p \circ g_j) = \nu$. Then, by Lemma 8.6.4, $\mu_j(p \circ g_j) = N$. Let $p \circ g_j = \sum q_{ab} x^a y^b$, S be the set of all (a, b) with $q_{ab} \ne 0$, and $T = \{(a, b) \in \mathbb{N}_0 \times \mathbb{N}_0 \mid k_j a + b \le N\}$. Then $S \subset T$ and ν is equal to the maximum of $a + k_{j-1} b$ with (a, b) varying over S. Since, again by

Lemma 8.6.4, there is i such that $0 \le i \le s-1$ and $(i, N - k_j i) \in S$, the a-coordinate of any point in $\{(a,b) \in S \mid a + k_{j-1}b = \nu\}$ is $\le s-1$. Therefore,

$$\#(\{(a,b) \mid a + k_{j-1}b = \nu\} \cap S)$$

$$\le \#(\{(a,b) \mid a + k_{j-1}b = \nu\} \cap T) \le 1 + \frac{s-1}{k_{j-1}}.$$

□

The next lemma is obvious.

Lemma 8.6.6 *Let p be a polynomial of degree k' (≥ 1) whose degree k' part consists of the unique term $cy^{k'}$ $(c \ne 0)$. Then the composition $p \circ g_j$ of p and the generalized Hénon map g_j is a polynomial of degree $k_j k'$ whose degree $k_j k'$ part consists of the unique term $cc_{0j}^{k'} y^{k_j k'}$.*

Proposition 8.6.7 *Let h be a non-constant polynomial. Then, there is an integer n_0 (≥ 0) such that, for any integer n with $n \ge n_0$, letting the degree of the polynomial $h \circ f^n$ be N, the degree N part of $h \circ f^n$ consists of the unique term cy^N.*

Proof In view of Lemma 8.6.6, it suffices to prove that there is $n \ge 0$ such that the degree N part of $h \circ f^n(x,y)$ consists of the unique term cy^N. Let s be the number of terms of the leading part of h in λ_m-weight. By Lemma 8.6.5, the number of terms of the leading part of $h \circ g_m$ in λ_{m-1}-weight is $\le 1 + (s-1)/k_{m-1}$. Applying Lemma 8.6.5 again, the number of terms of the leading part of $h \circ g_m \circ g_{m-1}$ in λ_{m-2}-weight is $\le 1 + (s-1)/(k_{m-1}k_{m-2})$. Continuing this process and using the notation $k = k_1 \cdots k_m$ as before, the number of terms of the leading part of $h \circ f^{n-1} \circ g_m \circ \cdots \circ g_2$ in λ_1-weight is $\le 1 + (s-1)k^{-n}k_m$. Take n so that $(s-1)k^{-n}k_m < 1$. Then the number of terms of the leading part of $h \circ f^{n-1} \circ g_m \circ \cdots \circ g_2$ in λ_1-weight is equal to 1. Let $N = \lambda_1(h \circ f^{n-1} \circ g_m \circ \cdots \circ g_2)$. Applying Lemma 8.6.4 for $s = 1$, $\mu_1(h \circ f^n) = N$ and $h \circ f^n$ has the term cy^N $(c \ne 0)$. So the degree of $h \circ f^n$ is N and the degree N part is cy^N. □

Theorem 8.6.8 *Let $h(x,y)$ be a non-constant polynomial defining the one-dimensional algebraic set M in $\mathbb{C}^2$ in the sense that $M = \{h = 0\}$ and that the zero multiplicity of h of each reducible component of M is equal to 1. Take an integer $n_0 \ge 0$ so that, letting N be the degree of $h \circ f^{n_0}$, the degree N part of $h \circ f^{n_0}(x,y)$ consists of the unique term $cy^N (c \ne 0)$.*

Then, $\int_{\mathbb{C}^2}[M] \wedge \mu^- = k^{-n_0} 2\pi N$ *and*

$$\lim_{n\to\infty} \frac{1}{k^n} f^{n*}[M] = \frac{N}{2\pi k^{n_0}} \mu^+,$$

where $[M]$ *denotes the current of integration defined by* M*.*

Proof Let $\varphi(x, y) = \frac{1}{N} \log|h \circ f^{n_0}(x, y)|$ and

$$\Phi_n(x, y) = \frac{1}{Nk^{n-n_0}} \log|h \circ f^n(x, y)| = \varphi^+_{n-n_0}.$$

Then Φ_n does not depend on the choice of n_0 in the following sense. For $n_1 > n_0$, letting $K = k^{n_1-n_0}$, the degree of $h \circ f^{n_1}$ is KN with the degree KN part equal to $c'y^{KN}(c' \neq 0)$ due to Lemma 8.6.6. Writing $\tilde{\varphi} = \frac{1}{KN} \log|h \circ f^{n_1}|$, we have

$$\Phi_n = \frac{1}{NKk^{n-n_1}} \log|h \circ f^n(x, y)| = \tilde{\varphi}^+_{n-n_1}.$$

Let $h \circ f^{n_0}(x, y) = cy^N + h_1(x, y)$, where $h_1(x, y)$ is the polynomial of degree $\leq N - 1$. Take V^+_R so that $|h_1(x, y)| \leq \frac{1}{2}|c|\,|y|^N$ for $(x, y) \in V^+_R$. Then, for $(x, y) \in V^+_R$, we have

$$\frac{1}{2}|c|\,|y|^N \leq |h \circ f^{n_0}(x, y)| \leq \frac{3}{2}|c|\,|y|^N, \text{ i.e.,}$$

$$\log|y| + \frac{1}{N} \log\left(\frac{1}{2}|c|\right) \leq \varphi(x, y) \leq \log|y| + \frac{1}{N} \log\left(\frac{3}{2}|c|\right). \tag{8.20}$$

By the Poincaré–Lelong formula to which we referred in §8.1, letting $T = dd^c\varphi$ we obtain $T = (2\pi/N) f^{n_0*}[M] \in \mathcal{K}_y$. Let $\nu = \lim_{n\to\infty} dd^c\varphi^+_n$. Then in view of the definition of c_ν in formula (8.15), formula (8.20) and Lemma 8.3.8 yield $c_\nu = 1$. So by Lemma 8.5.5, we have

$$c(T) = \frac{1}{4\pi^2} \frac{2\pi}{N} \int_{\mathbb{C}^2} f^{n_0*}[M] \wedge \mu^- = 1,$$

hence Theorem 8.5.14 yields

$$\lim_{n\to\infty} \frac{2\pi}{N} (f^{n_0*}[M])^+_n = \mu^+.$$

Now we have

$$\int_{\mathbb{C}^2} [M] \wedge \mu^- = \int_{\mathbb{C}^2} f^{n_0*}[M] \wedge f^{n_0*}\mu^- = \frac{1}{k^{n_0}} \int_{\mathbb{C}^2} f^{n_0*}[M] \wedge \mu^- = \frac{2\pi N}{k^{n_0}},$$

$$[M]^+_n = \frac{1}{k^{n_0}} (f^{n_0*}([M]))^+_{n-n_0} \to \frac{N}{2\pi k^{n_0}} \mu^+.$$

□

The next theorem is a direct translation of Theorem 8.5.14, which we need later.

Theorem 8.6.9 *Let Ω be an open set of $\mathbb{C}^2$, M be a purely one-dimensional closed submanifold in Ω and χ be a C^∞ function in $\mathbb{C}^2$ satisfying $\chi \geq 0$ and* supp$(\chi) \subset\subset \Omega$. *Then,*

$$\lim_{n\to\infty} \frac{1}{k^n} f^{n*}(\chi[M]) = \frac{1}{4\pi^2}\left(\int_M \chi \mu^-{}_{|M}\right)\mu^+.$$

9

Dynamics of polynomial automorphisms

The main purpose of this chapter is to explain some basic results on the dynamics of a polynomial automorphism f which are proved by Bedford and Smillie (1991a, 1991c, 1992) using the method of pluripotential theory.

In §9.1, from the viewpoint of dynamical systems, some kinds of the stable and unstable sets of f are studied. Specifically, in the case where the Jacobian δ of f satisfies $|\delta| < 1$, the Siegel disks and the Herman rings are discussed and the recurrent connected components of Int K^+ are characterized.

The remaining two sections of this chapter are devoted to giving some applications of the convergence theorems in Chapter 8 to the dynamics of f. The general case is treated in §9.2 and the hyperbolic case is studied in §9.3.

As a special C^∞ dynamical system, the study of dynamics of f inherits the vast quantity of results in the field of C^∞ dynamical system. In §§9.2 and 9.3, we use some of them without giving any proof. All the results we need are found in the book by Shub (1987).

9.1 The stable and unstable sets

In this section, concerning the dynamics of f, the stable set $W^s(X)$ and the unstable set $W^u(X)$ of a set $X \subset \mathbb{C}^2$ are defined. We shall show that $W^s(K) = K^+$ and $W^u(K) = K^-$. Next, in the case when the Jacobian δ of f satisfies $|\delta| < 1$, we investigate the stable sets $W^s(a)$ of an attracting periodic point a, $W^s(\mathcal{D})$ of a Siegel disk $\mathcal{D}$, and $W^s(\mathcal{H})$ of a Herman ring $\mathcal{H}$. Any one of these $W^s(a)$, $W^s(\mathcal{D})$ and $W^s(\mathcal{H})$ is shown to be a connected component of Int K^+. Further, we shall introduce the notion of a recurrent connected component of K^+ and show that a recurrent component of Int K^+ must be one of $W^s(a)$, $W^s(\mathcal{D})$ and $W^s(\mathcal{H})$.

Definition For a subset X of $\mathbb{C}^2$, *the stable set* $W^s(X)$ and *the unstable set* $W^u(X)$ of X are defined by

$$W^s(X) = \left\{ z \in \mathbb{C}^2 \,\middle|\, \lim_{n\to\infty} \mathrm{d}(f^n(z), f^n(X)) = 0 \right\},$$

$$W^u(X) = \left\{ z \in \mathbb{C}^2 \,\middle|\, \lim_{n\to\infty} \mathrm{d}(f^{-n}(z), f^{-n}(X)) = 0 \right\},$$

where d(,) denotes the Euclidean distance of sets in $\mathbb{C}^2$.

Proposition 9.1.1 (*a*) $W^s(K) = K^+$, (*b*) $W^u(K) = K^-$.

Proof We show (a) only. Since K is compact by Proposition 8.3.2, the inclusion $W^s(K) \subset K^+$ is obvious. We prove the converse inclusion. In view of Lemma 7.2.4, take $R > 0$ such that $K \subset\subset D_R$ and $K^+ \subset D_R \cup V_R^-$. Take $z \in K^+$ and an arbitrary open set $\tilde{U}$ with $K \subset \tilde{U} \subset\subset D_R$, and set $U = \tilde{U} \cap K^+$. It is enough to prove that there is a positive integer n_0 such that

$$f^n(z) \in U \qquad (n > n_0). \tag{9.1}$$

Set $z = (x_0, x_1)$. Take $R_1 > 0$ as in Lemma 8.3.4 and let $L_1 = K^+ \cap \{(x, y) \mid |x| \le \max\{R_1, |x_0|\}\}$. Then, by using the notation in Lemma 8.3.4, we have $(x_i, x_{i+1}) \in K^+ \subset D_R \cup V_R^-$ $(i \ge 0)$, hence $f^n(z) \in L_1$ $(n \ge 0)$. Since every $w \in L_1 \cap (K^+ - U)$ satisfies $w \notin K^-$, $\{f^{-n}(w)\}$ is not bounded, so $f^{-n_1}(w) \notin L_1$ for sufficiently large n_1. There is an open neighborhood Ω of w such that $f^{-n_1}(\Omega) \cap L_1 = \emptyset$. Then, $f^{-n}(\Omega) \cap L_1 = \emptyset$ for all $n > n_1$. Since $L_1 \cap (K^+ - U)$ is compact, we can choose a positive integer n_0 so that

$$f^{-n}(L_1 \cap (K^+ - U)) \cap L_1 = \emptyset \quad (n > n_0). \tag{9.2}$$

Now, if $n > n_0$, then $z = f^{-n} \circ f^n(z) \in L_1$ and (9.2) imply $f^n(z) \notin L_1 \cap (K^+ - U)$. Hence $f^n(z) \in U$ because $f^n(z) \in L_1 \cap K^+$, which proves (9.1). □

Let a be an attracting periodic point of f of period p. Then the stable set $W^s(a)$ is the basin of attraction of a for f^p, which we have seen in Theorem 6.3.1.

Proposition 9.1.2 *Let a be an attracting periodic point of f. Then, the stable set $W^s(a)$ is the connected component containing a of* Int K^+.

Proof Let p be the period of a, and let U be the connected component of Int K^+ containing a. Then, setting B to be a neighborhood of a such that $f^p(B) \subset\subset B$, we have $W^s(a) = \bigcup_{n=1}^{\infty} f^{-np}(B)$ and hence $W^s(a) \subset U$. On the other hand, by Proposition 8.3.3, $\{f^{np} \mid n \in \mathbb{N}\}$ is normal in U. So by Vitali's theorem, $\{f^{np}\}$ converges to the constant map a uniformly on every compact set in U, hence $U \subset W^s(a)$. □

A similar result holds for the unstable set $W^u(a)$ of a repelling periodic point a. For a periodic point of saddle type, the stable set $W^s(a)$ and the unstable set $W^u(a)$ were studied in §6.4. In §9.3, we will study the stable set $W^s(a)$ and the unstable set $W^u(a)$ of a point $a \in J$ in the case where f is hyperbolic.

Next, we study a stable set of f of another special type.

Definition Let κ be an irrational number, $b = e^{i\pi\kappa}$, and $\Delta = \{\zeta \in \mathbb{C} \mid |\zeta| < r\}$ $(r > 0)$. If there are an injective holomorphic map $\varphi : \Delta \to \mathbb{C}^2$ and a positive integer p such that

$$f^p(\varphi(\zeta)) = \varphi(b\zeta) \quad (\zeta \in \Delta),$$

then $\mathcal{D} = \varphi(\Delta)$ is called a *Siegel disk*. Here, as a Siegel disk, we only refer to the maximal $\mathcal{D}$ with respect to inclusion.

$\mathcal{D}_0 = \mathcal{D}, \mathcal{D}_1 = f(\mathcal{D}), \ldots, \mathcal{D}_{p-1} = f^{p-1}(\mathcal{D})$ are all Siegel disks. By the definition of K, we have $\mathcal{D} \subset K$, and by Proposition 8.3.2, K is compact. So by the Schwarz lemma, it is impossible that the radius r of a Siegel disk $\mathcal{D}$ equals ∞.

Definition Let κ be an irrational number and let $b = e^{i\pi\kappa}$. If there are an annulus $A = \{\zeta \in \mathbb{C} \mid r_1 < |\zeta| < r_2\}$ $(0 < r_1)$, an injective holomorphic map $\varphi : A \to \mathbb{C}^2$ and a positive integer p such that

$$f^p(\varphi(\zeta)) = \varphi(b\zeta) \quad (\zeta \in A),$$

then $\mathcal{H} = \varphi(A)$ is called a *Herman ring*. We only refer to that $\mathcal{H}$ which is maximal with respect to inclusion.

$\mathcal{H}_0 = \mathcal{H}, \mathcal{H}_1 = f(\mathcal{H}), \ldots, \mathcal{H}_{p-1} = f^{p-1}(\mathcal{H})$ are all Herman rings. When the inner radius r_1 of A equals 0, due to the Riemann continuation theorem, the holomorphic map $\varphi : \{0 < |\zeta| < r_2\} \to \mathcal{H}$ is continued holomorphically to $\tilde{\varphi} : \{|\zeta| < r_2\} \to \mathbb{C}^2$ because $\mathcal{H} \subset K$ and K is compact. So this case brings a Siegel disk $\tilde{\varphi}(\{|\zeta| < r_2\})$; therefore we exclude this case from the Herman rings. Similarly the case that the outer radius r_2 of the annulus A equals ∞ is not a Herman ring.

Remark In both cases, since $f^p \circ \varphi(\zeta) = \varphi(b\zeta)$, we can see that the derivative φ' satisfies $\varphi' \neq 0$ in A or in $\Delta - \{0\}$. Therefore, $\mathcal{H}$ and $\mathcal{D} - \{\varphi(0)\}$ are one-dimensional complex submanifolds immersed in $\mathbb{C}^2$.

In the rest of this section, we always assume that the Jacobian δ of f satisfies $|\delta| < 1$. In the next two lemmas, let A denote either $\{r_1 < |\zeta| < r_2\}$ in the case of a Herman ring or $\Delta - \{0\}$ in the case of a Siegel disk, and φ the map in the above definitions.

Lemma 9.1.3 *There are an open neighborhood $\mathcal{N}$ of $A \times \{0\}$ in $A \times \mathbb{C}$ and an injective holomorphic map $S : \mathcal{N} \to \mathbb{C}^2$ satisfying $S(\zeta, 0) = \varphi(\zeta)$ such that $F = S^{-1} \circ f^p \circ S : \mathcal{N} \to \mathbb{C}^2$ has the form*

$$F(\zeta, \eta) = (b\zeta + \eta^2 f_1, \delta^p b^{-1}\eta + \eta^2 f_2), \quad f_1, f_2 \in \mathcal{O}(\mathcal{N}),$$

where $\mathcal{O}(\mathcal{N})$ is the set of all holomorphic functions in $\mathcal{N}$.

Proof Let $\varphi = (\varphi_1, \varphi_2)$ with $\varphi_1, \varphi_2 \in \mathcal{O}(A)$. Since $(\varphi_1', \varphi_2') \neq (0, 0)$ in A, we can take $\psi_1, \psi_2 \in \mathcal{O}(A)$ such that $-\psi_1\varphi_2' + \psi_2\varphi_1' = 1$. Define $T : A \times \mathbb{C} \to \mathbb{C}^2$ by $T(\zeta, \eta) = (\varphi_1(\zeta) + \eta\psi_1(\zeta), \varphi_2(\zeta) + \eta\psi_2(\zeta))$. Then $T(\zeta, 0) = \varphi(\zeta)$ and $\det(T') = 1$ on $A \times \{0\}$.

Take two sufficiently small neighborhoods $\mathcal{N} \subset \mathcal{N}_1$ of $A \times \{0\}$ in $A \times \mathbb{C}$ so that $T_{|\mathcal{N}_1}$ is injective and $F_1 = (T_{|\mathcal{N}_1})^{-1} \circ f^p \circ T_{|\mathcal{N}}$ is well defined in $\mathcal{N}$. Since $F_1(\zeta, 0) = (b\zeta, 0)$, so $F_1'(\zeta, 0) = \begin{pmatrix} b & \lambda(\zeta) \\ 0 & \delta^p/b \end{pmatrix}$, hence

$$F_1(\zeta, \eta) = (b\zeta + \lambda(\zeta)\eta + \cdots, \delta^p b^{-1}\eta + \cdots),$$

where $\lambda \in \mathcal{O}(A)$ and $\cdots$ denotes the terms of degree greater than 1 in η. Let $g(\zeta) = -b^{-1}\sum_{j=0}^{\infty} \delta^{jp} b^{-2j} \lambda(b^j\zeta) \in \mathcal{O}(A)$. In view of $|\delta| < 1$, this series converges uniformly on every compact in A and g satisfies $\delta^p b^{-1} g(b\zeta) = bg(\zeta) + \lambda(\zeta)$.

Let $R(\zeta, \eta) = (\zeta + \eta g(\zeta), \eta)$. We replace the $\mathcal{N}$ with a smaller one so that $R^{-1}(\zeta, \eta) = (\zeta - g(\zeta)\eta + \cdots, \eta)$ and $R^{-1} \circ F_1 \circ R$ is well defined in $\mathcal{N}$. Then,

$$\begin{aligned} R^{-1} \circ F_1 \circ R(\zeta, \eta) & \\ = \; & (b(\zeta + \eta g(\zeta)) + \lambda(\zeta + \eta g(\zeta))\eta \\ & -g(b(\zeta + \eta g(\zeta)) \\ & +\lambda(\zeta + \eta g(\zeta))\eta + \cdots)\delta^p b^{-1}\eta + \cdots, \delta^p b^{-1}\eta + \cdots) \\ = \; & (b\zeta + (bg(\zeta) + \lambda(\zeta) - \delta^p b^{-1} g(b\zeta))\eta + \cdots, \delta^p b^{-1}\eta + \cdots) \\ = \; & (b\zeta + \eta^2 f_1(\zeta, \eta), \delta^p b^{-1}\eta + \eta^2 f_2(\zeta, \eta)). \end{aligned}$$

Therefore $S = T \circ R$ is the desired function. □

Lemma 9.1.4 *Fix two concentric open annuli A_0 and A_1 such that $A_0 \subset\subset A_1 \subset\subset A$. Then we can take $\varepsilon > 0$ so small that for any $(\zeta, \eta) \in A_0 \times \{|\eta| < \varepsilon\}$ and any positive integer n,*

$$(\zeta_n, \eta_n) = F^n(\zeta, \eta) \in A_1 \times \{|\eta| < \varepsilon\}$$

and $|\eta_n| \to 0$ as $n \to \infty$.

Proof Take $\varepsilon > 0$ such that $A_1 \times \{|\eta| < \varepsilon\} \subset\subset \mathcal{N}$, and $M > 0$ such that $|f_1| \leq M$, $|f_2| \leq M$ on $A_1 \times \{|\eta| < \varepsilon\}$. We make ε smaller so that $\tau = |\delta|^p + M\varepsilon < 1$ and the distance between A_0 and ∂A_1 in $\mathbb{C}$ is greater than $M\varepsilon^2/(1 - \tau^2)$. Let $(\zeta, \eta) \in A_0 \times \{|\eta| < \varepsilon\}$. We can prove inductively on n that $(\zeta_n, \eta_n) = F^n(\zeta, \eta)$ satisfy $(\zeta_n, \eta_n) \in A_1 \times \{|\eta| < \varepsilon\}$ and the following estimates:

$$\begin{aligned}
|\zeta_1 - b\zeta| &\leq M\varepsilon^2, & |\eta_1| &\leq |\delta|^p|\eta| + M|\eta|^2 \leq \tau\varepsilon,\\
|\zeta_2 - b\zeta_1| &\leq M\tau^2\varepsilon^2, & |\eta_2| &\leq \tau|\eta_1| \leq \tau^2\varepsilon,\\
&\ldots, & &\ldots,\\
|\zeta_{n+1} - b\zeta_n| &\leq M\tau^{2n}\varepsilon^2, & |\eta_{n+1}| &\leq \tau|\eta_n| \leq \tau^{n+1}\varepsilon.
\end{aligned}$$

□

Lemma 9.1.5 *Any Siegel disk $\mathcal{D}$ and any Herman ring $\mathcal{H}$ are subsets of* Int K^+.

Proof In the case of $\mathcal{H}$, since we can take arbitrary $A_0 \subset\subset A_1 \subset\subset A$ in Lemma 9.1.4, it can be seen that a neighborhood of $\mathcal{H}$ is contained in K^+. So $\mathcal{H} \subset \operatorname{Int} K^+$. Similarly, in the case of $\mathcal{D} = \varphi(\Delta)$, we have $\varphi(\Delta - \{0\}) \subset \operatorname{Int} K^+$.

It remains to show that $\varphi(0) \in \operatorname{Int} K^+$. Let $\mathcal{D}' = \varphi(\{|\zeta| < r/2\})$ and $\partial\mathcal{D}' = \varphi(\{|\zeta| = r/2\})$. Take a neighborhood B of $z = 0$ in $\mathbb{C}^2$ so that $\bigcup_{z \in B}(z + \partial\mathcal{D}') \subset\subset \operatorname{Int} K^+$, where $z + \partial\mathcal{D}'$ denotes the translate of $\partial\mathcal{D}'$.

Since $\{f^{np}\}$ is bounded on $\bigcup_{z \in B}(z + \partial\mathcal{D}')$, applying the maximum modulus principle on the analytic subset $(z + \mathcal{D}')$, $\{f^{np}\}$ is bounded on $\bigcup_{z \in B}(z + \mathcal{D}')$, which implies $\bigcup_{z \in B}(z + \mathcal{D}') \subset\subset K^+$.

Since $\varphi(0) \in \operatorname{Int}(\bigcup_{z \in B}(z + \mathcal{D}'))$, we obtain $\varphi(0) \in \operatorname{Int} K^+$. □

Lemma 9.1.6 *Let U be a connected component of* Int K^+ *containing a Siegel disk $\mathcal{D}$. Then, an appropriate subsequence $\{f^{n_j p}\}$ of $\{f^{np} \mid n \in \mathbb{N}\}$ converges on every compact set in U to a holomorphic map $\Phi : U \to \mathcal{D}$.*

Moreover, there are a holomorphic retraction $\rho : U \to \mathcal{D}$ *and* $\phi \in \mathrm{Aut}(\mathcal{D})$ *which is conjugate via* φ *to a rotation of the disk* Δ *such that* $\phi \circ \Phi = \rho$ *and* $\rho \circ f^p = f^p \circ \rho$ *on* U.

The same statement holds for a Hermann ring $\mathcal{H}$.

Proof By Proposition 8.3.3, there is a subsequence $\{n_j\}$ such that $f^{n_j p}$ converges uniformly on every compact set in U to a holomorphic map $\Phi : U \to \bar{U}$ and that $b^{n_j} \to b_1$, with $|b_1| = 1$. Then, $\Phi \circ f^p = f^p \circ \Phi$ and $f^p \circ \varphi(\zeta) = \varphi(b\zeta)$ imply $\Phi \circ \varphi(\zeta) = \varphi(b_1\zeta)$.

Let $\Phi(U) = M \subset \overline{U}$. Then, by Proposition 9.1.1, $M \subset K$. If $u = \Phi(v)$ with $v \in U$, then $f^p(u) = f^p \circ \Phi(v) = \Phi \circ f^p(v) \in M$, which proves $M \subset f^p(M)$. Similarly, by $f^{-p} \circ \Phi = \Phi \circ f^{-p}$, $M \subset f^{-p}(M)$. So combining these two yields $M = f^p(M)$.

Since $\mathcal{D} \subset U$ and $\Phi(\mathcal{D}) = \mathcal{D}$, we have $\mathcal{D} \subset M$. By Lemma 9.1.4, there is a neighborhood $\omega \subset \mathbb{C}^2$ of $\mathcal{D} - \{\varphi(0)\}$ such that $\Phi_{|\omega} : \omega \to \mathcal{D}$, so the rank of Φ' is equal to 1. Therefore $M = \Phi(U)$ is a one-dimensional locally closed analytic subset of $\mathbb{C}^2$, so there is a Riemann surface $\tilde{M}$ which is the desingularization of M and $\tilde{M}$ is hyperbolic, because $M \subset K$. Let $\tilde{f^p} : \tilde{M} \to \tilde{M}$ be the holomorphic automorphism induced by f^p. Then $\tilde{f^p}$ falls in case (iii) of Theorem 2.1.11 (the Denjoy–Wolff theorem), because cases (i), (ii) and (iv) do not occur owing to $f^p \circ \varphi(\zeta) = \varphi(b\zeta)$. Since $\mathcal{D} \subset M$, $\tilde{M}$ is biholomorphic to a disk and f^p is conjugate to the rotation $\zeta \to b\zeta$ via φ.

The maximality of $\mathcal{D}$ yields $\mathcal{D} = M$, so $\Phi : U \to \mathcal{D}$. Set $\phi \in \mathrm{Aut}(\mathcal{D})$ by $\phi(\varphi(b_1\zeta)) = \varphi(\zeta)$ and the restriction of the map $\rho = \phi \circ \Phi$ to $\mathcal{D}$ is the identity map, which implies that ρ is a retraction. Since both f^p and ϕ are conjugate to rotations via φ, $f^p \circ \phi = \phi \circ f^p$ on $\mathcal{D}$, hence

$$\rho \circ f^p = \phi \circ \Phi \circ f^p = \phi \circ f^p \circ \Phi = f^p \circ \phi \circ \Phi = f^p \circ \rho \text{ on } U.$$

□

Lemma 9.1.7 *For a Siegel disk* $\varphi : \Delta \to \mathcal{D}$, $\varphi(\mathcal{D})$ *is non-singular at* $\varphi(0) \in \mathcal{D}$.

Proof Since the restriction to $\mathcal{D}$ of ρ in Lemma 9.1.6 is the identity, the rank of the differential $\rho'(\varphi(0))$ is equal to 1. Therefore, $\mathcal{D}$ is non-singular at $\varphi(0)$. □

As a result of this, Lemma 9.1.7, for a Siegel disk $\mathcal{D}$, we can take concentric open disks $\Delta_0 \subset\subset \Delta_1 \subset\subset \Delta$ and can modify Lemma 9.1.4 so that the statement holds for $\Delta_0 \subset\subset \Delta_1 \subset\subset \Delta$ in place of $A_0 \subset\subset A_1 \subset\subset A$.

Theorem 9.1.8 *Suppose that the Jacobian δ of f satisfies $|\delta| < 1$. Then the stable set $W^s(\mathcal{D})$ of a Siegel disk $\mathcal{D}$ is a connected component of* $\operatorname{Int} K^+$ *containing $\mathcal{D}$.*

The same result holds also for a Herman ring.

Proof Let U be a connected component of $\operatorname{Int} K^+$ containing $\mathcal{D}$. By Lemmas 9.1.3 and 9.1.4, $W^s(\mathcal{D})$ contains a neighborhood Ω in $\operatorname{Int} K^+$ of $\mathcal{D}$. Therefore, $W^s(\mathcal{D}) = \bigcup_{n=1}^{\infty} f^{-np}(\Omega)$, so $W^s(\mathcal{D}) \subset U$. On the other hand, by the proof of Lemma 9.1.6, for an arbitrary subsequence of $\{f^{np}\}$, we can extract a further subsequence $\{f^{n_j p}\}$ which converges to a holomorphic map $\Phi : U \to \mathcal{D}$ of rank 1. So $U \subset W^s(\mathcal{D})$. □

We shall show that in the case of $|\delta| < 1$, f^p can be linearized in $W^s(\mathcal{D})$ and $W^s(\mathcal{H})$.

Theorem 9.1.9 *Assume that the Jacobian δ of f satisfies $|\delta| < 1$, and let $\mathcal{H}$ be a Herman ring. Set*

$$L(\zeta, \eta) = (b\zeta, \delta^p b^{-1}\eta).$$

Then there is a surjective biholomophic map $G : A \times \mathbb{C} \to W^s(\mathcal{H})$ such that $G(A \times \{0\}) = \mathcal{H}$ and $f^p \circ G = G \circ L$.

The same assertion holds for a Siegel disk.

Proof Let $\mathcal{N}$, S and $F(\zeta, \eta) = (b\zeta + \eta^2 f_1, \delta^p b^{-1}\eta + \eta^2 f_2)$ be as in Lemma 9.1.3. In particular, $\mathcal{N}$ is an open neighborhood of $A \times \{\eta = 0\}$ in $A \times \mathbb{C}$ such that $S(\mathcal{N}) \subset W^s(\mathcal{H})$. Let $\psi : \mathcal{H} \to A \subset \mathbb{C}$ be the inverse of $\varphi : A \to \mathcal{H}$.

Define $\tilde{\rho} : \mathcal{N} \to \mathbb{C}$ by setting $\tilde{\rho} = \psi \circ \rho \circ S$, where $\rho : W^s(\mathcal{H}) \to \mathcal{H}$ is a retraction as in Lemma 9.1.6.

Consider $f_3 = \delta^p b^{-1}(1 + \delta^{-p} b\eta f_2)^2$ and set

$$h = \delta^{-p} b \left(f_2 + f_2 \circ F f_3 + \cdots + f_2 \circ F^n \prod_{j=0}^{n-1} f_3 \circ F^j + \cdots \right).$$

By Lemma 9.1.4 and by $|\delta| < 1$, h converges uniformly on every compact set in a neighborhood $\mathcal{N}_1 \subset \mathcal{N}$ of $A \times \{0\}$ in $A \times \mathbb{C}$, and satisfies

$$f_2 + \delta^p b^{-1} h \circ F f_3 = \delta^p b^{-1} h. \tag{9.3}$$

Define $H : \mathcal{N}_1 \to \mathbb{C}^2$ by $H(\zeta, \eta) = (\tilde{\rho}(\zeta, \eta), \eta + \eta^2 h(\zeta, \eta))$. We prove $H \circ F = L \circ H$ by showing that the two sides coincide. Since

$$\tilde{\rho} \circ F = \psi \circ \rho \circ S \circ F = \psi \circ \rho \circ f^p \circ S = \psi \circ f^p \circ \varphi \circ \psi \circ \rho \circ S = b\psi \circ \rho \circ S = b\tilde{\rho}$$

on $\mathcal{N}_1$, the first components of both sides coincide. By (9.3),

$$\delta^p b^{-1}\eta + \eta^2 f_2 + (\delta^p b^{-1}\eta + \eta^2 f_2)^2 h \circ F = \delta^p b^{-1}(\eta + \eta^2 h),$$

which implies that the second components of both sides coincide.

Let $G = S \circ H^{-1}$ in a neighborhood of $A \times \{0\}$. Then, $G(A \times \{0\}) = \mathcal{H}$ and $f^p \circ G = G \circ L$. Using this formula, we can continue G holomorphically to a map of $A \times \mathbb{C}$, which we denote by the same notation G. Then, $G(A \times \mathbb{C}) = W^s(\mathcal{H})$. □

Finally, we characterize a periodic recurrent component of $\operatorname{Int} K^+$ in the case of $|\delta| < 1$.

Definition Let U be a connected component of $\operatorname{Int} K^+$. If there is a positive integer p with $f^p(U) = U$, then as before U is called *periodic*. The minimum of such p is called the *period* of U. On the contrary, when $f^n(U) \cap U = \emptyset$ for any positive integer n, U is called *wandering*.

A periodic component U is called *recurrent* if there are a point $w \in U$ and a compact $L \subset\subset U$ such that $f^n(w) \in L$ for infinitely many $n > 0$.

Theorem 9.1.10 *Assume that the Jacobian δ of f satisfies $|\delta| < 1$. Then, a periodic recurrent component U of* $\operatorname{Int} K^+$ *is either the stable set $W^s(a)$ of an attracting periodic point, $W^s(\mathcal{D})$ of a Siegel disk $\mathcal{D}$ or $W^s(\mathcal{H})$ of a Herman ring $\mathcal{H}$.*

Proof Let p be the period of U. Then, since U is recurrent, there are a subsequence $\{n_j\}$ and two points $w, w_1 \in U$ such that $\lim_{j\to\infty} f^{n_j p}(w) = w_1$. Since $\{f^{np} \mid n \geq 0\}$ is a normal family by Proposition 8.3.3, we may assume, by passing to a further subsequence which we denote again by $\{n_j\}$, that $\Phi = \lim_{j\to\infty} f^{n_j p} : U \to \bar{U}$ exists and $m_j = n_{j+1} - n_j \to \infty$.

The holomorphic map Φ satisfies $\Phi(w) = w_1$, $\Phi \circ f^p = f^p \circ \Phi$ and $\det \Phi' = 0$. Since the normality of $\{f^{np} \mid n \in \mathbb{N}_0\}$ ensures the existence of a constant A such that $\sup_j |(f^{m_j p})'_z| \leq A$ when z moves over a compact convex neighborhood of w_1, where $|\cdot|$ denotes the operator norm induced by the Euclidean metric, we have, for sufficiently large j,

$$\|f^{n_{j+1}p}(w) - f^{m_j p}(w_1)\|$$

$$= \|f^{m_j p} \circ f^{n_j p}(w) - f^{m_j p}(w_1)\| \leq A \|f^{n_j p}(w) - w_1\|,$$

hence

$$\lim_{j\to\infty} f^{m_j p}(w_1) = w_1. \tag{9.4}$$

Set $M = \Phi(U)$. Then, $w_1 \in M$, and by Proposition 9.1.1, $M \subset K$. We can prove, as in the proof of Lemma 9.1.6, that $f^p(M) = M$.

First, we assume rank $\Phi = 1$. Then, there is a hyperbolic Riemann surface $\tilde{M}$ which is the desingularization of M and f^p induces the automorphism $\tilde{f^p}$ of $\tilde{M}$. The automorphism $\tilde{f^p}_{|\tilde{M}}$ is in case (iii) of Theorem 2.1.11, because (i) and (iv) do not occur owing to (9.4) and Proposition 8.3.1 respectively. $\tilde{M}$ is equal to either a disk or an annulus and $\tilde{f^p}$ is an irrational rotation. So M is either $\mathcal{D}$ or $\mathcal{H}$.

Next, we assume that rank $\Phi = 0$. Then, Φ is a constant map w_1. Since $f^p(w_1) = f^p \circ \Phi(w_1) = \Phi \circ f^p(w_1) = w_1$, w_1 is a fixed point of f^p. So w_1 is an attracting fixed point, and $U = W^s(w_1)$. □

9.2 Applications of the convergence theorems

In this section, we prove several dynamical properties of polynomial automorphisms.

Theorem 9.2.1 *Suppose that the holomorphic Jacobian δ of f satisfies $|\delta| < 1$. Let U be either $W^s(a)$ of an attracting periodic point a, $W^s(\mathcal{D})$ of a Siegel disk $\mathcal{D}$ or $W^s(\mathcal{H})$ of a Herman ring $\mathcal{H}$. Then, for any one-dimensional algebraic set M of $\mathbb{C}^2$, we have*

(a) $M \cap U \neq \varnothing$,
(b) $M \cap \overline{U}$ *is compact.*

Proof (a) According as U equals either $W^s(a)$, $W^s(\mathcal{D})$, or $W^s(\mathcal{H})$, we have $a \in K^-$, $\varphi(\{|\zeta| = r/2\}) \subset K^-$, or $\varphi(\{|\zeta| = (r_1 + r_2)/2\}) \subset K^-$, respectively, which implies $K^- \cap U \neq \varnothing$.

Let $[M]$ be the current of integration of M; then, $\mathrm{supp}([M]) = M$. Since $|\delta| < 1$, $K^- = J^-$ by Theorem 7.2.9 applied to f^{-1}, so combining with Theorem 8.3.15, $\mathrm{supp}(\mu^-) = K^-$. Now, assume $M \cap U = \varnothing$. Then, letting p be the period of U, $\mathrm{supp}(k^{-np} f^{-np*}[M]) \cap U = \varnothing$. So by Theorem 8.6.8 for f^{-1}, we have $\mathrm{supp}(\mu^-) \cap U = \varnothing$, or equivalently $K^- \cap U = \varnothing$. This is a contradiction, and we conclude that $M \cap U \neq \varnothing$.

(b) Let $h(x, y)$ be a defining polynomial of M. Then, by Proposition 8.6.7 exchanging the roles of x and y, there is a positive integer n_1 such that, letting k' be the degree of the polynomial $h \circ f^{-n_1 p}(x, y)$, the degree k' part of $h \circ f^{-n_1 p}(x, y)$ consists of only one term $cx^{k'}$ with $c \neq 0$. Therefore, for sufficiently large R, $f^{n_1}(M) \cap \overline{V_R^-} = \varnothing$. (See the proof of Theorem 8.6.8.) On the other hand, by Proposition 7.2.3, $U \subset \mathrm{Int}\, K^+ \subset (D_R \cup V_R^-)$, so

$\overline{U} \subset \overline{(D_R \cup V_R^-)}$. Hence, $\overline{U} \cap f^{n_1}(M) \subset \overline{D_R}$. Since $\overline{U}$ is invariant under f, $\overline{U} \cap M \subset f^{-n_1}(\overline{D_R})$, which proves that $\overline{U} \cap M$ is compact. □

Theorem 9.2.2 *Suppose that the holomorphic Jacobian δ of f satisfies $|\delta| < 1$. Let U be a connected periodic component of* Int K^+ *and $U \cap K^- \neq \varnothing$. Then, $\partial U = J^+$.*

Remark Keeping the assumption $|\delta| < 1$, let U be either $W^s(a)$ of an attracting periodic point a, $W^s(\mathcal{D})$ of a Siegel disk $\mathcal{D}$ or $W^s(\mathcal{H})$ of a Herman ring $\mathcal{H}$. Then, U is a periodic connected component of Int K^+ by Proposition 9.1.2 and Theorem 9.1.8, and satisfies $U \cap K^- \neq \varnothing$ by Theorem 9.2.1. Hence the above theorem implies that $\partial U = J^+$.

Proof of the theorem Since $\partial U \subset \partial K^+ = J^+$, we shall prove the opposite inclusion. Let $z \in U \cap K^-$. By Theorem 7.2.9, we have Int $K^- = \varnothing$, so we can take a complex line L passing through z and a disk D in L with center z such that $D \subset U$, and $D \not\subset K^-$. Consider the Green function G^- for K^-. Then, $G^-_{|D}$ is subharmonic, non-negative and takes the value 0 at z. If $G^-_{|D}$ were harmonic in D, the minimum principle yields that $G^-_{|D}$ should be identically equal to 0 in D, which contradicts Theorem 8.3.14 and the fact that $D \not\subset K^-$. Hence $\mu^-_{|D}(D) > 0$. Take a C^∞ function χ in $\mathbb{C}^2$ with $\chi \geq 0$, $\mathrm{supp}(\chi) \cap D \subset\subset D$ and $\mu^-_{|D}(\chi D) > 0$. Then, letting p be the period of U, by Theorem 8.6.9, $k^{-np} f^{np*}(\chi[D]) \to c_1 \mu^+$ with $c_1 > 0$ as $n \to \infty$. So by Theorem 8.3.15, $J^+ = \mathrm{supp}(\mu^+) \subset \overline{\bigcup_{n=1}^{\infty} f^{-np}(D)} \subset \overline{U}$. Since $J^+ \cap U = \varnothing$, we have $J^+ \subset \partial U$, which completes the proof. □

Theorem 9.2.3 *Let a be a periodic point of saddle type of f. Then,*

$$\overline{W^s(a)} = J^+ \quad \textit{and} \quad \overline{W^u(a)} = J^-.$$

Proof Let p be the period of a. Since the eigenvalue β of the differential $(f^p)'$ with $|\beta| > 1$ satisfies $\beta^n \to \infty \quad (n \to \infty)$, the family $\{f^{np} \mid n \geq 1\}$ is not normal in any neighborhood of a. So by Proposition 8.3.3, $a \notin$ Int K^+. On the other hand, since a is a periodic point, $a \in K^+$. Therefore, $a \in J^+$. Then, it follows easily that $\overline{W^s(a)} \subset J^+$.

Conversely, we shall show $\overline{W^s(a)} \supset J^+$. Let M be a locally closed submanifold with $a \in M \subset W^s(a)$. Since $\{f^{-np}_{|M}\}$ is not normal around a, $M \not\subset K^-$. If $G^-_{|M}$ were harmonic in M, taking the minimum 0 at an interior point a, $G^-_{|M}$ would have to be identically equal to 0, which is a contradiction to $M \not\subset K^-$. So $\mu^-_{|M} > 0$. Take a C^∞ function χ in $\mathbb{C}^2$ satisfying $\mathrm{supp}(\chi) \cap M \subset\subset M$ and

$\mu^-_{|M}(\chi M) > 0$. Then by the same argument as in the proof of Theorem 9.2.2, we can conclude that $J^+ = \operatorname{supp}(\mu^+) \subset \overline{\bigcup_{n=1}^{\infty} f^{-np}(M)} \subset \overline{W^s(a)}$. □

Remark If we apply the theorem of A. Katok (1980), which assures us that f has a saddle point, then Theorem 9.2.1 yields the following.

Theorem 9.2.4 *J^+ and J^- are connected subsets.*

See Bedford and Smillie (1992). Bedford and Smillie also gave another direct proof of Theorem 9.2.4 in Bedford and Smillie (1991).

Definition A point $z \in \mathbb{C}^2$ is called *wandering* if there is an open neighborhood U such that $U \cap f^n(U) = \varnothing$ for all $n \geq 1$. Otherwise, z is called *non-wandering*. The set of all non-wandering points is called the *non-wandering set* and is denoted by $\Omega(f)$.

A point $z \in \mathbb{C}^2$ is called *chain recurrent* if, for any $\varepsilon > 0$, we can choose N points $\{z_n\}_{n=1}^{N+1}$ with $z = z_1 = z_{N+1}$ and $\|z_{n+1} - f(z_n)\| < \varepsilon \quad (1 \leq n \leq N)$. The set of all chain recurrent points is called the *chain recurrent set* and is denoted by $R(f)$.

Then we know the following.

Proposition 9.2.5 (a) *$\Omega(f)$ is a closed f-invariant set satisfying $\Omega(f) = \Omega(f^{-1})$.*

(b) *$R(f)$ is a closed f-invariant set satisfying $R(f) = R(f^{-1})$.*

(c) *$\Omega(f) \subset R(f)$.*

Proposition 9.2.6 *$R(f) \subset K$.*

Proof First let $w \notin K^+$; then we can take $n \geq 0$ with $z = f^n(w) \notin D_R \cup V_R^-$ by Proposition 7.2.3. Then the proof of Lemma 7.2.4 implies that, for sufficiently small $\varepsilon > 0$, it is impossible to take a sequence $\{z_n\}_{n=1}^{N+1}$ with $z = z_1 = z_{N+1}$ and $\|z_{n+1} - f(z_n)\| < \varepsilon$ $(1 \leq n \leq N)$, which implies that $R(f) \subset K^+$. Similarly, we have $R(f^{-1}) \subset K^-$, and hence, $R(f) = R(f^{-1}) \subset K$. □

Theorem 9.2.7 *For any open neighborhoods U_1, U_2 of $z_1 \in J^+$ and $z_2 \in J^-$ respectively, there is $n \geq 1$ such that $f^n(U_1) \cap U_2 \neq \varnothing$.*

In particular,

$$J \subset \Omega(f).$$

Proof Take a complex line L passing through z_1 and an open ball B in $\mathbb{C}^2$ with center z_1 such that the open disk $D = L \cap B$ satisfies $D \subset U_1$ and $D \cap (\mathbb{C}^2 - K^+) \neq \varnothing$. If $G^+_{|D}$ were harmonic in D, then since G^+ takes the minimum 0 at the interior point z_1, $G^+_{|D}$ would be identically equal to 0, which is a contradiction to $D \cap (\mathbb{C}^2 - K^+) \neq \varnothing$. Therefore $\mu^+_{|D}(D) > 0$.

Now, take a C^∞ function χ in $\mathbb{C}^2$ such that $\chi(z_1) > 0$, $\int_D \chi \mu^+{}_{|D} > 0$, $\chi \geq 0$ and $\mathrm{supp}(\chi) \subset\subset B$, and consider the positive $(1,1)$-current $\chi[D]$ in $\mathbb{C}^2$. Note that $\mathrm{supp}(\chi[D]) \subset U_1$. By Theorem 8.6.9 for f^{-1}, $(\chi[D])^-_n \to c_1\mu^-$ with $c_1 > 0$. Since Theorem 8.3.15 states $\mathrm{supp}(c_1\mu^-) = J^-$, there is n such that $\mathrm{supp}((\chi[D])^-_n) \cap U_2 \neq \varnothing$. Then, this integer n has the desired property because

$$\mathrm{supp}((\chi[D])^-_n) = f^n(\mathrm{supp}(\chi[D])) \subset f^n(U_1).$$

□

9.3 Hyperbolic polynomial automorphisms

We shall define the notion of hyperbolicity for polynomial automorphisms, and show some basic results for a hyperbolic f.

Definition Let Λ denote an f-invariant compact set, and taking a C^∞ Hermitian metric ds^2 in an open neighborhood of Λ, we denote by $\mathrm{d}_h(\ ,\ )$ the distance induced by this C^∞ Hermitian metric.

Let $\varepsilon > 0$ be sufficiently small. For $z \in \Lambda$,

$$W^s_\varepsilon(z) = \Big\{ w \in \mathbb{C}^2 \ \Big|\ \mathrm{d}_h(f^n(w), f^n(z)) \leq \varepsilon\ (n \geq 0),\ \lim_{n\to\infty} \mathrm{d}_h(f^n(w), f^n(z)) = 0 \Big\},$$

$$W^u_\varepsilon(z) = \Big\{ w \in \mathbb{C}^2 \ \Big|\ \mathrm{d}_h(f^{-n}(w), f^{-n}(z)) \leq \varepsilon\ (n \geq 0),\ \lim_{n\to\infty} \mathrm{d}_h(f^{-n}(w), f^{-n}(z)) = 0 \Big\},$$

$$W^s(z) = \bigcup_{n=0}^{\infty} f^{-n}(W^s_\varepsilon(f^n(z))), \quad W^u(z) = \bigcup_{n=0}^{\infty} f^n(W^u_\varepsilon(f^{-n}(z))).$$

Here, the formulas above have meaning if $\varepsilon > 0$ is so small that the C^∞ Hermitian metric is defined in the ε-neighborhood of Λ. Since Λ is compact, the metric is equivalent to the Euclidean metric in a neighborhood of Λ. So $W^{s/u}(z)$ is independent of the choice of C^∞ Hermitian metrics. Hence, the above $W^{s/u}(z)$ and that defined in §9.1 coincide.

Definition An f-invariant compact $\Lambda \subset \mathbb{C}^2$ is called a *hyperbolic set for* f if the restriction $T\mathbb{C}^2_{|\Lambda}$ of $T\mathbb{C}^2$ to Λ is the direct sum of two f'-invariant complex continuous subbundles E^s and E^u which satisfy: there are $C > 0$ and $0 < \sigma < 1$ such that

$$|(f^n)'| < C\sigma^n \text{ on } E^s, \quad \text{and} \quad |(f^{-n})'| < C\sigma^n \text{ on } E^u$$

for all $n \geq 1$.

When the complex dimension of E^u_z is constant at each point of Λ, this number is called the *index* of Λ. Here $T\mathbb{C}^2 = \mathbb{C}^2 \times \mathbb{C}^2$ is the holomorphic tangent bundle of $\mathbb{C}^2$, $(f^n)' : T\mathbb{C}^2 \to T\mathbb{C}^2$ is the differential of f^n and $|\cdot|$ is the operator norm according to the C^∞ Hermitian metric.

Lemma 9.3.1 *There is a C^∞ Hermitian metric in a neighborhood of Λ and there is A, $0 < A < 1$, such that $|f'| < A$ in E^s and $|(f^{-1})'| < A$ in E^u.*

Such a metric is called an *adapted metric* to the hyperbolic set Λ for f.

Also, the next theorem holds in the category of differentiable manifolds. See Shub.

Theorem 9.3.2 *For a hyperbolic set Λ for f, we fix an adapted metric ds^2. Then, for sufficiently small $\varepsilon > 0$ and for $z \in \Lambda$, $W^s_\varepsilon(z)$ and $W^s(z)$ are complex manifolds diffeomorphic to an open ball of complex Euclidean space of dimension* $\dim E^s_z$, *and their holomorphic tangent spaces at z satisfy*

$$T_z W^s_\varepsilon(z) = T_z W^s(z) = E^s_z.$$

For all $n \geq 0$ and $w \in W^s_\varepsilon(z)$, $\mathrm{d}_h(f^n(w), f^n(z)) \leq A^n \mathrm{d}_h(w, z)$, where A is as in Lemma 9.3.1.

Similar statements hold also for $W^u_\varepsilon(z)$ and $W^u(z)$.

Definition An automorphism f is called *hyperbolic* if J is a hyperbolic set for f.

See Theorem 9.3.14 for some conditions on f equivalent to hyperbolicity.

Theorem 9.3.3 *If f is hyperbolic, then the index of J is equal to* 1 *and*

$$W^s(J) \subset J^+, \quad W^u(J) \subset J^-.$$

Proof To reach a contradiction, assume $\dim E^u_z = 0$ for some $z \in J$. Since $\{f^n(z)\} \subset J$ and J is compact, there are a subsequence $\{n_j\}$ and $w \in J$

such that $f^{n_j}(z) \to w$ as $j \to \infty$. Since $\dim E^u_w = 0$, by Theorem 9.3.2, $B = W^s_\varepsilon(w)$ is an open set of $\mathbb{C}^2$ containing w, and for every $t \in B$

$$d_h(f^n(w), f^n(t)) \le A^n d_h(w, t) \quad \text{and} \quad |(f^n)'_t| \le A^n.$$

Let $\xi = f^{n_j}(z)$ and $\eta = f^{n_{j+l}}(z) \in B$. Here we take n_j and n_{j+l} sufficiently large so that, fixing $q = n_{j+l} - n_j$, $A^q < \frac{1}{2}$ and that the ball of center ξ with radius $2(d_h(\eta, w) + d_h(\xi, w))$ is contained in the interior of B. Then $\{f^{nq}(\xi) \mid n \ge 1\}$ satisfies

$$d_h(f^{iq}(\xi), f^{(i-1)q}(\xi))$$
$$= d_h(f^{(i-1)q}(\eta), f^{(i-1)q}(\xi)) \le A^{(i-1)q}(d_h(\eta, w) + d_h(\xi, w))$$

and hence

$$\begin{aligned} d_h(f^{nq}(\xi), \xi) &\le \left(\sum_{i=1}^{n} A^{(i-1)q}\right)(d_h(\eta, w) + d_h(\xi, w)) \\ &\le 2(d_h(\eta, w) + d_h(\xi, w)). \end{aligned}$$

So $\{f^{nq}(\xi) \mid n \in \mathbb{N}\} \subset B$. Therefore $\{f^{nq}(\xi)\}$ converges to some $\zeta \in \overline{B}$ and $f^q(\zeta) = \zeta$ holds. Then, ζ is an attracting fixed point of f^q, so $\zeta \in \operatorname{Int} K^+$ and $\zeta \notin J^+$, which implies $\zeta \notin J$, a contradiction.

If $\dim E^s_z = 0$ for some z, then a similar argument on f^{-1} reaches a contradiction. Therefore the index of J is 1.

Let N be a neighborhood in $\mathbb{C}^2$ of J, and let C_z, D_z in $T_z\mathbb{C}^2$ be two cones with the origin as vertex and moving continuously in $z \in N$ satisfying $\overline{C_z} \cap \overline{D_z} = \{0\}$, $E^u_z \subset C_z$ and $E^s_z \subset D_z$. Take N small enough so that an integer $n_0 > 0$, C_z and D_z can be chosen such that, for all $z \in N$,

$$|(f^{n_0})'(v)| \ge 2|v| \quad (v \in C_z) \quad \text{and} \quad (f^{n_0})'(C_z) \subset C_{f^{n_0}(z)},$$

where $|\cdot|$ denotes the norm induced by the adapted C^∞ metric.

We shall show $W^s(J) \subset J^+$. Since we have already known $W^s(J) \subset K^+$ by Proposition 9.1.1, we derive a contradiction from the existence of a point $z \in W^s(J) \cap \operatorname{Int} K^+$. By Proposition 8.3.3, $\{|(f^n)'_z| \mid n \in \mathbb{N}\}$ is bounded. On the other hand, since $z \in W^s(J)$, there is $M > 0$ such that $f^n(z) \in N$ for $n \ge M$. Then for v with $(f^M)'_z(v) \in C_{f^M(z)}$, and $l \in \mathbb{N}$, we have

$$|(f^{M+ln_0})'_z(v)| \ge 2^l |(f^M)'_z(v)|,$$

which implies that $\{|(f^n)'_z| \mid n \in \mathbb{N}\}$ is not bounded, a contradiction. □

Theorem 9.3.4 *Suppose f is hyperbolic. Then, for $a \in J$, both $W^s(a)$ and $W^u(a)$ are biholomorphic to $\mathbb{C}$.*

Proof By Theorem 9.3.2, $W^u(a)$ is a one-dimensional complex manifold diffeomorphic to $\mathbb{R}^2$. So $W^u(a)$ is biholomorphic to either the unit disk Δ or $\mathbb{C}$. Using the notation of Theorem 9.3.2, for each $z \in J$, set $A(z) = W^u_\varepsilon(z) - \overline{W^u_{\varepsilon/2}(z)}$, which is biholomorphic to an annulus. The moduli of the annuli $A(z)$ are bounded below by a positive constant when z moves over J since J is compact. Fix $n > 0$ so that $A^n < 1/2$; then $f^{-n}(A(z)) \subset\subset W^u_{\varepsilon/2}(f^{-n}(z))$.

Now, let $A(l, a) = f^{nl}(A(f^{-nl}(a)))$ for $l \geq 1$. Then, $\{A(l, a)\}$ is an infinite family of disjoint annuli in $W^u(a)$ so that $A(l+1, a)$ surrounds $A(l, a)$. The existence of such a family implies that $W^u(a)$ is biholomorphic to $\mathbb{C}$ by the modulus inequality. (See §1.1.) □

Proposition 9.3.5 *If f is hyperbolic, $W^s(z_1) \cap W^u(z_2) \subset J$ whenever z_1, $z_2 \in J$.*

Proof Let $z_1, z_2 \in J$. Then, by Theorem 9.3.3, $W^s(z_1) \subset J^+$ and $W^u(z_2) \subset J^-$. Hence, $W^s(z_1) \cap W^u(z_2) \subset J$. □

The next proposition follows from the facts that J is hyperbolic and J has local product structure due to Proposition 9.3.5.

Proposition 9.3.6 *Suppose f is hyperbolic. Then, there is an open neighborhood N of J having the following property: any f-invariant set Ξ contained in N satisfies $\Xi \subset J$.*

When f is hyperbolic, every connected component of $\operatorname{Int} K^+$ is a stable set $W^s(a)$ of an attracting periodic point. To show this, we need Poincaré's recurrence theorem.

Theorem 9.3.7 *Let (M, m) be a measure space. Here m is assumed to be positive with finite total measure $m(M)$, and to be invariant by a bijection $S : M \to M$. Let $A \subset M$ be a measurable subset. Then, there is a measurable subset B with $B \subset A$ such that $m(B) = m(A)$ and that $\{n \geq 1 \mid S^n(p) \in A\}$ is always an infinite set for every $p \in B$.*

Proof For non-negative integers ν, let $A_\nu = S^{-\nu}(A)$ and $C_\nu = \bigcup_{j \geq \nu} A_j$. Then, $S(C_\nu) = C_{\nu-1} \supset C_\nu$ and $m(C_{\nu-1}) = m(S(C_\nu)) = m(C_\nu)$. Let $C_\infty = \bigcap_{\nu=0}^\infty C_\nu$; then, $C_0 - C_\infty = \bigcup_{\nu>0}(C_{\nu-1} - C_\nu)$. Therefore, $m(C_0 - C_\infty) = 0$. Let $B = A \cap C_\infty$. Since $A = A_0 \subset C_0$, we have

$$A = A \cap C_0 = B \cup (A \cap (C_0 - C_\infty)),$$

which proves $m(B) = m(A)$. Since $p \in C_\nu$ is equivalent to the existence of j with $j \geq \nu$ and $S^j(p) \in A$, $p \in C_\infty$ has the desired property. □

Theorem 9.3.8 *Let f be hyperbolic. Then there are no wandering connected components of* $\operatorname{Int} K^+$.

Proof By Theorem 7.2.9, we restrict ourselves to the case $|\delta| \leq 1$. Let U be a connected component of $\operatorname{Int} K^+$.

When $|\delta| = 1$, since K^+ has finite Lebesgue measure by Theorem 7.2.9, Theorem 9.3.7 assures us that U is not wandering.

Now, we assume $|\delta| < 1$ and take $z \in U$. Let L be the set of all limit points of all the convergent subsequences $\{f^{n_j}(z) \mid n_j > 0\}$. By Proposition 9.1.1 and Theorem 7.2.9, $L \subset K = K^+ \cap K^- = K^+ \cap J^-$.

First assume that there is a point $w \in L \cap \operatorname{Int} K^+$. Let U_0 be the connected component of $\operatorname{Int} K^+$ containing w. There are distinct positive integers n_1, n_2 with $f^{n_1}(z), f^{n_2}(z) \in U_0$. Therefore, $f^{n_1}(U) = U_0 = f^{n_2}(U)$, so U is non-wandering.

Next suppose that $L \subset J^+$. Then since $L \subset J$, $z \in W^s(J)$. Theorem 9.3.3 yields $W^s(J) \subset J^+$, where we use the hyperbolicity of f for the first time. Hence $z \in J^+$, which contradicts $z \in \operatorname{Int} K^+$. In conclusion, this case does not occur. □

Theorem 9.3.9 *Suppose that f is hyperbolic, and that the Jacobian δ satisfies* $|\delta| = 1$. *Then,*

$$\operatorname{Int} K^+ = \operatorname{Int} K^- = \varnothing.$$

Proof First, without using the hyperbolicity, we show $m(K^+) = m(K^-) = m(K)$, hence $\operatorname{Int} K^+ = \operatorname{Int} K^- = \operatorname{Int} K$. By Theorem 7.2.9, we know $m(K^+) < \infty$, so it suffices to show $m(K^+) = m(K)$. To reach a contradiction, suppose there is a compact set $A \subset (K^+ - K)$ with $m(A) > 0$. Take R and R_1 as in Lemma 8.3.4 such that furthermore we have $A \subset \big\{(x, y) \,\big|\, |x| \leq R_1\big\}$. Then, $f^n(A) \subset \{|x| \leq R_1\}$ for all $n \geq 0$. Since $A \cap K^- = \varnothing$, we can choose positive increasing integers $\{n_j\}$ such that $\{f^{-n_j}(A)\}$ are pairwise disjoint. Take a $q \in \mathbb{N}$ such that $qm(A) > m(\{K^+ \cap \{|x| \leq R_1\})$ and let $T = f^{n_q}(\bigcup_{j=1}^{q} f^{-n_j}(A))$. Then, $T \subset K^+ \cap \{|x| \leq R_1\}$ but, since $|\delta| = 1$ implies f is volume-preserving, we have $m(T) = qm(A)$, a contradiction.

Now, to the contrary of the statement of the theorem, let $U \neq \varnothing$ be a connected component of $\operatorname{Int} K$. Then $\partial U \subset J$. In view of Theorem 9.3.8, let p (≥ 1) be the period of U.

For any compact $L \subset\subset U$, let $\hat{L} = \overline{\{f^{np}(z) \mid z \in L, n \in \mathbb{Z}\}}$. Then $\hat{L}$ is an f^p-invariant compact set. We shall show $\hat{L} \cap \partial U = \varnothing$. Assume, for subsequences $\{n_j\}$ and $\{z_j\} \subset L$, $\{f^{n_j p}(z_j)\}$ converges to $w \in \overline{U}$. Taking further subsequences if necessary, we may assume that $z_j \to \zeta \in L$ and $\{f^{n_j p}\}$ converges uniformly on every compact set in U to a holomorphic map $h : U \to \overline{U}$. (See Proposition 8.3.3.) Since $|\delta| = 1$, $|\det(h')| = 1$, so h must be locally biholomorphic, hence $h : U \to U$. Therefore $w = h(\zeta) \in U$ which implies $\hat{L} \cap \partial U = \varnothing$.

Take an open neighborhood N of J having the property of Proposition 9.3.6 for f^p. Since $\partial U \subset J$, $L = U - N$ is compact satisfying $L \subset\subset U$ and $(U - \hat{L}) \subset N$. The above observation implies that $(U - \hat{L}) \cap \operatorname{Int} K^+ \neq \varnothing$. Therefore $(U - \hat{L})$ is an f^p-invariant set contained in N and not contained in J. This contradicts Proposition 9.3.6. □

Theorem 9.3.10 *Assume f is hyperbolic. Then, each connected component U of* Int K^+ *is the stable set $W^s(a)$ of an attracting periodic point a.*

Proof By Theorems 7.2.9 and 9.3.9, it can happen that Int $K^+ \neq \varnothing$ only when $|\delta| < 1$. Let $p \geq 1$ be the period of U.

First we show that U is recurrent. Take R sufficiently large so that $B = U \cap D_R \neq \varnothing$ and $U \subset \operatorname{Int}(D_R \cup V_R^-)$ by Lemma 7.2.4. Then, since $f^p(U) = U \subset D_R \cup V_R^-$ and, by Lemma 7.2.5, $f^p(D_R) \subset D_R \cup V_R^+$, so $f^p(B) \subset B$. Let L be the set of all limit points $w = \lim_{j\to\infty} f^{n_j p}(z)$, where z moves in B and $\{n_j > 0\}$ moves all subsequences such that $\{f^{n_j p}(z)\}$ is convergent. Then $L \subset \overline{B}$, and hence $L \neq \varnothing$. Since $f^{-p}(L) \subset L$, we have $L \subset K^- = J^-$. If $L \cap U = \varnothing$, then $L \subset J^+$, so $L \subset J$, which implies $B \subset W^s(J)$. Hence, by Theorem 9.3.3, $B \subset J^+$, which contradicts $B \subset \operatorname{Int} K^+$. Therefore $L \cap U \neq \varnothing$, which implies that U is recurrent.

Now, by Theorem 9.1.10, U is either $W^s(a)$ of an attracting periodic point a, $W^s(\mathcal{D})$ of a Siegel disk or $W^s(\mathcal{H})$ of a Herman ring. We will remove the possibilities of the latter two types. Let $U = W^s(\mathcal{D})$. Take an open neighborhood N of J satisfying the property of Proposition 9.3.6. On the other hand, $\partial\mathcal{D} \subset J$ holds and a circle near $\partial\mathcal{D}$ of the Siegel disk $\mathcal{D}$ is an f^p-invariant set which is contained in N and not contained in J. This is a contradiction to the property of N. Similarly, the case $U = W^s(\mathcal{H})$ does not happen. □

Theorem 9.3.11 *Assume that f is hyperbolic and the Jacobian of f satisfies $|\delta| < 1$. Then the number of the attracting periodic points is finite.*

Proof Denote by A the set of all attracting periodic points. We assume that A is an infinite set. Let L be the set of all points which are limit points of a sequence consisting of distinct points in A. Then $L \neq \varnothing$. By Theorem 9.3.10, we have $L \cap \text{Int } K^+ = \varnothing$, so $L \subset J^+$. On the other hand, $L \subset K \subset K^- = J^-$. Hence, $L \subset J$.

Take an open neighborhood N of J as in Proposition 9.3.6. Then the existence of infinitely many attracting periodic points z_j of period p_j with orbits satisfying $\{f^i(z_j) \mid 0 \le i \le p_j - 1\} \not\subset N$ leads to a contradiction, for $L \subset J$.

On the other hand, the existence of an attracting periodic point z of period p such that $\{f^i(z) \mid 0 \le i \le p-1\} \subset N$ gives an f-invariant set $\{f^i(z) \mid 0 \le i \le n-1\} \subset \text{Int } K^+$, which contradicts the property of N. □

Theorem 9.3.12 *Let f be hyperbolic, and δ the Jacobian of f.* (i) *If* $|\delta| = 1$, *then* $R(f) = J$. (ii) *If* $|\delta| < 1$, *letting S be the finite set of all attracting periodic points of f, $R(f) = J \cup S$.* (iii) *If* $|\delta| > 1$, *letting S_1 be the finite set of all repelling periodic points of f, $R(f) = J \cup S_1$.*

Proof (i) By Theorem 9.3.9, $K = K^+ \cap K^- = J^+ \cap J^- = J$. So by Proposition 9.2.6, $R(f) \subset J$. On the other hand, $J \subset R(f)$ by Proposition 9.2.5 and Theorem 9.2.7.

(ii) By Theorem 9.3.10, a connected component of Int K^+ is $W^s(a)$ of an attracting periodic point a, and clearly, $(W^s(a) - a) \cap R(f) = \varnothing$. So by Proposition 9.2.6 and Theorem 7.2.9,

$$\begin{aligned} R(f) &= R(f) \cap K = ((R(f) \cap J^+) \cup S) \cap K^- \\ &= (R(f) \cap J^+ \cap K^-) \cup (S \cap K^-) = (R(f) \cap J) \cup S = J \cup S. \end{aligned}$$

Part (iii) follows from (ii). □

Theorem 9.3.13 *Let f be hyperbolic, and δ the Jacobian of f.*

(i) *If* $|\delta| = 1$, $W^s(J) = J^+$ *and* $W^u(J) = J^-$.

(ii) *If* $|\delta| < 1$, *letting S be the finite set of all attracting periodic points of f, $W^s(J) = J^+$ and $W^u(J) = J^- - S$.*

(iii) *If* $|\delta| > 1$, *letting S_1 be the finite set of all repelling periodic points of f, $W^s(J) = J^+ - S_1$ and $W^u(J) = J^-$.*

Proof We prove the result only when $|\delta| \le 1$. By Theorem 9.3.9 when $|\delta| = 1$, and by Theorem 7.2.9 when $|\delta| < 1$, we have

$$K^- = J^-. \tag{9.5}$$

By Theorem 9.3.3, we know $W^s(J) \subset J^+$ and $W^u(J) \subset J^-$. And by Proposition 9.1.1, $W^s(K) = K^+$, and hence $J^+ \subset W^s(K)$. Since J^+ is f-invariant, $J^+ \subset W^s(K \cap J^+)$. By (9.5), $K \cap J^+ = J$, hence $J^+ \subset W^s(J)$. Thus $W^s(J) = J^+$.

Next, we show $J^- - \operatorname{Int} K^+ \subset W^u(J)$. By Proposition 9.1.1 and (9.5), $J^- = K^- = W^u(K)$. Since $J^- - \operatorname{Int} K^+$ is an f-invariant closed set, $J^- - \operatorname{Int} K^+ \subset W^u(K \cap (J^- - \operatorname{Int} K^+))$. Since $K \cap (J^- - \operatorname{Int} K^+) = K^+ \cap (J^- - \operatorname{Int} K^+) = J^+ \cap J^- = J$, we have $J^- - \operatorname{Int} K^+ \subset W^u(J)$.

Further, $J^- \cap (\operatorname{Int} K^+ - S) \subset W^u(J)$. Indeed, by Theorem 9.3.10, a connected component of $\operatorname{Int} K^+$ is $W^s(a)$ of an attracting periodic point a. If $w \in W^s(a) - \{a\}$, then $w \in W^u(J^+)$. Moreover if $w \in J^-$, $w \in W^u(J^-)$. Therefore $J^- \cap (\operatorname{Int} K^+ - S) \subset W^u(J^+ \cap J^-) = W^u(J)$. Thus, we have $J^- - S \subset W^u(J)$. Since clearly $W^u(J) \subset J^- - S$, we complete the proof of (i) and (ii). □

Theorem 9.3.14 *The following three conditions are equivalent.*

(i) *The chain recurrent set $R(f)$ is a hyperbolic set for f.*
(ii) *The non-wandering set $\Omega(f)$ is a hyperbolic set for f.*
(iii) *J is a hyperbolic set for f.*

Proof By Proposition 9.2.5 and Theorem 9.2.7, $J \subset \Omega(f) \subset R(f)$. So to show the implication (iii)⇒(i) is the only problem. Since both the set of all attracting periodic points and that of all repelling ones are hyperbolic sets, the assertion follows from Theorem 9.3.12. □

Bibliography

Abikoff W., The residual limit sets of Kleinian groups, *Acta Math.*, **130** (1973), 127–44.

Ahlfors L. V., *Complex Analysis*, McGraw-Hill, 1953.

Ahlfors L. V., *Lectures on Quasiconformal Mappings*, Van Nostrand, 1966.

Baker I. N., The existence of fixpoints of entire functions, *Math. Z.*, **73** (1960), 280–4.

Baker I. N., Multiply connected domains of normality in iteration theory, *Math. Z.*, **81** (1963), 206–14.

Baker I. N., Fixpoints of polynomials and rational functions, *J. London Math. Soc.*, **39** (1964), 615–622.

Baker I. N., Completely invariant domains of entire functions, *Mathematical Essays Dedicated to A. J. MacIntyre*, Ohio Univ. Press, 1970, 33–5.

Baker I. N., The domains of normality of an entire functions, *Ann. Acad. Sci. Fenn. Ser. A I Math.*, **1** (1975), 277–283.

Baker I. N., An entire function which has wandering domains, *J. Austral. Math. Soc.*, **22** (1976), 173–6.

Baker I. N., Wandering domains in the iteration of entire functions, *Proc. London Math. Soc. (3)*, **49** (1984), 563–76.

Baker I. N., Some entire functions with multiply-connected wandering domains, *Ergod. Th. & Dynam. Sys.*, **5** (1985), 163–9.

Baker I. N. & Rippon P. J., Iteration of exponential functions, *Ann. Acad. Sci. Fenn. Ser. A I Math.*, **9** (1984), 49–77.

Banks J., Brooks G., Cains G., Davis G. & Stacey P., On Devaney's definition of chaos, *Amer. Math. Monthly* **99** (1992), 332–334.

Beardon A. F., The exponent of convergence of Poincaré series, *Proc. London Math. Soc. (3)*, **18** (1968), 461–83.

Beardon A. F., *Iteration of Rational Functions*, Grad. Texts Math. 132, Springer, 1991.

Beardon A. F. & Pommerenke Ch., The Poincaré metric of plane domains, *J. London Math. Soc. (2)*, **18** (1978), 475–483.

Bedford E., Iteration of polynomial automorphisms of $\mathbb{C}^2$, *Proceedings of the ICM, 1990, Kyoto, Japan*, 1991, 847–58.

Bedford E., Lyubich M. & Smillie J., Polynomial diffeomorphisms of $\mathbb{C}^2$ IV: The measure of maximal entropy and laminar currents, *Invent. Math.*, **112** (1993), 77–125.

Bedford E., Lyubich M. & Smillie J., Distribution of periodic points of polynomial diffeomorphisms of $\mathbb{C}^2$, *Invent. Math.*, **114** (1993), 277–88.

Bedford E. & Smillie J., Polynomial diffeomorphisms of $\mathbf{C}^2$: Currents, equilibrium measure and hyperbolicity, *Invent. Math.*, **103** (1991), 69–99.

Bedford E. & Smillie J., Fatou–Bieberbach domains arising from polynomial automorphisms, *Indiana U. Math. J.*, **40** (1991), 789–92.

Bedford E. & Smillie J., Polynomial diffeomorphisms of $\mathbf{C}^2$ II: Stable manifolds and recurrence, *J. Amer. Math. Soc.*, **4** (1991), 657–679.

Bedford E. & Smillie J., Polynomial diffeomorphisms of $\mathbf{C}^2$ III: Ergodicity, exponents and entropy of the equilibrium measure, *Math. Ann.*, 294 (1992), 395–420.

Bergweiler W., Periodic points of entire functions: proof of a conjecture of Baker, *Comp. Vari.*, **18** (1991), 57–72.

Bergweiler W., Iteration of meromorphic functions, *Bull. Amer. Math. Soc.*, **29** (1993), 151–88.

Bergweiler W., Newton's method and a class of meromorphic functions without wandering domains, *Ergod. Th. & Dynam. Sys.*, **13** (1993), 231–47.

Bergweiler W., On the Julia set of analytic self-maps of the punctured plane, *Analysis*, **15** (1995), 251–6.

Bergweiler W., Invariant domains and singularities, *Math. Proc. Camb. Phil. Soc.*, **117** (1995), 525–32.

Bergweiler W., Haruta M., Kriete H., Meier H.-G. & Terglane N., On the limit functions of iterates in wandering domains, *Ann. Acad. Sci. Fenn. Ser. A I Math.*, **18** (1993), 369–75.

Bergweiler W. & Terglane N., Weakly repelling fixpoints and connectivity of wandering domains, *Trans. Amer. Math. Soc.*, **348** (1996), 1–12.

Bers L. & Royden H. L., Holomorphic families of injections, *Acta Math.*, **157** (1986), 259–86.

Bishop C. J. & Jones P. W., Hausdorff dimension and Kleinian groups, *Acta Math.*, **179** (1997), 1–39.

Blanchard P., Complex analytic dynamics on the Riemann sphere, *Bull. Amer. Math. Soc.*, **11** (1984), 85–141.

Blum L., Shub M. & Smale S., On a theory of computation and complexity over the real numbers: NP-completeness, recursive functions and universal machines, *Bull. Amer. Math. Soc.*, **21** (1989), 1–46.

Blum L. & Smale S., The Gödel incompleteness theorem and decidability over a ring, *From Topology to Computation: Proceeding of Smalefest* (Eds. Hirsch M. W., Marsden J. E. & Shub M.), Springer-Verlag, 1993, 321–39.

Brolin H., Invariant sets under iteration of rational functions, *Ark. Mat.*, **6** (1965), 103–44.

Carleson L. & Gamelin T. W., *Complex Dynamics*, Universitext, Springer, 1993.

Cowen C., Iteration and the solution of functional equations for functions analytic in the unit disk, *Trans. Amer. Math. Soc.*, **265** (1981), 69–95.

Denker M. & Urabański M., Hausdorff measures on the Julia sets of subexpanding rational maps, *Israel J. Math.*, **76** (1991), 193–214.

Devaney R. L., *A First Course in Chaotic Dynamical Systems*, Addison-Wesley, 1993.

Devaney R. L. & Goldberg L. R., Uniformization of attracting basins for exponential maps, *Duke Math. J.*, **55** (1987), 253–66.

Devaney R. L. & Krych M., Dynamics of exp(z), *Ergod. Th. & Dynam. Sys.*, **4** (1984), 35–52.

Devaney R. L. & Nitecki Z., Shift automorphisms in the Hénon mapping, *Comm. Math. Phys.*, **67** (1979), 181–207.

Devaney R. L. & Tangerman F., Dynamics of entire functions near the essential singularity, *Ergod. Th. & Dynam. Sys.*, **6** (1986), 489–503.

Dominguez P., Connectedness properties of Julia sets of transcendental entire functions, *Complex Variables*, **32** (1997), 199–215.

Douady A. & Hubbard J. H., On the dynamics of polynomial-like mappings, *Ann. scient. Éc. Norm. Sup.*, **18** (1985), 287–343.

Douady A. & Hubbard J. H., A proof of Thurston's topological characterization of rational functions, *Acta Math.*, **171** (1993), 263–97.

Eremenko A. E., On the iteration of entire functions, *Dynamical Systems and Ergodic Theory*, Banach center publication, vol. **23** (1989), 339–44.

Eremenko A. E. & Lyubich M. Yu., Examples of entire functions with pathological dynamics, *J. London Math. Soc.*, **36** (1987), 458–68.

Eremenko A. E. & Lyubich M. Yu., The dynamics of analytic transformations, *Leningrad Math. J.*, **1** (1990), 563–634.

Eremenko A. E. & Lyubich M. Yu., Dynamical properties of some classes of entire functions, *Ann. Inst. Fourier Grenoble*, **42** (1992), 989–1020.

Falconer K. J., *The Geometry of Fractal Sets*, Cambridge Univ. Press, 1985.

Fatou P., Sur l'itérations des fonctions transcendentes entières, *Acta Math.*, **47** (1926), 337–70.

Fornaess J. E., *Dynamics in Several Complex Variables*, CBMS Regional Conference Series in Mathematics, 87, American Mathematical Society, 1996.

Fornaess J. & Sibony N., Complex Hénon mappings in $\mathbf{C}^2$ and Fatou–Bieberbach domains, *Duke Math. J.*, **65** (1992), 345–80.

Fornaess J. E. & Sibony N., Complex dynamics in higher dimensions. I., Complex analytic methods in dynamical systems, *Astérisque*, **222** (1994), 201–31.

Fornaess J. E. & Sibony N., Complex dynamics in higher dimensions. II., Modern methods in complex analysis, *Ann. of Math. Stud.*, **137** (1995), 135–82.

Friedland S. & Milnor J., Dynamical properties of plane polynomial automorphisms, *Ergod. Th. & Dynam. Sys.*, **9** (1989), 67–99.

Garber V. L., On the iteration of rational functions, *Math. Proc. Camb. Phil. Soc.*, **84** (1978), 497–505.

Goldberg L. R. & Keen L., A finiteness theorem for a dynamical class of entire functions, *Ergod. Th. & Dynam. Sys.*, **6** (1986), 183–92.

Graczyk J. & Swiatek G., *The Real Fatou Conjecture*, Ann. Math. Studies 144, Princeton Univ. Press, 1998.

Hamilton D. H., Length of Julia curves, *Pacific J. of Math.*, **169** (1995), 75–93.

Hayman W. K., *Meromorphic Functions*, Oxford Univ. Press, 1964.

Hayman W. K., The local growth of power series: a survey of the Wiman–Valiron method, *Can. Math. Bull.*, **17** (1974), 317–58.

Hénon M., A two dimensional mapping with a strange attractor, *Comm. Math. Phys.*, **50** (1976), 69–77.

Herman M. R., Are there critical points on the boundaries of singular domains?, *Comm. Math. Phys.*, **99** (1985), 593–612.

Hinkkanen A., Julia sets of rational functions are uniformly perfect, *Math. Proc. Cambr. Phil. Soc.*, **113** (1993), 543–59

Hua X. H. & Yang C. C., *Dynamics of Transcendental Functions*, Asian Mathematics Series, Gordon and Breach, 1998.

Hubbard J., The Hénon mapping in the complex domain, *Chaotic Dynamics and Fractals* (Eds. Bansley M. *et al.*), Academic Press, 1986, 101–111.

Hubbard J. H. & Oberste-Vorth R. W., Hénon mappings in the complex domain Part I: the global topology of dynamical space, *Publ. Math. IHES*, **79** (1994), 5–46

Hubbard J. H. & Oberste-Vorth R. W., Hénon mappings in the complex domain Part II: Projective and inductive limits of polynomials, *Real and Complex Dynamical*

Systems (Eds. Branner B. and Hjorth P.), Kluwer Academic Publishers, 1995, 89–132

Hubbard J. H. & Papadopol P., Superattractive fixed points in $\mathbf{C}^n$, *Indiana Univ. Math. J.*, **43** (1994), 321–65.

Hubbard J. H., Schleicher D. & Sutherland S., How to find roots of polynomials by Newton's method, preprint.

Hurley M., Multiple attractors in Newton's method, *Ergod. Th. & Dynam. Sys.*, **6** (1986), 561–9.

Imayoshi Y. and Taniguchi M., *An Introduciton to Teichmüller Spaces*, Springer, 1992.

Jungreis I., The uniformization of the complement of the Mandelbrot set, *Duke Math. J.*, **52** (1985), 935–8.

Katok A., Lyapunov exponents, entropy and periodic orbits for diffeomorphisms, *Publ. Math. IHES.*, **51** (1980), 137–74

Kisaka M., On some exceptional maps, *Proc. Japan Acad., Ser. A*, **71** (1995), 35–8.

Klimek M., *Pluripotential Theory*, Clarendon Press, 1991.

Lattès S., Sur les formes réduites des transformations ponctuelles à deux variables, *Comptes Rendus*, **152** (1911), 1566–9.

Lelong P., Fonctions entières (n variables) et fonctions plurisousharmoniques d'ordre fini dans $\mathbb{C}^n$, *J. Analyse Math. Jérusalem*, **12** (1964), 365–407.

Lelong P. & Gruman L., *Entire Functions of Several Complex Variables*, Grund. der Math. Wiss., 282, Springer, 1986.

Lyubich M., Dynamics of quadratic polynomials I/II, *Acta Math.*, **178** (1997), 185–297.

Mañé R., On the instability of Herman rings, *Inv. Math.*, **81** (1985), 459–471.

Mañé R. & Rocha da L. F., Julia sets are uniformly perfect, *Proc. Amer. Math. Soc.*, **116** (1992), 251–7.

Mañé R., Sad P. & Sullivan D., On the dynamics of rational maps, *Ann. Sci. Éc. Norm. Sup.*, **16** (1983), 193–217.

Matsuzaki K. and Taniguchi M., *Hyperbolic Manifolds and Kleinian Groups*, Oxford Univ. Press, 1998.

Mayer J. C., An explosion point for the set of endpoints of the Julia set of $\lambda \exp(z)$, *Ergod. Th. & Dynam. Sys.*, **10** (1990), 177–83.

McMullen C., Area and Hausdorff dimension of Julia sets of entire functions, *Trans. Amer. Math. Soc.*, **300** (1987), 329–42.

McMullen C., Families of rational maps and iterative root-finding algorithms, *Ann. of Math.*, **125** (1987), 467–93.

McMullen C., *Complex Dynamics and Renormalization*, Annals of Math. Studies 135, Princeton Univ. Press, 1994.

McMullen C., Frontiers in complex dynamics, *Bull. Amer. Math. Soc.*, **31** (1994), 155–72.

McMullen C., *Renormalization and 3-manifolds which Fiber over the Circle*, Annals of Math. Studies 142, Princeton Univ. Press, 1996.

McMullen C., The Mandelbrot set is universal, preprint, 1997.

McMullen C., Hausdorff dimension and conformal dynamics II: geometrically finite rational maps, preprint.

McMullen C. & Sullivan D., Quasiconformal homeomorphisms and dynamics III: the Teichmüller space of a holomorphic dynamical system, preprint.

de Melo W. & van Strien S., *One-Dimensional Dynamics*, Springer, 1993.

Milnor J., Dynamics in one complex variable: introductory lectures, SUNY Stony Brook Preprint #1990/5.

Milnor J., Geometry and dynamics of quadratic rational maps, *Experimental Math.*, **2** (1993), 37–83.

Misiurewicz M., On iterations of e^z, *Ergod. Th. & Dynam. Sys.*, **1** (1981), 103–6.

Morosawa S., Note on infinitely connected immediate basins of attracting cycles for iteration of rational functions, *Nonlinearity*, **8** (1995), 107–9.

Morosawa S., On the residual Julia sets of rational functions, *Ergod. Th. & Dynam. Sys.*, **17** (1997), 205–10.

Morosawa S., An example of cyclic Baker domains, *Mem. Fac. Sci. Kochi Univ. (Math.)*, **20** (1999), 123–6.

Nevanlinna R., *Analytic Functions*, Springer, 1970.

Patterson S. J., The limit set of a Fuchsian group, *Acta Math.*, **136** (1976), 241–273.

Pommerenke Ch., On uniformly perfect sets and Fuchsian groups, *Analysis* **4** (1984), 299–321.

Ransford T., Variation of Hausdorff dimension of Julia sets, *Ergod. Th. & Dynam. Sys.*, **13** (1993), 167–74.

Rosay J. P. & Rudin W., Holomorphic maps from $\mathbf{C}^n$ to $\mathbf{C}^n$, *Trans. Amer. Math. Soc.*, **310** (1988), 47–86.

Sario L. and Nakai M., *Classification Theory of Riemann Surfaces*, Springer, 1970.

Schiff J. L., *Normal Families*, Universitext, Springer, 1993.

Schwick W., Repelling periodic points in the Julia set, *Bull. London Math. Soc.*, **29** (1997), 314–6.

Shishikura M., On the quasiconformal surgery of rational functions, *Ann. Sci. Ecole Norm. Sup.*, **20** (1987), 1–29.

Shishikura M., The connectivity of the Julia set and fixed point, IHES preprint.

Shishikura M., The boundary of the Mandelbrot set has Hausdorff dimension two, *Asterisque*, **222** (1994), 389–405.

Shishikura M., The Hausdorff dimension of the boundary of the Mandelbrot set and Julia sets, *Ann. of Math.*, **147** (1998), 225–67.

Shub M., *Global Stability of Dynamical Systems*, Springer, 1987.

Shub M. & Smale S., On the existence of generally convergent algorithms, *J. Complexity*, **2** (1986), 2–11.

Steinmetz N., *Rational Iteration*, de Gruyter Stud. Math. 16, Walter de Gruyter, 1993.

Sugawa T., Various domain constants related to uniform perfectness, *Comp. Vari.*, **36** (1998), 311–45.

Sullivan D., The density at infinity of a discrete group of hyperbolic motions, *Publ. Math. IHES*, **50** (1979), 171–202.

Sullivan D., Conformal dynamical system, *Geometric Dynamics*, Springer Lecture Notes, **1007** (1983), 725–52.

Sullivan D., Quasiconformal homeomorphism and dynamics I: solution of Fatou–Julia problem on wandering domains, *Ann. of Math. (2)*, **122** (1985), 401–18.

Sullivan D., Quasiconformal homeomorphism and dynamics II: structural stability implies hyperbolicity for Kleinian groups, *Acta Math.*, **155** (1985), 243–60.

Sullivan D., Bounds, quadratic differentials and renormalization conjectures, *Proc. of the AMS Centennial Symposium*, 1992, 417–66.

Sullivan D., Linking the universalities of Milnor–Thurston, Feigenbaum and Ahlfors–Bers, *Toplogical Methods in Modern Mathematics: a Symposium in Honor of John Milnor's Sixtieth Birthday* (Eds. Goldberg L. & Phillips A. V.), Publish or Perish, 1993, 543–64.

Tan Lei, Similarity between the Mandelbrot set and Julia sets, *Comm. Math. Phys.*, **122** (1990), 401–18.

Ueda T., Local structure of analytic transformations of two complex variables I; II, *J. Math. Kyoto Univ.*, **26** (1986), 233–61, **31** (1991), 695–711.
Ueda T., Complex dynamical systems on complex projective spaces, *Chaotic Dynamical Systems*, (Ed. Ushiki S.), World Scientific, 1993, 120–38.
Ueda T., Fatou sets in the complex dynamics on projective spaces, *J. Math. Soc. Japan.*, **46** (1994), 545–55.
Wernstein L., The Bieberbach conjecture, *Intern. Math. Res. Notices*, **5** (1991), 61–4.
Whyburn G. T., *Analytic Topology*, Amer. Math. Soc. Coll. Publ. 28, 1942.
Yoccoz J. C., Théorème de Siegel, nombres de Bruno et polynômes quadratiques, *Astérisque*, **231** (1996), 3–88.

Index